ABOUT ISLAND PRESS

Since 1984, the nonprofit Island Press has been stimulating, shaping, and communicating the ideas that are essential for solving environmental problems worldwide. With more than 800 titles in print and some 40 new releases each year, we are the nation's leading publisher on environmental issues. We identify innovative thinkers and emerging trends in the environmental field. We work with world-renowned experts and authors to develop cross-disciplinary solutions to environmental challenges.

Island Press designs and implements coordinated book publication campaigns in order to communicate our critical messages in print, in person, and online using the latest technologies, programs, and the media. Our goal: to reach targeted audiences—scientists, policymakers, environmental advocates, the media, and concerned citizens–who can and will take action to protect the plants and animals that enrich our world, the ecosystems we need to survive, the water we drink, and the air we breathe.

Island Press gratefully acknowledges the support of its work by the Agua Fund, Inc., Annenberg Foundation, The Christensen Fund, The Nathan Cummings Foundation, The Geraldine R. Dodge Foundation, Doris Duke Charitable Foundation, The Educational Foundation of America, Betsy and Jesse Fink Foundation, The William and Flora Hewlett Foundation, The Kendeda Fund, The Forrest and Frances Lattner Foundation, The Andrew W. Mellon Foundation, The Curtis and Edith Munson Foundation, Oak Foundation, The Overbrook Foundation, the David and Lucile Packard Foundation, The Summit Fund of Washington, Trust for Architectural Easements, Wallace Global Fund, The Winslow Foundation, and other generous donors.

The opinions expressed in this book are those of the author(s) and do not necessarily reflect the views of our donors.

ABOUT THE SOCIETY FOR ECOLOGICAL RESTORATION INTERNATIONAL

The Society for Ecological Restoration (SER) International is an international nonprofit organization comprising members who are actively engaged in ecologically sensitive repair and management of ecosystems through an unusually broad array of experience, knowledge sets, and cultural perspectives.

The mission of SER is to promote ecological restoration as a means of sustaining the diversity of life on Earth and reestablishing an ecologically healthy relationship between nature and culture.

The opinions expressed in this book are those of the author(s) and are not necessarily the same as those of SER International. Contact SER International at 285 W. 18th Street, #1, Tucson, AZ 85701. Tel. (520)622-5485, Fax (270)626-5485, e-mail, info@ser.org, www.ser.org.

LARGE-SCALE ECOSYSTEM RESTORATION

Society for Ecological Restoration International

The Science and Practice of Ecological Restoration

James Aronson, editor
Donald A. Falk, associate editor
Margaret A. Palmer
Richard J. Hobbs

Wildlife Restoration: Techniques for Habitat Analysis and Animal Monitoring,
by Michael L. Morrison

Ecological Restoration of Southwestern Ponderosa Pine Forests,
edited by Peter Friederici, Ecological Restoration Institute at Northern Arizona University

Ex Situ Plant Conservation: Supporting the Survival of Wild Populations,
edited by Edward O. Guerrant Jr., Kayri Havens, and Mike Maunder

Great Basin Riparian Areas: Ecology, Management, and Restoration,
edited by Jeanne C. Chambers and Jerry R. Miller

Assembly Rules and Restoration Ecology: Bridging the Gap between Theory and Practice,
edited by Vicky M. Temperton, Richard J. Hobbs, Tim Nuttle, and Stefan Halle

The Tallgrass Restoration Handbook: For Prairies, Savannas, and Woodlands,
edited by Stephen Packard and Cornelia F. Mutel

The Historical Ecology Handbook: A Restorationist's Guide to Reference Ecosystems,
edited by Dave Egan and Evelyn A. Howell

Foundations of Restoration Ecology,
edited by Donald A. Falk, Margaret A. Palmer, and Joy B. Zedler

Restoring the Pacific Northwest: The Art and Science of Ecological Restoration in Cascadia,
edited by Dean Apostol and Marcia Sinclair

A Guide for Desert and Dryland Restoration: New Hope for Arid Lands,
by David A. Bainbridge

Restoring Natural Capital: Science, Business, and Practice,
edited by James Aronson, Suzanne J. Milton, and James N. Blignaut

Old Fields: Dynamics and Restoration of Abandoned Farmland,
edited by Viki A. Cramer and Richard J. Hobbs

Ecological Restoration: Principles, Values, and Structure of an Emerging Profession,
by Andre F. Clewell and James Aronson

River Futures: An Integrative Scientific Approach to River Repair,
edited by Gary J. Brierley and Kirstie A. Fryirs

Large-Scale Ecosystem Restoration: Five Case Studies from the United States,
edited by Mary Doyle and Cynthia A. Drew

Large-Scale Ecosystem Restoration

Five Case Studies from the United States

Edited by

Mary Doyle and Cynthia A. Drew

Society for Ecological Restoration International

Large-scale ecosystem restoration : five case studies from the United States / edited by Mary Doyle and Cynthia A. Drew.
p. cm.
Includes index.
ISBN-13: 978-1-59726-025-1 (cloth : alk. paper)
ISBN-10: 1-59726-025-8 (cloth : alk. paper)
ISBN-13: 978-1-59726-026-8 (pbk. : alk. paper)
ISBN-10: 1-59726-026-6 (pbk. : alk. paper)
1. Restoration ecology—United States—Case studies. 2. Ecosystem management—United States—Case studies. 3. Watershed restoration—United States—Case studies. I. Doyle, Mary. II. Drew, Cynthia A.
QH104.L327 2008
639.9—dc22 2007046018

Printed on recycled, acid-free paper

Manufactured in the United States of America

10 9 8 7 6 5 4 3 2 1

Keywords: large-scale restoration, watershed restoration, wetland restoration, river restoration, Chesapeake Bay, the Everglades, California Bay Delta, the Platte River Basin, the Upper Mississippi River System

CONTENTS

INTRODUCTION

The Watershed-Wide, Science-Based Approach to Ecosystem Restoration

MARY DOYLE

Until about thirty years ago, environmental degradation and habitat loss were addressed, if at all, on a piecemeal basis, river segment by river segment, species by species. Over time, however, scientists working to resolve problems of species and habitat loss understood what now seems obvious: the crucial aspect of watersheds and other natural systems is the interconnectedness of their component parts. These became objects of study, revealing intricacies and synergies that scientists are compelled to seek to understand today. The advice of scientists to policy makers about how to restore damaged natural systems—areas in nature containing scenic, hydrologic, habitat, or other values that are in need of restoration and protection—was that the traditional, segmented approach was largely ineffective. The only way to proceed was to use the best available science to comprehend the interconnected problems and to work on all aspects of restoration simultaneously and comprehensively. As decision makers accepted this point of view, the concept of science-based, ecosystem- or watershed-wide restoration became the generally accepted model for dealing with large-scale natural resource degradation and loss.

This book presents case studies of five large ecosystem restoration projects in the United States: the Everglades, Platte River Basin, California Bay-Delta, Chesapeake Bay, and Upper Mississippi River Basin (UMRB). These projects focus on widespread improvement in water quality, quantity, and delivery to restore and protect natural systems while providing sustainable water supply. Besides being geographically diverse, they are outstanding examples of the challenges involved in bringing science to bear in watershed-wide efforts to halt and reverse ecosystem deterioration. Each case has an established history of achievements and disappointments, providing useful lessons on planning and implementing environmental policy in highly complex situations. Our purposes in examining the case studies of these fascinating projects side by side are to allow readers to draw comparisons, identifying points of similarity and divergence, and to advance awareness and understanding of the immensely difficult, important issues involved in restoring and protecting natural systems.

A brief comparative look at the projects will help set the stage. First, each covers a wide, sometimes vast, geographical area. The Everglades watershed encompasses 18,000 square miles of the Florida peninsula. Sacramento–San Joaquin Delta in California is the largest Pacific coast estuary in North and South America and contains over 1,156 square miles of valuable habitat and some of the richest farmland in the world. Chesapeake Bay is the largest estuary in the nation, covering 2,300 square miles; its watershed extends 64,000 square miles. The Platte River and adjacent lands in Nebraska that make up critical habitat of the endangered whooping crane and other species, the subjects of the Platte River Recovery Implementation Program, cover 10,000 acres (15.6 square miles). The Upper Mississippi River (UMR) flows 900 miles through five states and twenty-nine locks and dams, from Minneapolis to Cairo, Illinois, draining a watershed of more than 189,000 square miles. Its floodplain ecosystem alone contains 2.6 million acres (4,063 square miles) of diverse habitat. The Mississippi Flyway is used by over 40 percent of migratory waterfowl that cross America.

The five projects addressed in this book are watershed-wide, science-based, collaborative efforts to restore and protect water resources that have been degraded or depleted to crisis levels. The federal government was a precipitating force in launching the five restoration projects, playing one or more roles—as regulator of water quality under the Clean Water Act (CWA), enforcer of the Endangered Species Act (ESA), manager of a federal facility or national park, or operator of inland waterway systems. For example, a lawsuit brought by the United States against the state of Florida in 1988 over uncontrolled releases of phosphorous-laden runoff from agricultural lands into Everglades National Park stimulated creation of the federal–state effort to restore the Everglades. The process of negotiation with water interests and state representatives, at the heart of the Platte River Basin project, was initiated by the U.S. Department of the Interior (DOI) when a confrontation arose over regulating river flow to protect habitat of two endangered and one threatened species of birds and one endangered fish species, as required by the ESA, administered and enforced by the U.S. Fish and Wildlife Service (USFWS), a DOI agency.

The presence of endangered species also played a large role in the creation of the CALFED Bay-Delta Program, as DOI initiated the project in part to try to resolve ESA impacts on state and regional water supply. Another DOI agency, the U.S. Bureau of Reclamation (USBR), operates water storage and delivery facilities in both the Platte River Basin and the California Bay-Delta watershed. In the 1970s, the U.S. Environmental Protection Agency (EPA) undertook the first in-depth study of the Chesapeake Bay's health and led the effort to found the Chesapeake Bay Program in 1983. Exercising its congressionally mandated responsibility for navigation on the river, the U.S. Army Corps of Engineers (the Corps) is the sponsor of the *Upper Mississippi River–Illinois Waterway System Navigation Feasibility Study*. The Study began as a traditional Corps investigation of the feasibility of adding locks and dams and is now a proposed, dual-purpose, integrated navigation and ecosystem sustainability program. The Corps is also cosponsor of the Comprehensive Everglades Restoration Project because of its long-standing operation of water supply and flood control facilities in South Florida, now being reengineered to restore flows to the Everglades natural system.

The roles of the federal government in these projects—as enforcer, regulator, resource manager, project planner, funding source, operator of locks and dams, and partner or antagonist to state governments and stakeholders—can be conflicting. Although the federal impact may sometimes be coercive, particularly when ESA or CWA enforcement is involved, in most instances the ongoing federal presence has developed into a partnership with states, local governments, and stakeholders. Stresses and strains regularly occur in these partnerships, however, and how the players deal with the conflicts is a recurring theme in the case studies.

Because these five projects focus on the critical issues of water quality and quantity and protection of natural systems, the array of stakeholders is similar from project to project and includes agricultural representatives, advocating for a reliable water supply and flood control for farmers; urban users, interested in sustaining future supply and ensuring flood protection; environmental groups, seeking to assure protection of natural systems; fishers and other recreational users, interested in protection of the resource, including water quality, to support their sporting activities; and business interests, recognizing that a secure water supply or an effective water transportation system is critical to sustain a strong economy. Because stakeholder disputes are common and chronic, project managers must search constantly for effective ways to manage and resolve conflicts or face stalemate.

The wide geographical sweep of these projects and their concerns with water and natural systems bring them within the jurisdictions of a plethora of federal, state, local, and tribal government entities. The Chesapeake Bay, Platte River, and Upper Mississippi River projects, unlike those of the Everglades and California Bay-Delta, must face the complexities that necessarily attend multistate partnerships. The UMR Basin drains portions of seven midwestern states; the Chesapeake Bay, portions of six eastern seaboard states; and the Platte River, portions of three western states.

Another difference among the projects concerns the participation of the federal government through its various resource agencies. In two of the projects—the Everglades and the Upper Mississippi—the Corps plays a dominant role in both funding and policy, giving these projects the key advantage of access to the massive Corps' budget and appropriations. Because the Everglades comprises two national parks and three national wildlife refuges, two DOI agencies, the National Park Service (NPS) and the USFWS, play prominent roles in its oversight and management. Under the federal legislation authorizing the Everglades restoration plan, on issues involving assurances of water deliveries to the natural system, DOI functions as an equal partner in decision making with the Corps. Similarly, in the UMR Basin, where two national wildlife and fish refuges and one national wildlife complex occupy nearly 285,000 acres (445 square miles), USFWS plays a key partnership role with the Corps. Only in the Chesapeake Bay is the EPA the lead federal agency. Through its constituent agencies, USBR and USFWS, DOI has the lead for the federal government in the Platte River Basin projects and the CALFED Bay-Delta Program and its successor, the California Bay-Delta Authority.

The ecosystem-wide approach to environmental problem solving is by definition comprehensive; large areas of land and water are in play. For this reason and because of the scientific complexities involved, the ecosystem-wide approach to restoration is expensive. The anticipated cost of completing the plan for Everglades restoration is around $20 billion. Estimates of the cost of the Chesapeake Bay cleanup approach $19 billion. The first eight-year phase of the CALFED program was estimated to cost around $8 billion, and the estimate for a multidecade plan to restore the UMR ecosystem is over $5 billion. The only entity in the picture financially capable of sustaining such high costs is the federal government, so securing and sustaining federal funding has become a central goal of most large environmental restoration projects. Federal funding, of course, necessitates active congressional involvement; thus the role of Congress in environmental restoration is another common theme in the chapters that follow.

Legislators, government managers, stakeholders, and policy makers responsible for complex ecosystem restoration and management seem to agree universally that policy decisions must be based on the best available scientific and technical knowledge. What is not yet fully understood or agreed upon is how best to integrate science into decision making. Questions about managing science abound:

- What subjects deserve research priority?
- Should research dollars be spent on abstract or applied questions?
- How is the scientific research produced from different sources and in different disciplines to be integrated and communicated to policy makers?
- What are the best approaches to dealing with scientific disagreement?
- How can decisions that cost large amounts of money be made in the face of acknowledged scientific or technical uncertainty?

- How are advancements in scientific understanding brought to bear, and how will decision makers respond if these advancements suggest a fundamental change in course with unknowable effects?
- Finally, what manner of peer review should be employed to assure the quality and objectivity of the underlying science research?

The issues relating to how to proceed in the face of scientific uncertainty and emerging scientific understanding are often referred to by the term "adaptive management" and are among the most fascinating presented by the case studies. The theory of adaptive management, widely endorsed by project planners and authorizers, is still largely untested. It assumes that ecosystem restoration planning and policy making will be flexible enough to change course if new scientific knowledge reveals that the previously established plan was misguided, is deficient, or needs to be adjusted. The assumption is based on a scenario whereby scientists working on a project and their peers come to a consensus opinion on the direction for restoration. However, consensus can be elusive in scientific inquiry. If project scientists cannot agree on whether or how to pursue a different course, this could leave policy makers stymied. In each of the five projects, a long, complex, painstaking process of negotiation among government representatives and stakeholders was required to reach agreement on a restoration plan. The plan forms the basis for cost estimates and funding commitments. If scientists were later to conclude that the plan should be rejected as scientifically untenable, these agreements would be undone, and the cost of the project and the way forward would be thrown into uncertainty. Adaptive management requires that the parties have an effective process for making changes in place, which, if followed, will set the project on a new, scientifically sound course in an expeditious way.

Among the five ecosystem restoration efforts chronicled here, only the Platte River so far has had to face the challenge of adaptive management as an issue central to the project plan. Project scientists agreed in 2000, after years of study and negotiation, that planned water releases could be counterproductive to the central goal of restoring stream bank habitat. Decision makers and stakeholders accepted this opinion and went back to the drawing board and negotiating table. Six more years of effort were required to find a scientifically sound and agreed-upon solution to the water release problem.

Adaptive management worked in the Platte River project to avoid implementing a plan that would have exacerbated the condition it was trying to remedy, though at a cost of years of uncertainty and delay. In contrast, the Everglades and UMR, like other projects, have struggled to incorporate adaptive management principles into their planning, funding, and implementation processes. While project managers are continuously adjusting their actions according to scientific inputs, adaptive management has not yet mandated a major course correction in planning, budgeting, or implementation.

The Five Case Studies

Each section of this book is devoted to one case study and consists of three chapters. The primary, most extensive chapter describes and analyzes the project as a whole. This overview presents the project: its location, the important physical characteristics of the area, the environmental harms that the project addresses, and their causes. The primary chapter analyzes the authorities and organizational approaches employed by the restoration effort to create intergovernmental and interagency processes, with emphases on how overarching leadership and priorities have been established and carried out and on how conflict management and consensus-building methods were employed. This main chapter describes the ongoing scientific effort at the center of each project, the means for regular peer review of the science, and how the project deals with the scientific uncertainties common to all such complex undertakings. Each primary chapter also analyzes issues involving stakeholders and the public, including information sharing and interest group participation in decision making.

Each of the five sections of this book includes two additional chapters, one addressing the ecological issues presented by the project and the other analyzing project-specific economic issues. The chapter on ecology sets out the natural elements of the ecosystem, challenges to the system resulting from hydrological alterations and other impacts of human activity on the food web and habitat of indigenous species, and the scientific basis of measures taken to redress the problems. The chapter on economic issues looks at the costs and benefits assigned in the planning of each restoration project. It notes the challenges of quantifying benefits associated with restoring and protecting natural values, especially in the face of predictably heavy costs.

For the readers' convenience, a list of Abbreviations and Acronyms frequently used throughout discussions of the ecosystem restoration projects has been included. All contributors to this volume hope that the case studies will demonstrate the transcendent value of the arduous, complex work undertaken in these projects. In some measure, people working to reestablish natural systems feel that they have been called upon to play the role of the Creator but without the knowledge, understanding, and confidence to pretend to greatness. This is humbling work. Though the course is often uncertain and conflicted, the crucial importance of seeking to restore and protect the water, land, and habitat of our most precious ecosystems is never debated or in doubt. Our tireless pursuit of this goal is a foundational societal task for this and future generations.

PART I

The Everglades

One of the most diverse, challenged bioregions on Earth, the Everglades is rich in 2,000 plant species, 45 mammal and 50 reptile species, 20 species of amphibians, hundreds of fish species, and 350 species of birds. For the past five decades, the most powerful force shaping the destiny of South Florida, home to the Everglades, has been population growth. While the Everglades merits preservation for its natural beauty and unique ecosystem, it must also be protected because it is a primary source of water for the fast-growing region. Development has caused South Florida to become one of the United States' greatest ecological problem zones. Stephen Polasky's chapter 3 discusses the economic perspective of the resultant competition for water allocation.

Half of the historic "river of grass" (Douglas 1947) has been lost, and the remaining natural areas are highly fragmented by the Central and South Florida Project, built and operated by the U.S. Army Corps of Engineers (USACE or the Corps) to drain large quantities of freshwater from the cities and farms of South Florida. Water quality is also an issue, due largely to phosphorous discharges from sugarcane farms upstream. Thomas L. Crisman's chapter 2 explores the impact of hydrologic alterations and restoration plans critical to the landscape of the Everglades proper.

In the Water Resources Development Act (WRDA 2000), Congress authorized the $7.8 billion Comprehensive Everglades Restoration Plan (CERP), the largest effort of its kind ever undertaken. The CERP sets up a partnership among the Corps, the U.S. Department of the Interior (manager of two national parks and five national wildlife refuges in South Florida), and the state of Florida with the goal of "getting the water right" in the Everglades. This means that the quantity, quality, timing, and distribution of the water have to be altered so as to mimic as closely as possible predrainage conditions, a highly ambitious undertaking. Since CERP was authorized, despite prodigious efforts by the federal and state partners, its progress has been slow. Delays in federal funding have frustrated state officials, who have launched their own, self-funded effort to build CERP projects. Disputes on the methods and schedule for meeting water quality standards for the Everglades have also strained the federal–state partnership. The South Florida Ecosystem Restoration Task Force, established by Congress (WRDA 1996) to coordinate agency efforts in implementing CERP, has grappled with key issues, such as coordinating and integrating science and providing independent, scientific peer review. Those involved in this vast, complex undertaking have learned valuable lessons about effective collaboration, the role of leadership in achieving success, the crucial need for organizational adaptability, and the continuing importance of effective communication with shareholders and the public.

References

Douglas, M. S. [1947] 1974. *The Everglades: River of Grass*. Sarasota, FL: Pineapple Press.

Water Resources Development Act (WRDA) 1996. 110 Stat. 3658, Public Law No. 104-303. October 12. At http://www.fws.gov/habitatconservation/Omnibus/WRDA1996.pdf.

WRDA 2000. Comprehensive Everglades Restoration Plan (CERP). 114 Stat. 2687, Sec. 601 (j), (h). Public Law No. 106-541. December 11. At http://www.fws.gov/ habitatconservation/Omnibus/WRDA2000.pdf.

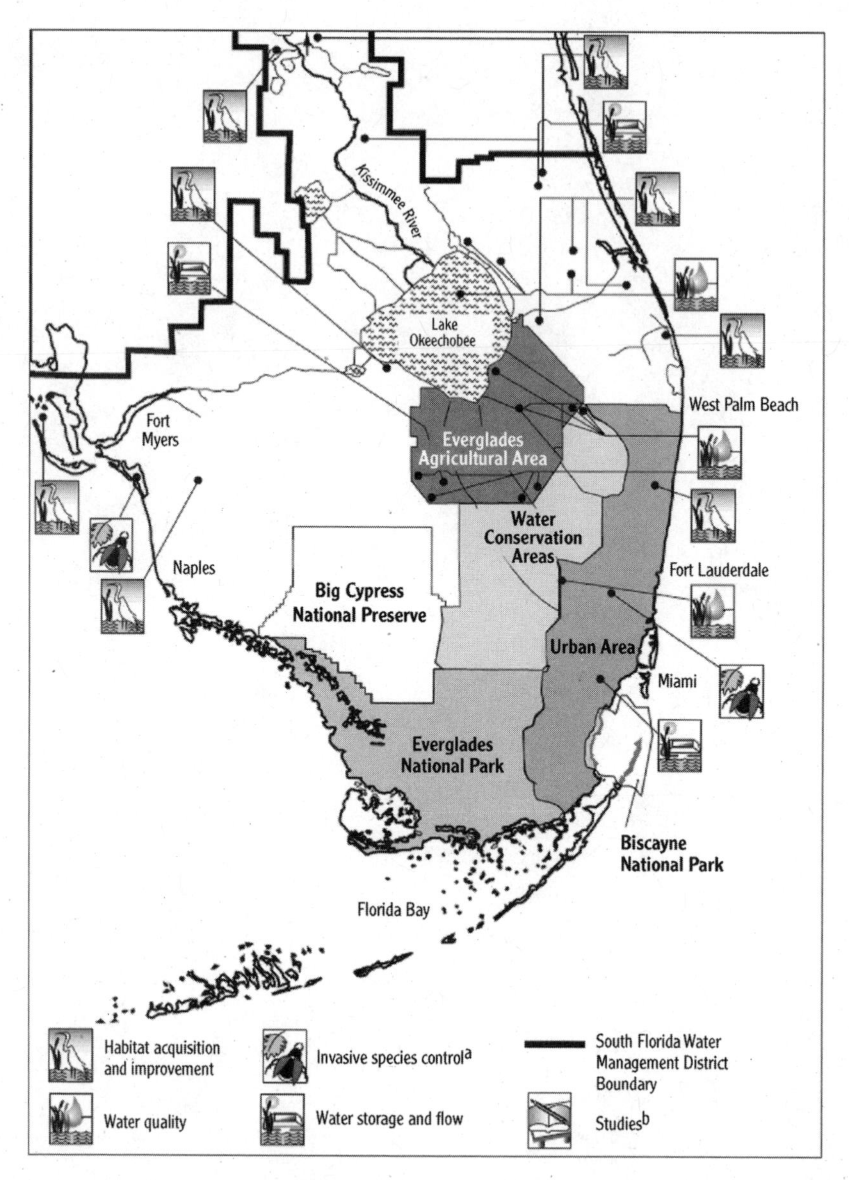

FIGURE 1.1. Everglades Restoration Areas (USGAO 2007).

Chapter 1

The Challenges of Restoring the Everglades Ecosystem

TERRENCE "ROCK" SALT,
STUART LANGTON, AND MARY DOYLE

One of the most complex, challenged bioregions on Earth, South Florida covers the lower third of the Florida peninsula, encompassing the Everglades, Lake Okeechobee, and the Florida Keys. A rare combination of elements—slow and vast water flow, high annual rainfall, low elevation, underlying limestone configuration, and proximity to the ocean—has made the Everglades a diverse, unique ecosystem, rich in 2,000 plant species, 45 mammal species, 50 reptile species, 20 species of amphibians, hundreds of fish species, and 350 species of birds.

To get a glimpse of the Everglades, try to imagine this kaleidoscope of flora and fauna: Caribbean pine, palmetto, yellow tea bush, tiny wild poinsettia, live oak, resurrection, maidenhair, and Boston fern (among 50 species of ferns), strangler fig, red-brown gumbo limbo, ilex, Eugenia, satinwood, cherry laurel, Florida boxwood, more wild orchid varieties than any other place in the United States, brown Florida deer, wildcat, diamondback rattler, otter, Florida panther, alligator, and crocodile. Add flocks of water fowl and birds: least and great blue heron, glossy, Louisiana, and great white heron, snowy egret, the warblers (pine, myrtle, black-throated, blue, and redstart), brown and white pelican, roseate spoonbill, and white ibis.

As pioneering environmentalist Marjory Stoneman Douglas observed, "There are no other Everglades in the world" (Douglas, [1947] 1974, 5). Despite being diked, dammed, and polluted, the Everglades is like no other place on Earth. In recognition of its rare beauty, the United Nations has designated Everglades National Park (ENP or the Park) as a World Heritage Site and a World Biosphere Reserve.

Quantity and Quality of Water

While the Everglades clearly merits preservation for its natural beauty and unique ecosystem, it must also be protected because it is a primary source of the region's water. As a result of the natural system's deterioration, water shortages are not uncommon for South Florida's communities. In the past, the limestone rock beneath the soil absorbed rainwater like a sponge, replenishing the natural aquifer, but human-made changes now divert the flow of freshwater before it soaks into the ground. Miles of paving, brought on by development of the human habitat, also prevent rainwater from penetrating the soil and entering the aquifer. Restoration efforts require more effective diversions and conservation of water to make it available to the ecosystem and for local water supplies.

Quantity is not the only issue: water quality is also of great concern. If less freshwater enters the aquifer, the probability of saltwater intrusion is increased, and agricultural runoff, including varying levels of pesticides, herbicides, fungicides, nitrogen, and phosphorous, contribute to the degradation of the water's purity. The fate of South Florida's water supply is directly related to the quantity and quality of water in the natural Everglades. Thus the long-term survival of all species in the region depends upon successful Everglades restoration.

The Larger South Florida Ecosystem

South Florida covers 18,000 square miles and supports a rapidly growing population. The region's landscape contains significant protected natural areas, surrounded by a large rim of urban, suburban, and agricultural development. This rim includes the Florida Keys, urban east coast, agricultural lands south and west of Lake Okeechobee, and lower west coast. Over 98 percent of South Florida's population lives along this 600-mile swath of land that is seldom wider than 20 miles. Inside the rim are several million acres of protected land. One of the paradoxes of South Florida is that large segments of the population live in congested areas, and yet the same region contains the largest remaining wilderness east of the Mississippi River.

Diverse land areas and habitats surround the Everglades. The slightly elevated Big Cypress Swamp, located to the northwest, serves as a natural levee. The Big Cypress region, 2,800 square miles in area, contains the largest variety and concentration of wildlife in South Florida, including endangered species such as the large black bear, Florida panther, and West Indian manatee. In 1974, the federal government established the Big Cypress National Preserve to protect the area.

South of the Everglades, at the point where its freshwater mingles with the saltwater of Florida Bay and the Gulf of Mexico, is located the largest strand of red, black, white, and buttonwood mangroves in America. Stretching for several hundred miles around the tip of the Florida peninsula, this mangrove wilderness, rich in bird life with an abundant fishery, is interspersed with bays and rivers that continue inland for many miles. Fallen mangrove leaves nourish young shrimp and other estuarine life. Although much has been done to protect this area, the Florida Keys and Florida Bay are under considerable environmental stress. Florida Bay has suffered algae blooms, hypersalinity, loss of sea grass, and reduction in pink shrimp and fish stocks. On the Atlantic side of the Keys, coral reefs show signs of degradation, though restoration efforts under way since 1992 appear to have stabilized some of them.

East of the Everglades, the coastal ridge originally contained all of the upland plant communities found in South Florida, including some that are indigenous only to the ridge. Miles of attractive sandy beaches border the ridge, as does Biscayne Bay, around which the cities of Miami and Miami Beach are built—the most intensely developed area of Florida, home to nearly one-third of the state's population.

The headwaters of the Everglades lie in the Kissimmee River valley, a 3,000-square-mile area that includes the Chain of Lakes region south of Orlando. The lakes flow into the Kissimmee River, which runs for some 90 miles south and flows into Lake Okeechobee. In 1962, the U.S. Army Corps of Engineers (USACE or the Corps) began to drain and convert the Kissimmee River for flood control purposes into a canal 56 miles long (named the C-38 Canal), which includes six water control structures (SFWMD n.d.). Much of the floodplain of the Kissimmee was destroyed by this project and replaced by large-scale dairy and cattle farms. One of the project's unintended, deleterious consequences was the introduction of phosphorous-laden runoff from these farms into Lake Okeechobee.

Located directly north of the Everglades and surrounded by agricultural lands, Lake Okeechobee is the only significant water storage area in the South Florida ecosystem, and the Kissimmee River provides the largest influx of water to the lake. Other tributaries include Nubbin Slough and Taylor Creek; Fisheating Creek is the only major unchanneled source of water flowing into the lake.

While Lake Okeechobee stores a vast amount of freshwater, the greatest source of South Florida's water supply is rainfall. The region receives an average of 40 to 60 inches of rain per year, about 75 percent between May and October (Lake Okeechobee.org n.d.). Historically, the weather pattern is marked by periods of flooding, caused by hurricanes and heavy rains, and periods of drought so severe that lightning often ignites vast fires in the dried-up muck of the Everglades.

Damage Done to the Natural System

The Everglades is particularly vulnerable to damage from a variety of human activities, including population growth, drainage, and agricultural practices.

Population Growth

For the last five decades, the most powerful force shaping the destiny of South Florida has been population growth. The population has grown from 3,283,712 in 1980 to an estimated 5,538,594 people in 2006 in Palm Beach, Miami-Dade, Monroe, and Broward counties (USCB 2007). This is an increase of 59 percent or 2,254,882 people. By 2020, the population is projected to increase by 50 percent and by 2050, 100 percent. One observer has noted that South Florida's population is growing faster than Haiti's and India's (Grunwald 2002). South Florida—with a substantial economy, diverse communities, and a vibrant cultural life—includes sixteen counties, 122 cities, two Indian tribal nations, five regional planning agencies, and scores of state and national parks and preserves. The only government agency that exclusively serves the region is the South Florida Water Management District (SFWMD or District), one of five such districts created by the Florida legislature in 1972. Supported in part by an *ad valorem* tax, the SFWMD had an annual budget of $1.1 billion in fiscal year (FY) 2006.

South Florida's rapid growth has significantly damaged the natural environment. Prompted by long-standing, urgent requests from Florida leaders, in 1947 the USACE began construction of a massive drainage and flood control project, called the Central and South Florida Project (C&SF Project) to allow increased development of South Florida. Operation of the C&SF Project has caused over half the original South Florida wetlands to be lost to urban and agricultural uses in the past fifty years. As a result, the hydrology of the Everglades has been seriously degraded. Today, the Everglades and other areas of South Florida are crisscrossed with 1,800 miles of canals and levees, 200 large and 2,000 small water control structures, and 25 pumping stations. Five large canals drain water into the Gulf of Mexico and the Atlantic Ocean to control flooding; an average of 1.7 billion gallons of freshwater are pumped into the gulf and the ocean every day. The result is that the Everglades receives only half its original water flow (CERP 2000). Development has caused South Florida to become one of America's greatest ecological problem zones. From the headwaters of Lake Okeechobee in the Kissimmee River valley to the north, to the coral reefs lying off the Florida

Keys, there are hundreds of instances of environmental disturbances and deterioration. The most serious include the following:

- Excessive amounts of phosphorous in Lake Okeechobee and in water flowing into the Everglades (EPA 2003)
- Significantly decreased wading bird populations (Gawlik and Sklar 2000)
- Sixty-eight threatened and endangered species, identified under the Endangered Species Act (ESA) by the U.S. Fish and Wildlife Service (USFWS) (McIntosh 2002)
- Invasive exotic plants that infest 1.5 million acres of land in South Florida (FDEP 2006)
- A decline of living corals in the reefs off the Florida Keys despite restoration efforts (Hu et al. 2003)
- Lesions on fish in the St. Lucie estuary, where water is flushed from Lake Okeechobee (DOI 1999)

Draining the Everglades

In fewer than one hundred years, South Florida has witnessed profound environmental change. Half of the historic Everglades has been lost, and the remaining natural areas are highly fragmented. Fragmentation started over a century ago, when government policies began to be directed toward draining large parts of the Everglades wetlands. By the 1920s, hundreds of miles of drainage canals had been built, and in 1930, a two-lane road, the Tamiami Trail, was completed across the Everglades. By the early 1960s, the C&SF Project had created 1,800 miles of levees and canals, and Alligator Alley, an interstate highway, traversed the South Florida peninsula.

In 1881, Hamilton Disston, an engineer, bought 4 million acres from the state of Florida and agreed to drain much of the land for development. He arranged for a canal to be excavated from Lake Okeechobee to the Caloosahatchee River to the west in 1882, allowing steamboats to penetrate South Florida. Within a few years, a commercial fishing industry started, tourism began to grow, and a real estate boom ensued.

Rapid growth of agriculture in the central region of South Florida was matched by growing urbanization on the East Coast. Henry Flagler, a partner of John D. Rockefeller, extended his Florida East Coast Railroad to Miami in 1896 and to Key West in 1912.

In the wake of devastating hurricanes and extensive flooding suffered by South Floridians in 1926 and 1928, state officials urgently requested federal flood control assistance. The federal government agreed to share with the state the cost of a project to control flooding around Lake Okeechobee: a large levee south of the lake and an expanded system of floodgates and canals were constructed. In spite of these efforts, severe flooding continued, and state officials repeatedly asked for increased federal flood protection. In 1947, Congress authorized the C&SF Project, which called for local flood control districts. Twenty-five years later, in 1972, the SFWMD was authorized as the nonfederal sponsor of the C&SF project and assumed responsibility for operating and maintaining the enormous flood control and water supply project constructed by the Corps.

To fulfill the flood-control purpose of the 1947 C&SF Project, the Corps built levees and canals to drain large quantities of freshwater away from farms and cities. Completion of the project in 1962 marked the demise of the Everglades natural system, with its vast unspoiled beauty, labeled by Stoneman Douglas as the "river of grass" (Douglas 1947). The C&SF Project radically altered natural flow through the Everglades, subjecting water to regulation schedules for movement and distribution controlled by the Corps (Dewitt 1994, 133–134). The project drastically reduced the quantity of water entering ENP through its major natural canals, Shark River and Taylor Sloughs. The C&SF Project disrupted the natural sheet flow of water from a flooding Lake Okeechobee into the Everglades, which caused massive damage to the natural system south of the lake, including loss of much natural habitat.

A perimeter levee was designed and built by the Corps to protect coastal areas from flooding. State-owned lands west of this levee were designated as water conservation areas (WCAs), undeveloped areas maintained by the SFWMD for the purposes of retaining floodwaters and replenishing the Biscayne aquifer that underlies South Florida. The District leased one of the WCAs to the U.S. Department of the Interior (DOI or Interior) to create the Loxahatchee National Wildlife Refuge. This project also created the Everglades Agricultural Area (EAA) to the south of Lake Okeechobee.

Farming in the Everglades Agricultural Area

In the fifteen years after the C&SF Project went into operation in 1947, sugar production in the United States increased sixfold. This was the result of Congress's embargo on sugar from Cuba and its imposition of quotas to protect the domestic sugar industry. Today, three-quarters of the EAA's more than 1,100 square miles is devoted to producing sugarcane; the Florida crop accounts for slightly more than half of annual U.S. sugarcane production. The advent of sugar production in the EAA brought a chronic problem of phosphorous pollution from farm runoff, carried by what was left of water flow to the Everglades.

The Evolution of an Environmental Approach to the Everglades

The evolution of environmental action to preserve and protect what remained of the natural Everglades began with important steps taken by the federal government and the state of Florida. Perhaps the most important of these steps was creation of the 1.3-million-acre ENP in 1947. Almost twenty-five years later, in 1970, Congress recognized the detrimental effects of massive drainage operations in South Florida and mandated a guaranteed minimum flow of water into the Park. In 1989, Congress added 107,600 acres to the ENP in the Everglades Expansion and Protection Act (Alvarez et al. 2002). The act also directed the Corps to construct modifications to its drainage project to improve water flows into the park and to take steps to restore natural hydrological conditions there.

The state's efforts to correct damage to the Everglades began in earnest in 1983, when the legislature passed the Save Our Everglades Act under the leadership of Governor Bob Graham. The Save Our Everglades (SOE) program introduced two approaches that proved to be fundamental to future efforts to save the Everglades. First, SOE was comprehensive; the program recognized that the entire South Florida ecosystem, not just parts of it, needed to be restored. It called for restoration of the Everglades and Florida Bay and protection of Lake Okeechobee, the WCAs, Big Cypress Swamp, and the habitats of the Florida panther and other endangered species. Second, the program was collaborative. It required that state agencies, the SFWMD, federal agencies, the Florida congressional delegation, and the Florida legislature cooperate to tackle comprehensive restoration. Governor Graham's Republican successor, Bob Martinez, continued SOE, initiating bipartisan support for Everglades restoration, which became a long-standing tradition in Florida.

The state took another important step in 1987 when it adopted the Surface Water Improvement and Management Act (SWIM), requiring Florida's water management districts to develop plans to clean up and preserve the rivers, lakes, estuaries, and bays within their jurisdictions. The SFWMD completed and set up implementation of SWIM plans for Lake Okeechobee and Biscayne Bay. The SWIM plan for the Everglades was more controversial; its development and implementation were later subject to litigation.

Federal Litigation and the Everglades Forever Act

In 1988, the federal government took another pivotal step affecting the future of the Everglades. In Miami, acting U.S. Attorney Dexter Lehtinen sued the state of Florida and the SFWMD, alleging their failure to protect federally owned and leased lands in the Everglades from excessive amounts of phosphorous being discharged from sugar farms located in the EAA south of Lake Okeechobee. Though the phosphorous was produced almost entirely by the sugar industry, the sugar companies were not named as defendants. Still, the defendants, the state, and the SFWMD believed the suit was politically motivated and amounted to an attack on sugar. Ten environmental groups and the Miccosukee Tribe of Indians joined the lawsuit.

One year after the suit was filed, the SFWMD developed a draft SWIM plan for the Everglades. Plaintiffs were critical of the draft plan because it did not establish standards limiting phosphorous in runoff from the EAA. The draft proposed creation of a 40,000-acre artificial wetland (the Everglades nutrient removal project) within the EAA to treat the runoff to reduce its phosphorous content. The sugar industry was concerned at the large acreage of the proposed artificial marsh and asked who would pay for it.

Unsuccessful negotiations among the parties dragged on for several years, costing the SFWMD and the federal government $15 million in attorneys' fees and other litigation expenses. In 1991, newly elected Florida Governor Lawton Chiles conceded publicly that the water was polluted and forced the parties into serious settlement talks. Six months later, in July 1991, the parties signed a settlement agreement, and a consent decree was entered in 1992. The settlement agreement and the consent decree (U.S. 1992, 1567) established interim and long-term total phosphorous concentration limits for Loxahatchee National Wildlife Refuge and ENP. To reduce phosphorous by filtering polluted agricultural runoff, a series of artificial wetlands, called storm water treatment areas (STAs), were created.

Although agricultural interests challenged the settlement and related actions by the state to implement it, in 1993, the largest sugar growers agreed to pay one-third of the more than $800 million cost of the STAs over 20 years. They also agreed to adopt best management practices to reduce phosphorous runoff from their fields. In 1994, the terms of the settlement were incorporated by the Florida legislature in the Everglades Forever Act (EFA). The EFA established a program to improve water quantity and quality in the Everglades, including a schedule for design and construction of the STAs, regulation of phosphorous in the Everglades by Florida's Department of Environmental Protection (DEP), a tax on farmers operating in the EAA to create a fund to help finance the cleanup, incentives for farmers to adopt best management practices to reduce phosphorous outputs, and steps to control the infestation of invasive exotic species. The EFA, together with the federal Water Resources Development Acts (WRDAs) of 1992, 1996, and 2000, form much of the legal basis for current Everglades restoration.

A New Era in Restoration

The settlement of the water quality litigation between the United States and the state of Florida and the SFWMD cleared the way for the parties and other stakeholders to consider the wider issues of Everglades restoration. In 1993, Bruce Babbitt, newly appointed secretary of the U.S. DOI, spoke to the annual meeting of the Everglades Coalition, a consortium of environmental groups. In this speech, Secretary Babbitt laid out a philosophy for the restoration of the environment in South Florida, which formed the framework for much of what followed. He emphasized that restoration should deal with the entire South Florida ecosystem, be science based, and include strong working partnerships among federal, state, local, and tribal government agencies and stakeholders. He proposed, for the first time, that the federal government closely coordinate the work of its agencies and collaborate with the state, the tribes, and other interested parties.

Later in 1993, Secretary Babbitt's proposal came to life, as six federal agencies established an interagency task force, now called the South Florida Ecosystem Restoration Task Force (SFERTF or Task Force), to coordinate federal restoration efforts in the area. In WRDA 1996, Congress formally constituted the SFERTF and added members representing state, local, and tribal governments. While the Task Force was given the authority to coordinate the efforts of its constituent agencies and entities, the lead role for planning and implementation of the restoration fell to the Corps.

By 1994, political leaders in Florida, along with environmentalists and others, agreed that the need for improved water quality and quantity in the Everglades was urgent. At their urging, the Corps carried out a reconnaissance study of how the C&SF Project could be modified to deliver these improvements. That same year, Governor Chiles appointed the Governor's Commission for a Sustainable South Florida (Governor's Commission). After the Commission issued its first report in 1995, the Corps requested its help to design a conceptual plan to improve the water supply and control system of South Florida.

The Comprehensive Everglades Restoration Plan

The Corps and the Governor's Commission created a conceptual plan that the Corps submitted to Congress in August 1996, which still serves as the framework and vision for "getting the water right" in Everglades restoration. The Water Resources Development Act (WRDA 1996) directed the Corps, working with its nonfederal partner, the SFWMD, to use this conceptual plan as the basis for development of a comprehensive restoration plan. In 1999, the Corps and the SFWMD completed the Comprehensive Everglades Review Study (called the

Restudy), the centerpiece of the strategy to restore the Everglades. The Restudy, authorized by Congress in WRDA 2000, is referred to as the Comprehensive Everglades Restoration Plan (CERP or the Plan). CERP is designed (1) to meet the ecosystem's restoration needs relating to the (a) quantity, (b) timing, and (c) delivery of water and (2) to make water available to meet urban and agricultural demands. In 2000, planners estimated that CERP's projects would cost $7.8 billion and that the total cost of restoration—including restoration of water flows, habitat protections, and other components—in South Florida would reach $14.8 billion. The total restoration/water supply effort comprises over two hundred projects, sixty-eight of which are water projects in CERP. Dozens of federal and state agencies are charged with various aspects of the restoration. Designed to take place over several decades, CERP is scheduled for completion by 2050.

Working Together

The individuals who have worked for ecosystem restoration in South Florida have faced much controversy and conflict. They have built a tradition of cooperation over the years, though at times, and particularly during the years since passage of WRDA 2000, restoration partnerships have experienced strains and rifts. Since 1990, two organizations, the SFERTF and the Governor's Commission, have helped the many groups engaged in ecosystem restoration to seek cooperation. The history of these entities provides an instructive view of the challenges in making collaboration work in the context of complex, expensive, long-term projects such as Everglades restoration.

The South Florida Ecosystem Restoration Task Force

When the SFERTF was created in the early 1990s, its members consisted of assistant secretaries of four federal departments—DOI, the Departments of Justice and Commerce, and the U.S. Army—and an assistant administrator of EPA. The federal officials had intended to include representatives of the state and SFWMD but were barred from doing so by the strictures of the Federal Advisory Committee Act (FACA). In WRDA 1996, Congress created a new Task Force (still called SFERTF), which was exempt from FACA's mandates, and expanded membership to include state, local, and tribal representatives. The chair of SFERTF has always been a senior official of the DOI. In its early meetings, the expanded Task Force agreed that it would concern itself with general policy issues and oversight and meet quarterly. It established a management and coordination team based in Florida, called the Working Group, to deal with issues assigned by the Task Force.

The Governor's Commission for a Sustainable South Florida

The Governor's Commission represented the entire panoply of stakeholders in Everglades restoration, including two members from the state legislature, six from various state agencies and organizations, four from regional groups, seven from local government, ten from the business community, and eleven from environmental and public interest groups. In addition, the Commission had five nonvoting members representing the U.S. Department of Commerce's National Oceanic and Atmospheric Administration, the Environmental Protection Agency (EPA), the Corps, the DOI, and the SFERTF. The forty-member Commission shared membership with and interacted effectively with the Working Group. Late in 1994, the executive director of the SFERTF was appointed to the Commission and was able to communicate regularly with its chair and executive director. To maximize interaction among members of the Commission and the Working Group, monthly meetings were often scheduled back-to-back at the same site. The Governor's Commission remained in operation from March 1994 until June 1999, when it was succeeded by the now defunct Governor's Commission for the Everglades.

Between 1994 and 1999, the Commission produced a series of reports addressing restoration issues and providing input from various stakeholders. In 1996, it published *The Conceptual Plan for the C&SF Project Restudy*, containing proposals for improving and expediting the Restudy, based on three categories of objectives: ecological, hydrologic, and socioeconomic. The plan included detailed, ongoing water resource projects, divided into thirteen thematic concepts, to help guide the restoration to success. Additionally, the plan contained several specific recommendations: to expedite the Restudy process; reinforce and broaden its scope and participants; and infuse the process with diversity, complexity, and clarity. The Conceptual Plan called for an improved and more efficient partnership between the federal and state governments. Other reports followed the initial Conceptual Plan with further recommendations, including the *Interim Report on the C&SF Project Restudy* in 1998, the *Restudy Plan Report* in 1999, and the *Report on the Draft January 1, 1999 Implementation Plan of the C&SF Project Restudy* in 1999.

WRDA 1996 vindicated the work of the Governor's Commission because Congress expressly directed the Corps to consider the Commission's *Conceptual Restoration Plan* in its restudy of the C&SF Project. Passage of the WRDA bills with authorization of substantial funding for Everglades restoration would not have been possible without the unanimous support of the large, often contentious, Florida congressional delegation. That unanimity is generally considered to be attributable to the hard-won consensus among government entities and stakeholders achieved through the painstaking work of the Governor's Commission.

Building an Ecosystem Restoration Plan for South Florida

As CERP was being authorized, the SFERTF took up the difficult issue of setting goals for the restoration effort as a whole. The size and complexity of the project, together with its long-term nature, meant that it was crucial for the Task Force to set objectives against which the multiplicity of actions necessary to implement restoration would be planned and assessed. For this reason, in its 2000 report to Congress, the Task Force adopted three broad goals, together with several subgoals, to guide the planning, measuring, and tracking of the project: "The three goals recognize that water, habitat and species, and the built environment must be addressed simultaneously if the ecosystem is to be restored and preserved over the long term" (SFERTF 2000, 14).

Get the Water Right

The first goal of the restoration, in the terms adopted by the SFERTF, is to "Get the water right . . . by restoring natural hydrologic functions and water quality in wetland, estuarine, marine, and groundwater systems, while also providing for the water resource needs of the urban and agricultural landscapes" (SFERTF 2000, 14). This means that the quantity, quality, timing, and distribution of water to the natural system have to mimic as closely as possible pre-drainage conditions (SFERTF 2000, 15). Getting the water right requires getting the hydrology right by (1) capturing freshwater now being drained to the oceans, (2) storing this water, (3) redirecting the water to the natural system, and (4) supplementing urban and agricultural water needs without increased incidence of flooding. The distribution of water to the natural areas must follow the natural hydropattern of extensive flooding in summer and fall, followed by a dry season through winter and spring. Getting the hydrology right is CERP's purpose.

The goal of getting the water right also requires getting the quality of the water right by (1) reducing phosphorous concentration in the runoff from the EAA, (2) treating wastewater, and (3) reducing saltwater intrusion into the estuaries. The agreement has been that water quality improvements would be the responsibility of the state of Florida.

Restore, Preserve, and Protect Natural Habitats and Species

The SFERTF's second goal for Everglades restoration is "to restore, preserve, and protect natural habitats and species." Fifty percent of the natural areas of South Florida have been lost to development in fewer than one hundred years, with the result that a large number of animal and plant species indigenous to the Everglades are currently designated by the USFWS as threatened or endangered under the ESA (SFERTF 2000, 22). One of two subgoals in this category is to "restore, preserve, and protect natural habitats." This is the purpose of the mammoth land acquisition programs undertaken by the state and federal governments in South Florida. A second subgoal is to "control invasive exotic plants," which have spread throughout the Everglades to the extent that some native species could become extinct. This goal will be achieved "when the diversity, abundance, and behavior of native South Florida animals and plants in terrestrial and aquatic environs are characteristic of pre-drainage conditions" (SFERTF 2000, 14).

Foster Compatibility of the Built and Natural Systems

The third goal for Everglades restoration articulated by the Task Force is "to foster compatibility of the built and natural systems" (SFERTF 2000, 24). Unmanaged population growth and rapid urban development threaten to defeat gains in the restoration effort and must be addressed by state and local governments in ways that support achievement and sustainability of the other goals. "This goal will be realized when the built environment is compatible with ecosystem restoration and preservation goals" (SFERTF 2000, 29). Of the three comprehensive restoration goals, this one has received the least attention.

Scope and Diversity of Restoration Projects

Ecosystem restoration in South Florida is not a single undertaking; it is a collection of 226 projects—some started decades ago and some not yet begun. Dozens of agencies are engaged in planning, implementing, and monitoring these restoration projects.

The scope and diversity of ecosystem restoration in South Florida were described in a report by the DOI to Congress in March 2000, entitled *Report to Congress on Restoring the South Florida Ecosystem*, which set the total estimated cost of Everglades restoration at $14.8 billion. The Task Force's report broke down costs according to its three goals for restoration. The cost of reaching goal one, getting the water right, was estimated at $10.4 billion. This cost estimate included nine projects in addition to the $7.8 billion CERP. The largest of these were modified water delivery, then estimated to cost $212 million, and completion of the Kissimmee River restoration at $518 million. The cost of goal two, habitat and species protection, was stated as $4.7 billion; this category consisted largely of land acquisition. The costs attributed to goal three, relating to the built environment, were negligible.

The costs of CERP and the Kissimmee River restoration are to be shared equally by the federal and state governments. Three-quarters of the land acquisition costs are borne by the state of Florida.

Everglades Restoration Science

The Corps and the SFWMD established an interdisciplinary, interagency group of scientists, called the Restoration, Coordination, and Verification (RECOVER) team, to bring science into the planning, assessment, and integration of activities in implementing CERP. The RECOVER team has become the primary entity responsible for the application of scientific knowledge to planning and implementation of CERP projects.

The role of the RECOVER team is twofold: (1) to promote an integrated, systemwide perspective on all essential science and technical issues and (2) to make sure that the best available scientific and technical information and judgments are put forward in development, implementation, and evaluation of the restoration. RECOVER is responsible for scientific activities related to CERP; its work is directed toward the first two of SFERTF's three restoration goals—restoring the flow of water in the natural areas and restoring wetland habitats affected by water management.

In issuing the programmatic regulations for CERP in 2003, the Corps made clear that RECOVER, overseen by the Corps and the SFWMD, is not an independent agency but a group of individuals with scientific and technical expertise who represent various government agencies. To assuage concerns of other governmental stakeholders and to promote confidence in the RECOVER team, the Corps and the SFWMD created the RECOVER Leadership Group consisting of a representative of each of ten state and federal agencies and tribes "to assist the program managers in coordinating and managing the activities of RECOVER" (CERP 2003, 64231). The science assessment function is central to the work of the RECOVER team, which establishes performance measures for assessing progress toward interim and final goals for the hydrologic and biological aspects of CERP. The RECOVER team is constantly updating and improving hydrologic models and other tools and applying them, with revised data and planning assumptions, to predict whether CERP will meet its performance measures in later decades.

Adaptive Management

In authorizing the complex and costly conceptual Everglades restoration plan, Congress understood that it was acting in the face of imperfect scientific information. The Senate committee report accompanying the CERP authorization in WRDA 2000 stated the following:

> One of the key components of the CERP is the inherent flexibility provided by adaptive assessment. Under the adaptive assessment approach, the Plan can be modified, based on any new and improved information or modeling. With a project of this size and duration, it is inevitable that new technologies will emerge, modeling systems will be perfected, and monitoring of the ecosystem will continue to provide up-to-the-minute data on the effectiveness of project components. It is important that these factors be incorporated into the Plan, when the new and improved information will enhance the restoration effort. (CEPW 2000)

To manage uncertainties, scientists use adaptive management—an approach that continuously assesses the ecosystem's responses to changes effected by the Plan and recognizes that the Plan will be modified to accommodate new information, improved modeling, new technologies, and changed circumstances. The Corps' programmatic regulations, which serve as the roadmap for CERP implementation, lay out an elaborate "adaptive management program" (CERP 2003).

Based on scientific and technical information from the RECOVER team, the Corps and SFWMD in consultation with DOI and others are required to develop externally peer-reviewed, periodic reports, assessing responses of the natural system to implementation of the Plan. When these responses fall short of expected results, the programmatic regulations require the agencies to consider whether "the Plan['s] . . . system or project operations or the sequence and schedule of projects should be modified to achieve the goals and purposes of the Plan, or to increase net benefits, or to improve cost effectiveness" (CERP 2003). The regulations further provide that CERP itself be updated on the basis of the most recent scientific and technical information at least every five years (CERP 2003).

Independent Scientific Peer Review

The Task Force has always advocated the view that science must drive policy decisions in Everglades restoration and that policy makers, stakeholders, and citizens must have confidence in the quality and credibility of that science. The hopes for adaptive management in CERP implementation cannot be fulfilled without widespread assurance that Everglades science is the best available. With this in mind, in 1999, the Task Force established a panel of independent scientists, called the Committee for the Restoration of the Greater Everglades Ecosystem (CROGEE), through a contract with the National Academy of Sciences (NAS). With its members designated by the NAS, CROGEE offered credible scientific peer review of CERP during the initial phase of its implementation. CROGEE held its first meeting in December 1999 and drew up a work plan to address five issues: (1) aquifer storage and recovery (ASR) technology, (2) identification of ecological indicators of restoration, (3) the effects of CERP on marine ecosystems, (4) water storage options, and (5) the human dynamics of the restoration (CROGEE 2000). The Task Force approved the first four of the proposed issues but did not agree to the fifth. Disagreement about control of the CROGEE work plan created a lasting strain between the Task Force and the science review panel, and the parties were unable to find a process to agree on the issues to be scrutinized by CROGEE.

In WRDA 2000, Congress required the federal agencies and the governor, in consultation with the Task Force, to "establish an independent scientific review panel convened by a body such as the National Academy of Sciences to review [and report on] the Plan's progress toward achieving the natural system restoration goals" (WRDA 2000, 601 (j)). In 2004, the federal agencies and the state contracted with the NAS to provide an independent scientific review panel to report biennially to Congress on CERP progress. The panel, which has met four times a year beginning in November 2004, reviews progress toward the restoration goals of CERP and publishes a report every other year that assesses "progress in restoring the natural system," discusses "significant accomplishments" and "specific scientific and engineering issues" impacting restoration goals, and provides an "independent review of monitoring and assessment protocols to be used for evaluation of CERP progress" (CISRERP 2004).

The Corps' programmatic regulations for CERP contain several provisions designed to ensure the independence of the scientific review panel, including a requirement that the principal federal agencies, the Task Force and its members, and the governor will not influence the panel in any way (CERP 2003, 385.22(5)). In May 2006, the Scientific Review Panel published its first report (SRP 2006).

Task Force Efforts to Coordinate Science in the Everglades

From its inception, the SFERTF has been concerned with the coordination of science programs and projects connected with the restoration effort. At its first meeting in September 1993, the Task Force created a Science Sub-Group, now called the Science Coordination Group (SCG), made up of sixteen scientists from nine federal agencies. Its November 1993 report provided a conceptual foundation for the restoration and helped to educate policy makers and legislators on the need for adequate funding for Everglades science (Science Sub-Group 1993).

Since the Science Sub-Group's creation, scientists and others associated with the project have been searching for an integrated rationale and strategy to guide Everglades restoration science. Efforts to articulate a unifying theme were included in the 1993, 1996, and 1997 public reports of the Task Force. In 2002, the Task Force issued its most comprehensive formulation of an integrated approach to Everglades restoration science. Appendix E of the 2002 Strategic Plan (SFERTF 2002) proffered an Integrated Science Plan (ISP), defined as "an organizing framework of scientific information and knowledge needed by managers and policy makers" engaged in the restoration.

In its 2003 report, the U.S. Government Accountability Office (GAO) evaluated the SFERTF's efforts at coordinating Everglades science. In testimony before the House Subcommittee on Interior and Related Agencies, Director of Natural Resources and Environment Barry Hill concluded, "The Task Force has yet to establish an effective means of coordination." The report cited three factors limiting the Task Force's effort to carry out its coordination responsibilities: "(i) Lack of clear direction from the Task Force to the SCG on what it was expected to accomplish; (ii) Lack of established processes to ensure that key management issues in science planning are identified or that critical science issues are prioritized; and (iii) Inadequacy of the resources allotted to the effort" (GAO 2003, 2–3).

Late in 2004, the SFERTF, still struggling with the difficult challenge of integrating and prioritizing restoration science, announced that it was using its SCG to develop a plan for coordinating science in two phases. Phase I would focus on (1) creating an approach for identifying programmatic-level science needs and gaps and (2) stimulating and coordinating efforts to bring high quality science to bear to fill those needs and gaps.

Critical Issues

While the public and governmental officials generally support ecosystem restoration in South Florida, serious questions have been raised about some aspects of CERP and wider restoration. These questions have focused largely on five areas: (1) assurances that sufficient water is allocated to the natural system and that other water-related needs are met, (2) water quality issues, (3) reliance of CERP on unproven technologies, (4) the modified water delivery project to enhance flows to ENP, and (5) the viability of the state–federal partnership, the engine of restoration.

Assurances of Water Deliveries to the Natural System

In authorizing CERP, Congress makes clear in WRDA 2000 that the goal of environmental restoration is the primary, though not the only, purpose of the project and will remain so throughout its construction and operation: "The overarching objective of the Plan is the restoration, preservation, and protection of the South Florida ecosystem, while providing for the other water-related needs of the region, including water supply and flood protection" (WRDA 2000, 601(h)).

The issue of assurances in the allocation of water between the needs of the natural system and the demands of urban users and those seeking flood protection was central to the negotiations leading to congressional authorization of CERP and shadows the project still. The DOI, the manager of the national parks and refuges in the Everglades, and environmental groups were concerned that water salvaged by CERP would be diverted from the natural areas to support the relentless urban growth that characterized South Florida. Urban and agricultural interests and the state of Florida made clear that their support of CERP was premised on the agreement that some portion of the water would be made available to meet "the other water-related needs of the region." Designed to provide assurances to urban and agricultural users that their water supply and flood protection levels will be undisturbed by CERP, WRDA 2000 also includes a water savings clause (WRDA 2000, 601(h)(5)).

The state of Florida was adamant that water from CERP be subject to allocation and permits under state law. To assure those concerned about meeting the needs of the natural system, WRDA 2000 provides that, although the water will be allocated under state law, as a condition of future CERP appropriations, the president of the United States and the governor of Florida must enter a "binding agreement," by which "the State shall ensure, by regulation or other appropriate means, that water made available by each project in the Plan shall not be permitted for a consumptive use or otherwise made unavailable by the State until such time as sufficient reservations of water for the restoration of the natural system are made under State law" (WRDA 2000, 601(h)(2)). President George W. Bush and Governor Jeb Bush signed the binding agreement in January 2002.

Frustrated by the slow pace of the federal agencies and Congress in implementing and funding CERP, in October 2004, Governor Bush approved an

SFWMD initiative called Acceler8. This plan targets eight CERP projects for expeditious construction by the SFWMD, including some for water storage, at a cost to the state of $1 billion. Reaction to Acceler8 from environmental groups and DOI has been positive but guarded. Those whose primary interest in CERP is restoring water flows to the natural system want to see the project implemented on a fast track, and the water storage elements are critical to achieving that goal. Others are concerned that the advent of Acceler8 means the federal–state partnership, on which CERP is based, may be unraveling, together with the primacy of the overarching goal of restoring and protecting the natural system.

Water Quality Issues

Water quality was the issue in the 1988 lawsuit brought by the United States against the state of Florida and the SFWMD. The consent decree of 1992 that settled the lawsuit makes the issue of reducing phosphorous loading in agricultural runoff subject to the continuing jurisdiction of the federal district court. Although sometimes a source of friction between the state and federal governments, the consent decree is still in place more than fifteen years later. The state has made substantial progress on this issue, but controversy remains on how and when to take final steps to achieve the agreed-upon standard of no more than 10 parts per billion (ppb) of phosphorous in water entering ENP. Although Governor Bush endorsed the 10 ppb standard, in 2004, the Florida legislature, at the governor's initiative, amended the EFA to roll back the deadline for achieving the 10 ppb standard by 20 years, from 2006 to 2026. This led to enormous outcry from environmentalists and others, and Republicans and Democrats in Congress questioned the state's commitment to CERP and to Everglades restoration itself.

In June 2005, on a motion of the Miccosukee Tribe, the federal district court found the state had violated interim measures to control phosphorous levels mandated by the consent decree. As a result, the court appointed a special master to hold hearings to determine appropriate remedies for the violations.

Phosphorous runoff from the EAA is not the only water quality concern in the region. For example, water entering Lake Okeechobee is contaminated by urban storm water and runoff from cattle ranches. The policy debate centers on the extent to which pollution will be contained at and by the source or cleaned at public expense.

Unproven Technologies Adopted by CERP

When CERP was developed in the late 1990s, engineers and ecologists could not predict how various approaches to restoration would affect the natural system. Acknowledging its inherent uncertainties, CERP was termed a conceptual plan, far less detailed than a typical Corps feasibility proposal. To maintain some level of control in the face of uncertainty, scientists use adaptive management: they continuously monitor and assess the ecosystem's responses to changes and modify their plans to accommodate new data and conditions.

CERP calls for construction of 333 ASR wells to provide massive capacity for water storage underground. Because ASR has never been employed on such a grand scale, serious questions have arisen regarding (1) whether the water recovery rate can be achieved at a sufficiently high level and (2) how water quality and the structural integrity of the aquifers will be affected. CERP also contemplates converting several large limestone quarries into reservoirs by lining them to limit leakage; however, no one can say for sure whether the reservoirs will live up to their planned holding capacities, and water quality questions will have to be answered. The plan also proposes to limit seepage along the eastern edge of the Everglades by various engineering approaches, each of which is attended by still unresolved technical issues.

Modified Water Delivery

The unfulfilled Modified Water Delivery Project (Mod Water) has been such a stubborn obstacle to progress in the Everglades that it has acquired iconic status as a symbol of frustration, delay, and discord among the federal partners and between them and the stakeholders. Responding to this frustration, Congress expressly stated in WRDA 2000 that appropriations would not be made on a number of critical CERP components in the east Everglades until completion of Mod Water, which had been under way for eleven years at that time (WRDA 2000, 601(b)(2)(D)(iv)).

Congress first authorized Mod Water in the Everglades National Park Protection and Expansion Act of 1989 by directing the Corps to modify operation of the C&SF Project to improve water deliveries and restore natural hydrological conditions in the Shark River Slough flow way in the northeast portion of ENP (ENPPEA 1989). Increasing water flow to this area of the Park had the potential to exacerbate natural flooding conditions on some neighboring agricultural and residential areas. Having been given this information, Congress directed the Corps to determine whether the adjacent areas would be adversely affected, and, if so, to undertake mitigation measures to prevent damage from implementing the Mod Water Project. The Corps estimated that the Project would be completed by 1997, a prediction that proved to be unrealistic in the extreme.

For almost two decades, planners for the Corps, DOI, and SFWMD have struggled without success to develop an acceptable plan for Mod Water. Beset by numerous, conflicting demands, they had to find a cost-effective way to provide flood mitigation to the adjacent property owners by means that also permitted increased water flow to the Park, did not impede existing natural water flow elsewhere, and did not threaten or harm the nesting habitat of the endangered Cape Sable seaside sparrow. Residents of the East Everglades who lived adjacent to the Park in what was known as the 8.5 Square Mile Area (8.5 SMA) vocally insisted that Congress, by authorizing Mod Water, had obligated the Corps to provide them full flood protection. These residents had elected to live on the west side of the protective levee, so their lands were naturally flooded for several months during each year's rainy season. They fought implementation of Mod Water unless provisions were made to relieve them from the adverse effects of the project, as well as from their properties' naturally flooded conditions. A long dispute ensued about the level of flood protection the Corps was required and authorized to provide to the residents of the 8.5 SMA.

In the late 1990s, the SFWMD board of directors voted to condemn the 8.5 SMA. The residents fiercely opposed condemnation, aligning themselves with anti–eminent domain movements in other parts of the United States. Under pressure from the 8.5 SMA residents and their advocates, the SFWMD (whose board was reconstituted by Governor Bush when he took office in 1999) reversed itself and elected not to condemn the property. Six more years of struggle over the 8.5 SMA, marked by congressional hearings, litigation, and stalemate, finally ended in an agreement brokered by a local member of Congress, under which the resident holdouts consented to convey their property to the government on very favorable terms.

However, the resolution of the flooding issues in the 8.5 SMA did not lead to the expeditious implementation of Mod Waters. Other intractable issues have emerged to thwart progress. These include whether increased flows will meet state water quality standards and what approach to take in removing Tamiami Trail, a major east–west roadway and an obstacle to reestablishing the Shark River Slough flow way from the WCAs to Everglades National Park. Funding estimates for Mod Water have ballooned from $90 million in 1990 to almost $400 million fifteen years later. By FY 2005, $192 million had been appropriated for constructing and implementing Mod Water. The participants dispute the reasons for the surge in cost, though most agree that changes in the implementation plan and rising land acquisition expenses were factors. Whatever the reasons, the dramatic rise in cost has caused critics to question the viability of the Mod Water Project (Sheikh 2005). The Corps and DOI locked horns over which of them should bear what portion of the funding obligation going forward. Though federal and local officials continue to struggle to find a plan, the stalemate over plans and funding for Mod Water threatens to undermine Everglades restoration as a whole.

State–Federal Relations

With the authorization of CERP in WRDA 2000, the state of Florida and the federal government entered what was intended to be a fifty-fifty partnership. The cost of implementing CERP, then estimated at $7.8 billion, was to be shared equally, as were the responsibilities of planning, constructing, and operating the constituent elements of CERP. The partnership has faced hurdles and stresses in the years since CERP was authorized. One source of tension between the federal and state governments relates to the water quality issue and the ongoing supervision by a federal judge, pursuant to the consent decree discussed earlier, of the state's continuing effort to reach the goal of 10 ppb of phosphorus in water entering the Everglades natural system. Lack of follow-on federal funding has been another source of frustration since Congress authorized CERP.

At the request of the U.S. House Committee on Transportation and Infrastructure, in May 2007, the GAO issued a report on CERP progress (GAO 2007). The report offered four findings of particular importance. First, previous cost estimates for CERP and the wider restoration are shown to be unrealistically low. CERP cost estimates have risen by 15 percent, from $8.8 billion to $10.1 billion, largely because of inflation and expansion in the scope of work for two CERP projects (GAO 2007, 35–36). In 2000, planners estimated that CERP's projects would cost $7.8 billion. The GAO's sole explanation for the $8.8 billion figure in 2007 was that it had "converted all funding data to constant 2006 dollars" (GAO 2007, 34).

The cost estimate for the overall restoration has increased by 28 percent, from $15.4 billion in 2000 to over $19.7 billion in 2007 (GAO 2007, 35). GAO reports that the costs of some components of the restoration, specifically nonconstruction projects related to water supply and water management planning, have not been estimated (GAO 2007, 36). In addition, the full costs of land acquisition for the project are unknown or are known but not reported: "Costs are not estimated due to price escalation and also to avoid adversely impacting ongoing negotiations of land acquisitions" (GAO 2007, 37).

The second finding of the GAO report relates to the comparative funding effort of the state and federal governments. In a significant departure from the idea of an equal partnership, from 1999 through 2006, of the total $7.1 billion tax dollars spent on Everglades restoration, Florida provided $4.8 billion, and the federal government $2.3 billion (GAO 2007, 29). In part, the lower federal funding is explained by the failure of the Bush administration and Congress to enact another WRDA bill since WRDA 2000.

Third, noting that the lack of federal funding and the complexities of planning have put Everglades restoration behind schedule, the report is critical of what it terms the Corps' failure to adopt "overall sequencing criteria [to] guide implementation" of the hundreds of components of the project (GAO 2007, 24). Calling the "correct sequencing of CERP projects . . . essential to the overall success of the restoration effort," the report recommends that the Corps gather the necessary data and then "comprehensively reassess its sequencing decisions to ensure that CERP projects have been appropriately sequenced to maximize the achievement of restoration goals" (GAO 2007, 42). The GAO report also criticizes the mathematical models developed by the Corps for restoration planning on grounds that they are insufficiently interactive with models used by other agencies: "[C]urrent agency efforts are focused on meeting the modeling needs of individual agencies, not on coordinating the efforts and needs of all the agencies involved in the restoration efforts" (GAO 2007, 38).

On a more positive note, the GAO report cites officials of the Corps, DOI, and SFWMD who, despite the delays in funding and implementation they have encountered, "believe that significant progress has been made in acquiring land, constructing water quality projects, and restoring a natural water flow to the Kissimmee River—the headwater of the ecosystem. In addition, many of the policies, strategies, and agreements required to guide restoration in the future are now in place" (GAO 2007 "Highlights").

Lessons Learned: Ten Elements of Success

From 1983, when Florida adopted the Everglades Forever Act, to 2050, the date projected for completion of the last of the CERP projects, sixty-seven years will have passed. Patience and persistence amid political changes and scientific advances are as difficult to sustain as they are essential to ultimate success. The scope and complexity of the restoration, interconnections among its component parts, and continuing scientific uncertainties are daunting. These conditions demand sound science and a strong will to succeed on the part of responsible governments, committed scientists and stakeholders, and informed citizens. Those who have served the Task Force over many years have identified ten elements that have contributed to the progress this ecosystem restoration project has so far achieved.

Collaborative Leadership

Productive periods in the history of ecosystem restoration in South Florida, such as the decade of the 1990s when Congress authorized CERP, have been characterized by high-quality leadership among responsible government and stakeholder entities. Successful leaders exhibit a strong orientation toward collaboration and consensus building. This is evident in their good listening skills, willingness to attend and participate in multiparty meetings, accessibility to colleagues and the public, readiness to ask for and offer help, capacity for dealing collectively with uncertainty, and positive approaches toward limiting and resolving conflict. Within their organizations, these leaders are known for their collaborative way of working, as they look to empower their subordinates by delegating authority and responsibility to them.

Achieving a Shared Vision

The responsible government agencies and major stakeholders must develop a shared vision of what they want the restoration project to achieve. The federal agencies engaged in Everglades restoration used the SFERTF as the means for reaching agreement on CERP. The Governor's Commission for a Sustainable South Florida achieved consensus among the many stakeholders, who often had competing interests.

The process of coming to agreement—meeting together regularly over a period of years and debating every disputed issue—changed attitudes. It drew those who were initially engaged in conflict away from short-term disagreements and established entirely new relationships among them. Participants eventually thought and acted as potential supporters of a common endeavor, as partners rather than adversaries. The debates taught those who had been in conflict to understand each other's concerns, values, and objectives. The process of creating a shared vision over several years resulted in lasting trust and even friendship among the participants, which helped them deal effectively with later conflicts.

Conflict Management

Addressing conflict is essential to progress in these endeavors. Without adequate means of timely conflict management, participants' disagreements can intensify and positions harden, causing stalemates and confidence in their commonalities to weaken. Of course, managing conflict is easier when the parties have a shared vision of what they want to achieve. Beyond that, it is important for agencies and stakeholder groups periodically to review and renew their commitment to that vision, to secure any needed changes, and to engage new leaders in the process. Once those involved in the restoration program agree on a vision of what they want to accomplish, they must create and agree upon multiple methods to address conflicts as they arise.

Making the Case for Restoration

Strong public support is essential to secure legislative funding. The professionals who plan and manage ecosystem projects must continuously make the case for restoration priorities and funding. They must communicate with the public and legislatures in ways that spell out the problems and alternative remedies, justify decisions, and present the material in an accessible style that is informative and convincing to uninitiated audiences. Planners and managers must be thoroughly prepared to answer questions in public forums and to be genuinely responsive to citizens' doubts and needs.

Implementation Capacity

The model for implementing restoration in South Florida distributes responsibilities among a number of government agencies and assigns to SFERTF the role of coordinating plans and activities. The Task Force's Working Group, its modest support staff, and volunteers representing member agencies, carry out most SFERTF functions. Ideally, as volunteers comprehend their agencies' work, they become committed to help achieve project goals. Long-term volunteers usually bring a high degree of expertise and experience to the Task Force. The drawbacks of this model are problems typical in reliance on volunteers: their levels of interest and participation may lag, their work priorities lie elsewhere, and they lack accountability to the Task Force. After years of experience with this model, those in charge of the Task Force have learned that they cannot rely exclusively on volunteers and have moved to add support staff and paid contractors to their team.

Organizational Adaptability

The agencies responsible for implementation of the project over decades must respond to an ever-widening range of social, political, financial, science, and public policy issues. Their willingness, with the approval of their governing bodies, to reform organizational structures and add new ones has been essential for progress. For example, CERP implementation has led to the creation of new entities, such as the RECOVER team and the SCT.

Bipartisan Support

One reason that ecosystem restoration has progressed over the years in South Florida is that leaders of both political parties at the state and federal levels have supported restoration and have coordinated their efforts to shape and secure passage of legislation to authorize and fund restoration programs. Widespread popular support for restoration of the Everglades in Florida and throughout the United States has helped engender bipartisan agreement.

Bold Ideas

A number of bold ideas have inspired ecosystem restoration in South Florida and, in turn, have fueled legislation, restoration planning, and implementation structures and practices. The boldest, most crucial idea has been the necessity of an ecosystem-wide approach to restoration. By challenging the tradition of compartmentalized problem solving, this holistic approach has promoted systemic rather than isolated solutions, requiring that restoration practices be integrated to the greatest possible degree and that decisions be guided by the long view. The ecosystem-wide approach is necessarily based on sound science because it calls for ever greater understanding of the causes of environmental problems, analysis of how components and natural processes interact, and development of more imaginative options and wiser investments in solutions than those resulting from a scattered, piecemeal approach.

Science Coordination

Scientists working on the restoration project have learned that they must coordinate their activities to manage the interconnections within the natural system. This coordination involves communication among scientists and between scientists and nonscientist policy makers, agreement on research priorities and sharing of research, and collaboration on various programs and projects.

Coordination is necessary to help assure that a scientific solution crafted for one environmental problem will not aggravate another. When widespread problems have occurred throughout the ecosystem, information sharing has accelerated research. To assure quality and resolve important scientific disputes where they arise, independent scientific peer review has been useful.

When science coordination has worked well in South Florida, some or most of the following conditions have been present: research objectives are clearly stated and achievable; capable team leaders are assigned, and efforts are well organized; protocols are established and agreed to; team scientists have the appropriate expertise, experience, equipment, and staff assistance and are supported in their work by their agency employer; differences among scientists are handled effectively; and reasonable schedules have been established and met.

Stakeholder Relations

The agencies engaged in South Florida ecosystem restoration have placed a high priority on communication with stakeholders. The SFERTF and its Working Group have expanded their membership over time. The agencies generally operate in a transparent manner, are careful to provide information to the public on a continuous basis, and have cultivated relationships with stakeholder organizations. In the future, if conflicts intensify, confidence in the progress of the restoration declines, consensus on how to proceed weakens, or opposition arises, the agencies will need to strengthen efforts to engage stakeholders and the general public. In the meantime, the Task Force and its members must continue to attend to stakeholder and public attitudes and concerns.

References

Alvarez, C., M. Eng, and A. Mayes. 2002. *Assessment of Opportunities for Multi-Stakeholder Collaboration on the Environmental Impact Statement Process for the Combined Structural and Operational Plan for Modified Water Deliveries to Everglades National Park and C-111 Canal Projects.* November 12. Tucson: U.S. Institute for Environmental Conflict Resolution. Prepared for the U.S. Army Corps of Engineers, Everglades National Park, South Florida Water Management District, and U.S. Fish and Wildlife Service. At www.ecr.gov/pdf/everglades_final_report.pdf.

Committee on Environment and Public Works (CEPW). 2000. *Restoring the Everglades, An American Legacy Act. Report of the Committee on the Environment and Public Works United State Senate together with Minority Views to Accompany S. 2797.* July 27. Washington, DC: U.S. Government Printing Office.

Committee on Independent Scientific Review of Everglades Restoration Project (CISR-ERP). 2004. *National Academies Project: Independent Scientific Review of Everglades Restoration Project.* Washington, DC: National Research Council (NRC) of the National Academies Press. At www8.nationalacaemies.org/cp/Projectview.asp?key=95.

Committee on Restoration of the Greater Everglades Ecosystem (CROGEE). 2000. "Potential Work Plan Items for the Committee on Restoration of the Greater Everglades Ecosystem." Appendix E of letter from Stephen D. Parker, director, Water Science and Technology Board, The National Academies, to Col. Terrence "Rock" Salt, executive director of the South Florida Ecosystem Restoration Task Force. January 5. At www.sfrestore.org/crogee/jan500letter.pdf.

Comprehensive Everglades Restoration Plan (CERP). 2000. "Why Restore the Everglades? Part 4, Four Factors: Quality, Quantity, Timing, & Distribution." At www.evergladesplan.org/about/why_restore_pt_04.aspx.

CERP. 2003. "Programmatic Regulations." 33 CFR 385: 64231. At www.epa.gov/fedrgstr/EPA-IMPACT/2003/November/Day-12/;27968.htm.

DeWitt, J. 1994. *Civic Environmentalism: Alternatives to Regulation in States and Communities.* Washington, DC: Congressional Quarterly Press.

Douglas, M. S. [1947] 1974. *The Everglades: River of Grass.* Sarasota, FL: Pineapple Press.

Environmental Protection Agency (EPA). 2003. "Overview of Total Phosphorus TMDL Lake Okeechobee, Florida, 2003." At www.epa.gov/region04/water/tmdl/florida/lake_o/okee_phos.pdf.

Everglades National Park Protection and Expansion Act (ENPPEA) of 1989, P.L. 101-229. At http://planning.saj.usace.army.mil/envdocs_A-D/Dade_Co/Tamiami_Trail/B2B/AppendicesA-E.pdf.

Florida Department of Environmental Protection (FDEP). 2006. At www.dep.state.fl.us/lands/invaspec/2ndlevpgs/faq.htm.

Gawlik, D. E., and F. Sklar. 2000. *Effects of Hydrology on Wading Bird Foraging Parameters Report: Critical Ecosystems Studies Initiative Task* 3.Washington, DC: U.S. Geological Survey (USGS). At http://sofia.usgs.gov/publications/reports/wading_bird/.

Governor's Commission for a Sustainable South Florida. 1996. *The Conceptual Plan for the C&SF Project Restudy*. At www.evergladesplan.org/docs/comp_plan_apr99/sect6.pdf.

Governor's Commission for a Sustainable South Florida. 1998. *An Interim Report on the C&SF Project Restudy*. August 11. The report was forwarded to Governor Lawton Chiles by letter, dated July 27, 1998, from Richard Pettigrew, chairman of the Governor's Commission. Coral Gables, FL: Governor's Commission for a Sustainable South Florida.

Governor's Commission for a Sustainable South Florida. 1999. *Restudy Plan Report*. January 20. The report was forwarded to Governor Jeb Bush by letter, dated January 27, 1999, from Richard Pettigrew, chairman of the Governor's Commission. Coral Gables, FL: Governor's Commission for a Sustainable South Florida.

Governor's Commission for a Sustainable South Florida. 1999. *Report on the January 25, 1999 Draft Implementation Plan of the C&SF Project Restudy*. March 3. The report was forwarded to Governor Jeb Bush by letter, dated March 19, 1999, from Richard Pettigrew, chairman of the Governor's Commission. Coral Gables, FL: Governor's Commission for a Sustainable South Florida.

Grunwald, M. 2002. "An Environmental Reversal of Fortune: The Kissimmee's Revival Could Provide Lessons for Restoring the Everglades." *Washington Post*, June 22, A01. At www.everglades.org/washingtonpost_com%20An%20Environmental%20Reversal%20of%20Fortune.htm.

Hu, C., K. Hackett, M. K. Callahan, S. Andréfouët, J. L. Wheaton, J. W. Porter, and F. E. Muller-Karger. 2003. "The 2002 Ocean Color Anomaly in the Florida Bight: A Cause of Local Coral Reef Decline?" *Geophysical Research Letters* 30:3, 1151. USDOI. At http://imars.usf.edu/~hu/papers/black_water/Hu_GRL_ocean_color_anomaly.pdf.

Lake Okeechobee.org. n.d. "Geology and Hydrology." At www.lakeokeechobee.org/content.php?section=about_the_lake&page=about_the_lake/science.html.

McIntosh, P. 2002. "Reviving the Everglades." *National Parks Magazine*, January/February. Washington, DC: National Parks Conservation Association (NPCA). At www.npca.org/magazine/2002/january_february/everglades.html.

Scientific Review Panel (SRP). 2006. *Report of the Review Panel Concerning Indicators for Restoration: Report to the South Florida Ecosystem Restoration Task Force*. May 8. At ww.sfrestore.org/scg/documents/indicators%20REPORT%20final%20from%20Jordan%20MAY%208%202006_w_line%20numbers.pdf.

Science Sub-Group. 1993. The South Florida Management and Coordination Working Group. *Federal Objectives for the South Florida Restoration*. November 15. At www.sfrestore.org/crogee/ra2/p1_31.html.

Sheikh, P. A. 2005. *Everglades Restoration: Modified Water Deliveries Project*. August 23. Washington, DC: Library of Congress, Congressional Research Service. At www.cnie.org/nle/crsreports/05aug/RS21331.pdf.

South Florida Ecosystem Restoration Task Force (SFERTF). 1993. *Science Sub-Group Report: Federal Objectives for the South Florida Restoration*. November. Miami: SFERTF. At www.sfrestore.org/sct/docs/subgrouprpt/index.htm.

SFERTF. 1998. *Integrated Financial Plan*. June 24. Miami: SFERTF. At www.sfrestore.org/documents/work_products/ifp_1998.pdf.

SFERTF. 2000. *Coordinating Success: Strategy for Restoration in the South Florida Ecosystem. July* 31. At www.sfrestore.org/documents/isp/sfweb/sfindex.htm.

SFERTF. 2002. *Tracking Success: Biennial Report for FY 2001-2002 of the South Florida Ecosystem Restoration Task Force to the U.S. Congress, Florida Legislature, Seminole Tribe of Florida, and Miccosukee Tribe of Indians of Florida*. Appendix E. August. Miami: SFERTF. At www.sfrestore.org/documents/work_products/2002_strategic%20plan_volume%20II.pdf.

South Florida Water Management District (SFWMD). n.d. "Kissimmee River Restoration Past/Present." At www.sfwmd.gov/org/erd/krr/pastpres/krrchanl.html.

United States (U.S.) *v. South Florida Water Management District* (SFWM), 847 F. Supp. 1567 (S.D. Fla. 1992).

U.S. Census Bureau (USCB). 2007. State and County QuickFacts. Washington, DC: U.S. Census Bureau. At http://quickfacts.census.gov.

U.S. Department of the Interior (DOI) and Science Sub-Group of the South Florida Management and Coordination Working Group. 1993. *Federal (DOI)Objectives for the South Florida Restoration*. November 15. At www.sfrestore.org/crogee/index.html.

U.S. DOI. 1999. "Fish Health in the St. Lucie Estuarine System." Results of the Florida Marine Research Institute Study and NOAA Fisheries Study. Cited in South Florida Information Access-South Florida Restoration Science Forum Web site, May, hosted by the U.S. Geological Survey, South Florida Water Management District, and U.S. Army Corps of Engineers. Washington, DC: DOI, USGS, Center for Coastal Geology. At http://sofia.usgs.gov/sfrsf/rooms/coastal/stlucie/fish/index.html.

U.S. DOI. 2000. Report to Congress on Restoring the South Florida Ecosystem. Washington, DC: DOI.

U.S. General Accountability Office (GAO). 2003. *South Florida Ecosystem Restoration: Improved Science Coordination Needed to Increase Likelihood of Success*. GAO-03-518T. Washington, DC: U.S. General Accounting Office. Statement cited from the report in testimony by Barry Hill, director of Natural Resources and Environment, to the U.S. House of Representatives' Subcommittee on Interior and Related Agencies, Committee on Appropriations, March 26. At www.gao.gov/new.items/d03518t.pdf.

USGAO (GAO). 2007. *South Florida Ecosystem Restoration Is Moving Forward but Is Facing Significant Delays, Implementation Challenges, and Rising Costs*. Washington, DC: GAO. At www.gao.gov/new.items/d07520.pdf.

Water Resources Development Act (WRDA) of 1992. 106 Stat. 4863, Public Law No. 102-580. October 31. At www.usace.army.mil/cw/cecwp/leg_manage/pdf/wrda92.pdf.

WRDA 1996. 110 Stat. 3658, Public Law No. 104-303. October 12. At www.fws.gov/habitatconservation/Omnibus/WRDA1996.pdf.

WRDA 2000. 114 Stat. 2687, Sec. 601 (j), (h). Public Law No. 106-541. December 11. At www.fws.gov/habitatconservation/Omnibus/WRDA2000.pdf.

Chapter 2

Everglades Ecology

The Impacts of Altered Hydrology

THOMAS L. CRISMAN

In her work, *The Everglades: River of Grass* (1947), Marjory Stoneman Douglas painted a literary portrait of the Everglades that emblazoned images of this vast system into the public mind, which resulted in federal protection of significant sections of this unique area. Stoneman Douglas depicted the Everglades, beginning at the southern shore of Lake Okeechobee, as a vast river of grass, dominated by sawgrass (*Cladium jamaicense*), interspersed with a mosaic of islands and sloughs (linear swamps), flowing gently south to Florida Bay. Far too many people, however, locked onto the image of a river of grass and lost sight of the extreme habitat diversity of the system.

In reality, the Everglades is a series of interconnected ecosystems variously linked hydrologically over temporal scales from seasonal to decadal, punctuated by occasional catastrophic climatic events, including extended drought and hurricanes. The Everglades is the most extensively studied large wetland system in the world. A literature search in May 2006 identified 1,299 refereed papers for the Everglades, followed by 367 for Donana in Spain and 335 for the Pantanal of Brazil, the largest wetland in the world (Web of Science 2006).

Prior to human modifications over the past two centuries, the Everglades watershed comprised the Kissimmee River, Lake Okeechobee, and the Everglades wetland. From its origin just south of Orlando, the Kissimmee River meandered across a broad floodplain and, together with Fisheating Creek from the west, discharged into Lake Okeechobee from the north. Lake Okeechobee, the largest lake in Florida, is the third largest lake situated entirely within the United States. Consisting of a central, shallow, open water area surrounded by a broad, fringing wetland, Lake Okeechobee has likely been eutrophic (having excessive aquatic plant growth in response to particular nutrients) since its formation (Gleason and Stone

1976). Prior to human manipulation, a sill along the southern lake margin restricted surface water flow from the eutrophic lake southward to the oligotrophic (low in plant nutrients with high oxygen levels) Everglades wetland to periods of extremely high water, hurricanes, and a few vestigial creeks that breached the sill periodically (Brooks 1974; Gleason and Stone 1994). Although the lake and the Everglades were hydrologically connected via groundwater and periodically via surface water, the two systems were biologically distinct.

The northern margin of the Everglades wetland was the shoreline of Lake Okeechobee south of the sill, and the southern margin was the mangrove fringe on the northern shore of Florida Bay (Davis and Ogden 1994). The preimpact (prior to human alteration) Everglades was a complex of sloughs, tree islands, wet prairies, and alligator holes. Deeper water sloughs and alligator holes were characterized by several oligotrophic macrophytes (visible aquatic plants) along a depth gradient with sawgrass aerially dominant. Primary production in shallow, wet prairies was mainly from benthic algal mats, representing a gradient of community types from low pH soft water, to high pH alkaline water, the distribution of which reflected organic matter accumulation, hydroperiod, and direct interaction with bedrock limestone. There is no indication that phytoplankton (microscopic algae) production in the water column was significant in any portion of the preimpact Everglades. The Everglades displayed pronounced vertical and horizontal habitat heterogeneity in its preimpact state. Vertical relief to the system was provided by solution features in bedrock limestone and alligator holes and trails. The extent of the Everglades included the areas that increased and decreased in direct reflection of water input to the system. Duration (how long the water stayed at a given level) was controlled by seasonal wet and dry cycles and storm events, including hurricanes and fire.

The history of the Everglades for the past two centuries has been one of progressive size reduction, compartmentalization, fragmentation, and general abuse. Significant changes have occurred in the ecosystem's structure and function, associated with altered hydrology, nutrient loading, fire frequency and intensity, and introduction of biological and chemical contaminants. While impacts and restoration plans in the Kissimmee River and Lake Okeechobee are critical to landscape and water management of South Florida, this chapter focuses on the Everglades' wetlands proper, that portion of the system extending from the southern shore of Lake Okeechobee south to Florida Bay.

Hydrologic Alterations

Several major engineering actions have altered the hydrology of the Everglades significantly: channelization of the Kissimmee River, construction of the Herbert

Hoover Dike around Lake Okeechobee, building of roads across the Everglades, drainage of wetlands south of Lake Okeechobee to become the Everglades Agricultural Area (EAA), partitioning of the Everglades into diked water conservation areas (WCAs), and development of the interconnective, extensive canal system throughout the basin. Channelization of the Kissimmee River increased water flow and volume to Lake Okeechobee, while construction of the Hoover Dike decades earlier had stopped the lake from topping its natural berm during peak hydrologic events (Grunwald 2006). Within the Everglades proper, WCAs were constructed to impede water flow south and serve as storage basins, whereas the canal system throughout the basin was developed to speed delivery of water for flood protection and agricultural drainage.

Although ongoing research is discovering that Native Americans prior to European settlement constructed canals through the Everglades for transport (Grunwald 2006), this pales in comparison with the interconnected network of variously sized canals begun in the late 1800s and expanded exponentially throughout the twentieth century, which increased delivery of water to the sea east and west from Lake Okeechobee through the St. Lucie Canal and Caloosahatchee River, respectively (Light and Dineen 1994). Other engineering actions compartmentalized and increased water storage within the Everglades proper, backpumped excess agricultural drainage north into Lake Okeechobee and south into WCAs, and generally reduced water delivery to the southern terminus of the Everglades, Florida Bay.

While canals increase habitat heterogeneity and ecosystem connectivity, they profoundly alter the structure and function of the wetlands through which they pass. The larger, deeper canals serve as faunal refugia (places where organisms can exist or be protected) during drought periods and usually have significantly higher dissolved oxygen concentrations than surrounding wetlands (Rader and Richardson 1994). Such canals also serve as biotic conduits, providing habitat pathways along which exotic and invasive species of flora and fauna can disseminate over broad geographic regions. Attracted to the canals' food sources, animals fleeing desiccating wetlands may prey upon and become the prey of other species.

Canals receive elevated loadings of pesticides, herbicides, heavy metals, and nutrients from urban storm water and agricultural runoff (Miller and Mattraw 1982) that have the potential to alter algal and microbial communities significantly (Downing et al. 2004). Canals, in turn, can become significant point source contributors of these chemicals to Lake Okeechobee and Everglades wetlands, especially to the WCAs, where surface water is stored for long periods. Gradients of decreasing phosphorus in both surface water and sediments have been noted from canal entry points into receiving wetlands (DeBusk et al. 1994), and these are usually accompanied by profound changes in macrophyte and attached algal productivity and species composition, including a shift in plant dominance to dense, near monocultures of the invasive macrophyte, cattails (*Typha domingensis*) (Childers et al. 2003).

Human modification of hydrology within the Everglades wetlands has resulted in profound changes in the structure and function of the ecosystem. Construction of an extensive system of canals and levees has significantly altered hydroperiods and fire regimes (King et al. 2004). In addition to concern for deep, burning peat fires in drained lands (Smith et al. 2003), the dominant native macrophyte, sawgrass, is hindered by increased fire frequency, especially if followed by a period of reflooding (Herndon et al. 1991). However, cattails appear to flourish following fire, in part because of increased nutrient availability (Smith and Newman 2001), whereas the exotic invasive, Brazilian pepper (*Schinus terebinthifolius*), invades drained habitats regardless of fire regime (Whiteaker and LaRosa 1991). Altered hydroperiods in portions of the Everglades and increased cattail dominance are estimated to have significantly lowered fire frequency and the area burned during higher water years, but have significantly increased fire size during drought periods (Wu et al. 1996). Fire is of special concern during drought periods because of persistent peat fires, as well as its potential impact on wading bird reproduction (Smith et al. 2003). However, deaths of adult birds trapped by flames while foraging can be more profound than destruction of nests if the latter are built in trees in the remaining wet areas (Epanchin et al. 2002).

Altered fire regimes also partially affect how the ecosystem will respond to natural catastrophic events, such as hurricanes. Dry season fires strongly influence the response of pine to subsequent hurricanes (Platt et al. 2002). Hurricanes open patches in old growth stands of trees, but beneficial, high-intensity fires slow growth of young trees and favor larger trees that are less susceptible to hurricanes (Platt et al. 2000).

In addition to pronounced changes horizontally within the landscape, major alterations of vertical habitat heterogeneity and quality have resulted from human manipulation of hydrological regimes in the Everglades. While the extensive human-constructed canal system within the Everglades has provided both horizontal connectivity and potential animal refugia during droughts, the impact has been equally negative by facilitating broad distribution of exotic fish species in the system. Tree islands separated by deeper slough areas were a common feature that reflected natural hydroperiod and regional flow patterns of surface water to the south.

Areas subject to human hydrologic modification, promoting higher water and longer hydroperiods, have experienced a reduced size of tree islands, while areas shifting to shorter hydroperiods have seen increases in both the number and the size of islands (Brandt et al. 2000). Macrophyte communities in sloughs are a clear reflection of local hydroperiods, with shallow water communities dominated by emergent macrophytes and deep-water areas by floating leaved and submersed plant species. Fish assemblages display distinct differences between sloughs and shallow wet prairies, with densities higher in areas of greater plant biomass and lower in deep-water areas (Jordan et al. 1998). The main predators of such wetland fish, wading birds, display a clear high-water tolerance level, above which bird numbers decline (Bancroft et al. 2002). Wading bird populations are favored by reduced cattails and increased slough habitats with characteristic vegetation.

Throughout the Everglades landscape, solution holes are common where the karstic (degraded from dissolution in water) limestone is exposed. Most solution holes are less than 46 centimeters (cm) deep and serve as important drought refugia for native fish (Kobza et al. 2004). Any drop in groundwater levels below 46 cm will force small native fish to the few remaining deeper karstic holes that are favored by exotic fish species, thus increasing predation pressure on the native assemblage.

Alligators actively excavate depressions in the Everglades where extensive organic sediments have accumulated. Palmer and Mazzotti (2004) identified three types of alligator holes in the Everglades: shrub holes, dominated by shrub/tree vegetation; marsh holes, dominated by marsh plant species; and cattail holes, dominated by *Typha* species. With the exception of cattail holes, which tended to be the deepest, all hole types displayed higher plant diversity than the higher, surrounding marsh. Alligator holes have plant species that are often missing from the adjacent wetland (Campbell and Mazzotti 2004). Hole area and depth are positively correlated with more abundant small alligators in smaller holes to avoid predation from large alligators that frequent larger, deeper holes. Regardless of origin, vertical habitat heterogeneity must be an integral part of any wetland management plan (Crisman et al. 2005).

Hydroperiod alteration as a result of human manipulation has had a profound impact on all components of the Everglades food web. The extent of macrophyte cover in the southern Everglades is inversely related to water depth, with periphyton (algae attached to a surface, such as on plants) being favored in extremely shallow water and calcareous (containing calcium carbonate) substrates (Busch et al. 1998). Although desiccated periphytic mats (masses of algae attached to the bottom of a wetland) rapidly recover upon rehydration and can effectively strip phosphorus from the water column (Thomas et al. 2006), prolonged drying alters the composition of algal species and increases nutrient release upon rewetting (Gottlieb et al. 2005).

Sawgrass was the dominant native in oligotrophic, shallow water areas of the northern Everglades (Lorenzen et al. 2001). Sawgrass displays slow growth and phosphorus uptake and has been replaced by cattails as the dominant emergent macrophyte in portions of the northern Everglades that have experienced higher water levels and nutrient loadings. Sawgrass can sometimes outcompete cattails in nutrient-enriched areas if water levels are kept low (Urban et al. 1993), but cattails become dominant in dried areas with frequent muck fires (Smith and Newman 2001).

Benthic invertebrates (aquatic animals without backbones) are a key energy linkage in the aquatic food web of the Everglades, from primary production and detritus to fish and, in the case of crayfish, to birds, reptiles, and terrestrial fauna. Invertebrates are also the most poorly investigated component of the Everglades. Extremely limited data suggest that any future hydroperiod alterations that break the predation link between zooplankton (microscopic, free-swimming animals within the water column) and higher trophic levels could result in major structural and functional changes in the food web of the Everglades.

The response of birds to altered hydrology is perhaps the most complicated relationship within the Everglades food web. The number of nests of white ibis (*Eudocimus albus*) and wood stork (*Mycteria americana*) have decreased by 87 and 78 percent, respectively, since the 1930s, while the percentage of total wading bird nests in the southern Everglades has decreased by 61 percent and increased by 41 percent in the central and northern Everglades (Crozier and Gawlik 2003). In contrast, the numbers of nests of the great egret (*Ardea alba*) have increased during the period. Optimal hydrological conditions for wading birds' successful breeding include intermediate water levels at the beginning of the dry season, a rapid rate of drying, and no disruption in the drying process (Russell et al. 2002). Breeding success does not appear to be significantly affected by fire (Epanchin et al. 2002), but increased water levels during the dry season as a result of human manipulation can affect foraging abilities of birds and result in lowered success (Russell et al. 2002).

Select wading birds, however, can exhibit exceptionally high productivity for up to two years following a periodic severe drought period (Frederick and Ogden 2001). In addition to altered hydroperiod, wading bird breeding success is affected by loss of habitat heterogeneity, especially sloughs (Fleming et al. 1994), excessive cattails, trophic state (Crozier and Gawlik 2002), and prey availability (Fleming et al. 1994). Gawlik and Rocque (1998) also found that forest and marsh bird species richness varied significantly among vegetation types within relatively unimpacted wetlands but not for areas experiencing reduced hydroperiods.

The response of other wildlife components to altered hydrology is less understood, but a major modeling effort is integrating these components into both systemwide (DeAngelis et al. 1998) and ecosystem-specific models (Ogden 2005). Even at the southern extremity of the Everglades, the endangered American crocodile (*Crocodylus acutus*) has demonstrated a great sensitivity to high salinity when young; thus populations would benefit from a return to increased delivery of freshwater to Florida Bay (Richards et al. 2004).

In its preimpact state, the Everglades displayed a great deal of vertical and horizontal habitat heterogeneity that was reflected in a mosaic of variously connected food webs. Essential to maintenance of this complex ecosystem was a seasonally divergent hydroperiod of wet and dry cycles, punctuated periodically by extreme events of drought and hurricanes. Hydrology, in part, will determine the dominant primary producer in both the northern (sawgrass versus cattails) and southern (calcareous algal mats versus floating algal mats) Everglades, as well as the structure and function of invertebrate, fish, and bird communities. However, the secondary and at times primary control of nutrient loading over Everglades food webs cannot be discounted. Originally an oligotrophic system, cultural eutrophication via human modification of nutrient loadings from agricultural and urban sources has vastly changed the character of major portions of the Everglades.

Restoration versus Rehabilitation

Restoration implies a return to the structure and function of a previous condition. Given the spatial and temporal dynamics of the Everglades, this lofty goal might be approached more effectively through rehabilitation, establishing achievable goals for ecosystem structure and function within the context of a sliding scale of environmental dynamics and a realization that complete restoration of the Everglades is no longer possible.

Even if nutrient, pesticide, herbicide, and heavy metal loadings to the Everglades could be curtailed; cattails brought under control; and the hydrologic cycle returned to something more in line with preimpact conditions; an extended lag time would take place before the system could respond positively. As with spatial alterations, profound changes in temporal scale components affecting the Everglades have taken place. Sorting out natural and human-induced climate changes and their interrelationships is a field of study still in its infancy, but it is clear that Florida is undergoing climate change, with mean winter temperatures becoming warmer over at least the past thirty years. In addition, the sea level is rising progressively, and changes have occurred in the frequency and intensity of hurricanes. The long-term response of the hydrologic cycle is less clear, perhaps because of the scale of current models, but the reality remains that human demands upon a finite water resource in

South Florida will only increase through time, decreasing water availability for the Everglades.

The Comprehensive Everglades Restoration Plan (CERP) is a reasonable starting point for ecosystem rehabilitation (see chapter 1, this volume), one that recognizes the sliding temporal and urbanization scales and incorporates them conceptually as part of adaptive management. Living within the constraints of a finite water resource as human demands increase is the reality of South Florida, and it is essential for ecologists and ecosystem managers to determine minimum water requirements for maintaining the resistance and resilience of the Everglades' structure and function. To the trained eye, the Everglades has little resemblance to the system that Stoneman Douglas heralded in her influential book (1947) or what eventually was gazetted as the Everglades National Park.

People's perception of environmental baseline conditions changes with every human generation. What was considered the inviolate structure of the Everglades fifty years ago is beyond the collective memory of today's society. Restoration of ecosystem is much more difficult to achieve than function, which can be performed by a variety of species. Increasingly, management of the Everglades must focus on critical functional aspects that are obtainable under a variety of biotic structural scenarios. In this regard, embracing agriculture, as well as other land-use practices, as partners in the management equation is critical for success. As recognized by Weisskoff (2005), restoration of the Everglades must be viewed within a sound economic sustainability plan. A new mantra must be adopted that recognizes good ecology is good business.

References

Bancroft, G. T., D. E. Gawlik, and K. Rutchey. 2002. "Distribution of Wading Birds Relative to Vegetation and Water Depths in the Northern Everglades of Florida, USA." *Waterbirds* 25:265–77.

Brandt, L. A., K. M. Portier, and W. M. Kitchens. 2000. "Patterns of Change in Tree Islands in Arthur R. Marshall Loxahatchee National Wildlife Refuge from 1950 to 1991." *Wetlands* 20:1–14.

Brooks, H. K. 1974. "Lake Okeechobee." In *Environments of South Florida: Present and Past. Memoirs of the Miami Geological Society*, ed. P. J. Gleason. Miami: Miami Geological Society.

Busch, D. E., W. F. Loftus, and O. L. Bass. 1998. "Long Term Hydrologic Effects on Marsh Plant Community Structure in the Southern Everglades." *Wetlands* 18:230–41.

Campbell, M. R., and F. J. Mazzotti. 2004. "Characterization of Natural and Artificial Alligator Holes." *Southeastern Naturalist* 3:583–94.

Childers, D. L., R. F. Doren, R. Jones, G. B. Noe, M. Rugge, and L. J. Scinto. 2003. "Decadal Change in Vegetation and Soil Phosphorus Pattern across the Everglades Landscape." *Journal of Environmental Quality* 32:344–62.

Crisman, T. L., C. Mitraki, and G. Zalidis. 2005. "Integrating Vertical and Horizontal Approaches for Management of Shallow Lakes and Wetlands." *Ecological Engineering* 24:379–89.

Crozier, G. E., and D. E. Gawlik. 2002. "Avian Response to Nutrient Enrichment in an Oligotrophic Wetland, the Florida Everglades." *Condor* 104:631–42.

Crozier, G. E., and D. E. Gawlik. 2003. "Wading Bird Nesting Effort as an Index to Wetland Ecosystem Integrity." *Waterbirds* 26:303–24.

Davis, S. M., and J. C. Ogden. 1994. "Introduction." In *Everglades: The Ecosystem and Its Restoration*, ed. S. M. Davis and J. C. Ogden. Delray Beach, FL: St. Lucie Press.

DeAngelis, D. L., L. J. Gross, M. A. Huston, W. F. Wolff, D. M. Fleming, E. J. Comiskey, and S. M. Sylvester. 1998. "Landscape Modeling for Everglades Ecosystem Restoration." *Ecosystems* 1:64–75.

DeBusk, W. F., K. R. Reddy, M. S. Koch, and Y. Wang. 1994. "Spatial Distribution of Soil Nutrients in a Northern Everglades Marsh: Water Conservation Area 2A." *Soil Science Society of America Journal* 58:543–52.

Douglas, M. S. 1947. *The Everglades: River of Grass*. Atlanta: Mockingbird Books.

Downing, H.F., M.E. Delorenzo, M.H. Fulton, G.I. Scott, C.J. Madden, and J.R. Kucklick. 2004. "Effects of the Agricultural Pesticides Atrazine, Chlorothalonil and Endosulfan on South Florida Microbial Assemblages." *Ecotoxicology* 13: 245–60.

Epanchin, P. N., J. A. Heath, and P. C. Frederick. 2002. "Effects of Fires on Foraging and Breeding Wading Birds in the Everglades." *Wilson Bulletin* 114:139–41.

Fleming, D. M., W. F. Wolff, and D. L. DeAngelis. 1994. "Importance of Landscape Heterogeneity to Wood Storks in Florida Everglades." *Environmental Management* 18:743–57.

Frederick, P. C., and J. C. Ogden. 2001. "Pulsed Breeding of Long-Legged Wading Birds and the Importance of Infrequent Severe Drought Conditions in the Florida Everglades." *Wetlands* 21:484–91.

Gawlik, D. E., and D. A. Rocque. 1998. "Avian Communities in Bayheads, Willowheads and Sawgrass Marshes of the Central Everglades." *Wilson Bulletin* 110:45–55.

Gleason, P. J., and P. A. Stone. 1976. *Prehistoric Trophic Level Status and Possible Cultural Influences on the Enrichment of Lake Okeechobee*. West Palm Beach, FL: South Florida Water Management District.

Gleason, P. J., and P. A. Stone. 1994. "Age, Origin, and Landscape Evolution of the Everglades Peatland." In *Everglades: The Ecosystem and Its Restoration*, eds. S. M. Davis and J. C. Ogden. Delray Beach, FL: St. Lucie Press.

Gottlieb, A., J. Richards, and E. Gaiser. 2005. "Effects of Desiccation Duration on the Community Structure and Nutrient Retention of Short and Long Hydroperiod Everglades Periphyton Mats. *Aquatic Botany* 82:99–112.

Grunwald, M. 2006. *The Swamp: The Everglades, Florida and the Politics of Paradise*. New York: Simon and Schuster.

Herndon, A., L. Gunderson, and J. Stenberg. 1991. "Sawgrass (*Cladium jamaicense*) Survival in a Regime of Fire and Flooding." *Wetlands* 11:17–28.

Jordan, F., K. J. Babbitt, and C. C. McIvor. 1998. "Seasonal Variation in Habitat Use by Marsh Fishes." *Ecology of Freshwater Fish* 7:159–66.

King, R. S., C. J. Richardson, D. L. Urban, and E. S. Romanowicz. 2004. "Spatial Dependency of Vegetation–Environment Linkages in an Anthropologically Influenced Wetland Ecosystem." *Ecosystems* 7:75–97.

Kobza, R. M., J. C. Trexler, W. F. Loftus, and S. A. Perry. 2004. "Community Structure of Fishes Inhabiting Aquatic Refuges in a Threatened Karst Wetlands and Its Implications for Ecosystem Management." *Biological Conservation* 116:153–65.

Light, S. S., and J. W. Dineen. 1994. "Water Control in the Everglades: A Historical Perspective." In *Everglades: The Ecosystem and Its Restoration*, eds. S. M. Davis and J. C. Ogden. Delray Beach, FL: St. Lucie Press.

Lorenzen, B., H. Brix, I. A. Mendelssohn, K. L. McKee, and S. L. Miao. 2001. "Growth, Biomass Allocation and Nutrient Use Efficiency in *Cladium jamaicense* and *Typha domingensis* as Affected by Phosphorus and Oxygen Availability." *Aquatic Botany* 70:117–33.

Miller, R. A., and H. C. Mattraw, Jr. 1982. "Storm Water Runoff Quality from Three Land-Use Areas in South Florida." *Water Resources Bulletin* 18:513–19.

Ogden, J. C. 2005. "Everglades Ridge and Slough Conceptual Ecological Model." *Wetlands* 25:810–20.

Palmer, M. L., and F. J. Mazzotti. 2004. "Structure of Everglades Alligator Holes." *Wetlands* 24:115–22.

Platt, W. J., R. F. Doren, and T. V. Armentano. 2000. "Effects of Hurricane Andrew on Stands of Slash Pine (*Pinus elliottii* var. *densa*) in the Everglades Region of South Florida (USA)." *Plant Ecology* 146:43–60.

Platt, W. J., B. Beckage, R. F. Doren, and H. H. Slater. 2002. "Interactions of Large Scale Disturbances: Prior Fire Regimes and Hurricane Mortality of Savanna Pines." *Ecology* 83:1566–72.

Rader, R. B., and C. J. Richardson. 1994. "Response of Macroinvertebrates and Small Fish to Nutrient Enrichment in the Northern Everglades." *Wetlands* 14:134–46.

Richards, P. M., W. M. Mooij, and D. L. DeAngelis. 2004. "Evaluating the Effect of Salinity on a Simulated American Crocodile (*Crocodylus acutus*) Population with Applications to Conservation and Everglades Restoration." *Ecological Modelling* 180:371–94.

Russell, G. J., O. L. Bass, and S. L. Pimm. 2002. "The Effect of Hydrological Patterns and Breeding Season Flooding on the Numbers and Distribution of Wading Birds in Everglades National Park." *Animal Conservation* 5:185–99.

Smith, S. M., and S. Newman. 2001. "Growth of Southern Cattail (*Typha domingensis* pers.) Seedlings in Response to Fire-Related Soil Transformations in the Northern Florida Everglades." *Wetlands* 21:363–69.

Smith, S. M., D. E. Gawlik, K. Rutchey, G. E. Crozier, and S. Gray. 2003. "Assessing Drought-Related Ecological Risk in the Florida Everglades." *Journal of Environmental Management* 68:355–66.

Thomas, S., E. E. Gaiser, M. Gantar, and L. J. Scinto. 2006. "Quantifying the Responses of Calcareous Periphyton Crusts to Rehydration: A Microcosm Study (Florida Everglades)." *Aquatic Botany* 84:317–23.

Urban, N. H., S. M. Davis, and N. G. Aumen. 1993. "Fluctuations in Sawgrass and Cattail Densities in Everglades Water Conservation Area 2A under Varying Nutrient, Hydrologic and Fire Regimes." *Aquatic Botany* 46:203–23.

Web of Science. 2006. Database at http://wos.mimas.ac.uk/.

Weisskoff, R. 2005. *The Economics of Everglades Restoration: Missing Pieces in the Future of South Florida*. Cheltenham, UK: Edward Elgar.

Whiteaker, L. D., and A. M. LaRosa. 1991. "Evaluation of Fire as a Management Tool for Controlling *Schinus terebinthifolius* as Secondary Successional Growth on Abandoned Agricultural Land." *Environmental Management* 15:121–29.

Wu, Y. G, F. H. Sklar, K. Gopu, and K. Rutchey. 1996. "Fire Simulations in the Everglades Landscape Using Parallel Programming." *Ecological Modelling* 93:113–24.

Chapter 3

Rivers of Plans for the River of Grass

The Political Economy of Everglades Restoration

STEPHEN POLASKY

Only a few unique ecosystems in the world are extraordinary enough to be recognized instantly by name. The Everglades is one. As evidence of its status, Everglades National Park has been designated as a World Heritage Site and a Biosphere Reserve. The largest wetlands ecosystem in the United States, the Everglades comprises an interconnected web of sawgrass marsh, sloughs, and timber islands with an incredible diversity of plants and animals, sustained by slowly flowing water that moves almost imperceptibly southward from Lake Okeechobee to Florida Bay.

Over the past one hundred years, the Everglades has been modified extensively by human activities. Large metropolitan areas have sprung up along Florida's east coast, forming an almost continuous urban strip from Miami to West Palm Beach. Urban areas along the Gulf Coast and in the interior south from Orlando through the watershed of the Kissimmee River, whose waters flow into Lake Okeechobee, are also experiencing rapid population growth. From 2000 to 2006, Fort Myers and Naples, Florida, were the third and sixth fastest growing metropolitan areas in the United States (USCB 2007a).

In addition to expansion of urban areas, areas south of Lake Okeechobee were drained for the Everglades Agricultural Area to grow sugarcane and other agricultural commodities. To allow for agricultural expansion and urban growth and to provide increased water supplies and more effective flood control, wetlands were drained, canals were dug, dikes were constructed, and rivers were straightened. The entire water flow regime of southern Florida, the central lifeblood of the Everglades, was reengineered. Approximately 70 percent less water flows through the Everglades to Florida Bay than prior to the reengineering (CERP 2007). These changes have reduced the total area of the Everglades by 50 percent and caused large increases in the levels of nutrients in the water.

Phosphorous from agricultural operations has been a particular concern. Tight interconnections via water flow through the Everglades mean that disturbance in one place leads to disturbances elsewhere. Even Everglades National Park, spared from direct modification, has suffered serious effects from increased nutrient loadings from agricultural and urban sources and from changes in hydrology upstream.

Because the Everglades is so extraordinary and changes to it have been so extensive, restoration efforts are large and comprehensive. Current efforts and plans to restore the Everglades involve a multi-billion-dollar investment over decades and promise to be "America's largest civil-engineering project and the world's most expensive environmental clean-up" (*Economist* 2005).

Pressure to "save the Everglades" has been present for almost as long as large-scale efforts to develop Florida. Notable early successes to protect the Everglades came in 1947 with the establishment of the Everglades National Park, the same year Marjory Stoneman Douglas's *The Everglades: River of Grass* was published. Efforts to prevent further degradation and to restore the Everglades gained momentum in the 1980s and 1990s with a series of initiatives by the state of Florida, as well as an important federal court case to protect Everglades National Park, filed in 1988 (see chapter 1, this volume).

These efforts eventually led to the landmark announcement of the Comprehensive Everglades Restoration Plan (CERP) in 1999 by the U.S. Army Corps of Engineers (USACE) and the South Florida Water Management District (SFWMD). This ambitious plan aims for "the restoration, preservation and protection of the south Florida ecosystem while providing for other water related needs of the region" (CSFP 1999, ii). Restoration efforts are now well under way, though progress has been slower than first anticipated because of significant, unresolved scientific uncertainties regarding restoration, as well as a lack of adequate funding for restoration projects (NRC 2007).

Restoration of the Everglades is a noble yet daunting goal. Many of the large development and infrastructure projects largely responsible for changing the hydrology of South Florida were authorized in 1948 and carried out during the 1950s and 1960s to control floods and prevent saltwater intrusion. At that time, many of the destructive consequences of development of South Florida and alteration of water flow for the Everglades ecosystem were poorly understood. Over time, the importance of conserving the Everglades has grown, as has the understanding of some of the unintended negative consequences of development. Restoring the Everglades will require reversing many of development's negative consequences even in the face of ongoing development pressure.

For the past sixty years, Florida has been one of the fastest growing states in the country. In 1940, its total population was less than 2 million; as of 2006,

Miami-Dade County alone had a population of 2.4 million (USCB 2007a). In 2000, Florida's population had grown to nearly 16 million and is projected to rise to 28.6 million by 2030 (USCB 2007b). Additional people will demand clean drinking water and flood protection and will generate additional sewage outflow. Significant investment will be required to prevent further deterioration in the conditions of the Everglades in the face of such large population growth (Weisss-koff 2004).

The goal of restoring the Everglades raises three fundamental questions. First, can the Everglades actually be restored? This question, primarily an ecological one, is addressed by Thomas L. Crisman (see chapter 2, this volume) and provides the foundation upon which economic and political questions rest. Second, if restoration can be accomplished, will its benefits exceed the costs of accomplishment? Third, even if restoration could be done, will the outcome of the political process allow restoration to occur? These three questions are discussed in the following sections.

Restoration Ecology

South Florida has undergone radical alterations over the past century. Restoring the Everglades to pre-twentieth-century conditions will require extensive changes. The consequences of large-scale changes in the Everglades' complex ecosystem cannot be fully predicted or understood prior to those changes' occurring. New conditions and interactions will require continual monitoring and adjustment as restoration activities are undertaken—an ongoing process that ecologists call "adaptive management."

The central element in the restoration efforts is to "get the water right," by restoring the sheet flow as well as the volume of water running through the Everglades, and CERP projects are largely dedicated to this goal. Reducing excess nutrient flow into the Everglades is another major component of restoration and involves acquiring and retiring land from agriculture and using storm water treatment areas for water runoff from urban and agricultural areas. Even if these efforts are successful at restoring both the flow of water and water quality, it is not clear that the ecological character of the Everglades will be restored to conditions that existed prior to the twentieth century.

The Everglades is a highly interconnected and complex system. One feature of complex, interconnected systems is that disturbances, even temporary ones, can have long-lasting effects that are difficult to reverse. For example, addition of nutrients from agricultural runoff to shallow lakes can cause them to shift from oligotrophic (nutrient-poor with plenty of oxygen at deeper levels) conditions to eutrophic (nutrient-rich) conditions, where algal blooms and plant decomposition deplete oxygen, making it difficult for some fish species to survive. Even when nutrient inputs are reduced, the lake may remain eutrophic, sustained by new species composition and ecological processes brought about by the shift to the eutrophic state (Carpenter et al. 1999; see chapter 2, this volume). It is likely that such long-lasting effects of disturbance will occur in the Everglades. If so, restoring the initial set of hydrologic conditions ("getting the water right") will not be sufficient to restore the area's initial set of ecological conditions. Despite intensive research on the Everglades, a great deal of uncertainty remains about how the ecological system as a whole will respond to restoration projects.

Restoration of the Everglades is complicated further by global climate change, invasive species, and other major environmental changes. Climate change can affect the Everglades through changes in annual rainfall patterns and the frequency and intensity of hurricanes. The most profound change, however, will probably come from rising sea levels: 60 percent of Everglades National Park is less than three feet above sea level (Kimball 2007). Using the Intergovernmental Panel on Climate Change's (IPCC) projections from its 2007 report of sea-level rise of from 7 to 23 inches by 2100, "10% to 50% of the park's freshwater marsh would be transformed by salt water pushed landward by rising seas" (Kimball 2007). In addition to climate change, the spread of alien invasive species (Simberloff et al. 1997) and long-lasting contaminants (such as mercury) make it virtually impossible to restore the Everglades to pre-twentieth-century conditions.

Given the largely irreversible environmental changes, it is unrealistic to expect that the Everglades can be restored to pre-twentieth-century conditions. As ecologist Crisman (see chapter 2, this volume) concludes, rehabilitation of the Everglades to provide habitat and ecological functions similar to pre-twentieth-century conditions, rather than restoration to pre-twentieth-century conditions should be the real policy goal.

Restoration Economics

Economists approach virtually all policy questions through the lens of benefits and costs: will changes brought about by a policy or a project likely generate benefits that exceed costs? Addressing benefits and costs from restoration projects in the Everglades presents challenging and vexing issues, due to the underlying scientific uncertainties in how the ecosystem will respond to various interventions, as well as the difficulty of estimating the benefits to society of ecosystem restoration.

The cost side of the equation is easier to address than the benefit side. For budgetary purposes, cost estimates of restoration projects are prepared periodically.

Restoration projects in the Everglades are estimated to cost multiple billions of dollars. The initial cost estimate for restoration projects under CERP was $7.8 billion in year 2000 dollars. The current estimated cost for CERP projects has climbed to over $10 billion in 2006 dollars (GAO 2007). In addition to CERP projects, a number of other restoration projects in the Everglades have been approved, such as land acquisitions to retire land from agriculture and restore marshes and construction of storm water treatment areas to reduce the amount of nutrients flowing into the Everglades. Adding these costs to the cost of CERP projects generates estimates of total restoration costs of close to $20 billion in 2006 dollars (GAO 2007). Even this estimate is probably on the low side because it does not include some project costs (GAO 2007) and because it does not adequately factor in currently unplanned expenses (surprises) that will undoubtedly arise as restoration efforts evolve through time.

What does society get in return for its investment of $20 billion? Some of the benefits from Everglades restoration, such as increased water supply and improved navigation, recreation, and fishing, can be fairly readily translated into estimates of benefits in dollar terms. For example, estimates of the value of improved fisheries to the commercial fishing industry can be obtained by estimating the increase in returns to the fishery prior to ecological restoration versus after restoration. In 1999, the USACE and the SFWMD estimated the change in benefits for agricultural water supply, urban water supply, commercial navigation, flood control, recreation, and commercial and recreational fishing from Everglades restoration projects (CSFP 1999). The total of these benefits, while significant, fell well short of the costs for restoration (Milon and Hodges 2000). Restoration projects in the Everglades would fail a benefit–cost test if it included only those benefits that could be readily translated into dollar terms.

However, the major benefits of restoration are ecological and cultural, exceedingly difficult to translate into dollar terms. Ecological benefits of restoration include the improvement of habitat that supports a wide array of biodiversity throughout the Everglades plus an increased supply of freshwater to Florida Bay, which supports the rich biodiversity in and around the Bay. Restoration of the Everglades is also important to the people of the region, as well as to the rest of the world as a unique, irreplaceable ecosystem: "The Everglades is to south Florida what the Rockies are to many western states; the old growth forests are to the Pacific Northwest; the Adirondack, White and Green Mountains are to the Northeast; and the Mississippi River is to the nation's heartland. The Everglades epitomizes the region's sense of definition and place, both substantially and spiritually" (CSFP 1999, xvii).

Unlike more tangible benefits with fairly direct links to the production of marketed goods and services, ecological and cultural benefits typically fall well outside the realm of markets. In fact, ecological and cultural benefits are often what economists call "nonuse" values, that is, values that arise from the *existence* of an environment rather than from the *use* of it. Many people place great value on the potential for the Everglades to function as it did prior to major human interventions, even though they may never visit the Everglades or benefit directly from flood control or water provision. Nonuse benefits may be substantial, but attempts to quantify them in dollar terms are controversial.

The only economic valuation methods that can be used to estimate the dollar value of nonuse benefits are stated preference methods, such as contingent valuation. In a typical contingent valuation survey, respondents are asked whether they would be willing to pay a specified amount to receive an environmental benefit. The median "willingness-to-pay" for the environmental benefit can be estimated by varying the amount and observing the percentage of positive responses at each amount. Milon and Scroggins (2002) conducted a contingent valuation survey on a sample of households in Florida. They estimated that the average household in Florida would gain annual benefits of $58.79 from full restoration of the Everglades, which translates to a total benefit for all Florida households of $342.2 million annually (Milon and Scroggins 2002). Translating these annual figures to total value of benefits through time, as of 2007, generates estimates of $7.2 billion, using a 5 percent discount rate, to $17.5 billion, using a 2 percent discount rate. The latter estimate approaches the estimate for total cost of restoration projects of roughly $20 billion yet falls short by $2.5 billion.

The difficulties in accurately evaluating the benefits of Everglades restoration stem from the difficulties of measuring nonuse benefits and from uncertainties about the effects that restoration activities will have on the Everglades. If restoration activities fail or are only marginally successful, then few if any benefits will be realized. Although, in principle, it is appealing to try to measure all possible ecological benefits of restoration and to weigh those benefits against costs, the daunting challenges of scientific uncertainty and economic valuation preclude this. In summarizing attempts at evaluating all of the benefits derived from ecosystem restoration, such as in the Everglades, the National Research Council (NRC) concluded that

> It would certainly be advantageous to have evidence on *all* benefits and costs prior to decision-making as anything less will be partial and incomplete and risks giving incorrect advice to decision-makers. Yet, trying to attain the "value of everything" through a complete and reliable accounting of all ecosystem services cannot be done with current understanding and methods and is unlikely to be accomplished anytime soon. . . . In the case of the Exxon *Valdez* and the Florida Everglades restoration, how-

ever, many of the important values are linked to existence of species or the existence of the ecosystem itself in something akin to its original (pre-human-altered) condition. Valuing such services presents difficult challenges even when ecological knowledge is relatively complete. (NRC 2004, 189)

Restoration projects for the Everglades do not need to pass a benefit–cost test. The stated objective of the Comprehensive Everglades Restoration Plan is to restore, preserve, and protect the Everglades while providing water-related needs of South Florida (CSFP 1999). Current policy is to carry out projects necessary to accomplish these goals whether or not the benefits exceed the costs. As Milon and Hodges state, "The presumption is that restoring this area is a national priority, and economic measures of the value of the environment are either unreliable or unavailable" (2000, 13). In his study of the economics of Everglades restoration, Weisskoff asserts, "The value of a restored environment—saving species, preserving wetlands, recreating habitats—these are all unknown or inestimable, so that the need for a benefit–cost ratio has been suspended" (2004, 7).

Though it is not possible to conduct a formal, economic benefit–cost analysis at this time, decision makers responsible for guiding Everglades restoration policy remain subject to the pressures of informal, political benefit–cost tests. If the costs of a policy are viewed as being far greater than the benefits, at least by some significant or powerful sectors of society, then political pressure will mount to change policy. For the Everglades, political considerations are never far below the surface.

Restoration Politics

In 2000, the stars were aligned for moving toward restoration of the Everglades. After decades of studies, political wrangling, initiatives, legislation, and court cases, the federal government and the state of Florida agreed to share fifty-fifty the multi-billion-dollar costs of the CERP. However, whether this agreement was the "dawn of a new era in conservation" and a "model for the world, proof that man and nature could live in harmony," according to Governor Jeb Bush (Grunwald 2006, 6), or a brief aberration from politics as usual remains to be seen.

Reasons for optimism can be found. Restoring the Everglades is a goal with widespread support among the general public. Policy makers across the political spectrum are in favor of restoration. In 2000, Everglades restoration received bipartisan support in Congress as well as support from President Bill Clinton and Governor Bush (despite the bitter partisan contest over Florida's electoral votes between George W. Bush and Vice President Al Gore to decide the 2000 presidential election).

Reasons for pessimism also abound. Since 2000, the process toward Everglades restoration has not gone smoothly. By 2004, restoration projects were two years behind schedule and $1 billion over budget (*Economist* 2005). The federal government has lagged behind in its commitment to fund restoration of the Everglades, contributing $1.4 billion less than it had promised to give through 2006 (GAO 2007). The state of Florida has increased its funding, thereby partially rescuing restoration plans, but it has faced political problems as well. In 2003, Florida passed a law backed by the sugar industry that appeared to weaken commitments to providing clean water to the Everglades, to the consternation of environmentalists and members of Congress (Drew 2007).

With so much at stake directly, in terms of the multi-billion-dollar expenses from the restoration projects, and indirectly, through changes in water management and land management that could affect virtually all of South Florida, some political infighting was bound to occur as restoration planning moved beyond platitudes and into the hard work of implementation. After CERP was agreed upon, Grunwald, in his book analyzing Everglades politics, observed, "America's politicians would finally pass the Everglades test. It was a noble sentiment. But man had been flunking that test for a long time" (2006, 6). Whether sufficient goodwill and consensus remain to keep restoration projects on track and moving forward will be severely tested in the coming years.

Conclusion

Comprehensive restoration of an ecosystem as complex and interconnected as the Everglades requires a level of understanding of system dynamics that we currently lack. We do not know exactly how restoration activities that change one part of the system will affect other parts of the system. Will "getting the water right" mean that we will "get the ecology right," or will a number of further restoration efforts be required? What other surprises or unintended consequences will result from restoration actions? It is impossible to know now what eventually will be required to restore the Everglades. Ecosystem restoration planning will have to continue to evolve in light of constantly changing conditions and developing understanding of the ecosystem.

The NRC recommended the use of incremental adaptive restoration to invest in projects "large enough to provide some restoration benefits and address critical scientific uncertainties . . . to promote learning and that can guide the remainder of the project design " (NRC 2007, 11). Following this advice, restoration plans should be thought of as interrelated processes rather than a fixed plan.

Restoration will continually, slowly evolve, much like a virtual river of plans, mirroring the slow, continuous, yet ever changing flow of water through the river of grass.

In the near future, population growth, invasive species, and climate changes will create unpredictable conditions affecting the macro and micro ecosystems of the Everglades, making rehabilitation a more realistic goal than restoration. Rehabilitation, however, is a less clear-cut objective than restoration. For restoration, the target ecological conditions are those that prevailed in the nineteenth century, prior to major human interventions. For rehabilitation, a wider latitude exists for setting ecological target conditions.

The major test for the Everglades will be whether sufficient long-term political agreement can be maintained to make substantial progress toward ecosystem restoration or, at minimum, rehabilitation goals. The agreement between the federal government and the state of Florida to create the CERP was a remarkable achievement. Sustaining the agreement, so that restoration project funds continue to flow toward projects that satisfy ecological rehabilitation needs, rather than flowing toward projects that appease powerful interest groups, will be an equally remarkable achievement.

References

Carpenter, S. R., D. Ludwig, and W. A. Brock. 1999. "Management and Eutrophication for Lakes Subject to Potentially Irreversible Change." *Ecological Applications* 9:751–71.

Central and Southern Florida Project (CSFP). 1999. *Central and Southern Florida Project Comprehensive Review Study, Final Integrated Feasibility Report and Programmatic Environmental Impact Statement.* U.S. Army Corp of Engineers and South Florida Water Management District. At www.evergladesplan.org/docs/comp_plan_apr99/summary.pdf.

Comprehensive Everglades Restoration Plan (CERP). 2007. "Why Restore the Everglades?" At www.evergladesplan.org.

Douglas, M. S. [1947] 1974. *The Everglades: River of Grass.* Atlanta: Mockingbird Books.

Drew, C. A. 2007. "Storm Water and the Consent Decree: The Life or Death of the Everglades." *Natural Resources and Environment* 21 (4): 30–35.

"The Everglades: Water, Bird and Man." 2005. *Economist* 377 (8447): 29–30, 33.

Grunwald, M. 2006. *The Swamp: The Everglades, Florida and the Politics of Paradise.* New York: Simon and Schuster.

Kimball, D. 2007. Testimony on Climate Change to the Subcommittee on Interior, Environment and Related Agencies. House Appropriations Committee. U.S. House of Representatives. April 26. At www.nps.gov/ever/parknews/everclimatechangetestimony.htm.

Milon, J. W., and A. W. Hodges. 2000. "Who Wants to Pay for Everglades Restoration?" *Choices* 15 (2): 12–16.

Milon, J. W., and D. Scroggins. 2002. "Heterogeneous Preferences and Complex Environmental Goods: The Case of Ecosystem Restoration." In *Recent Advances in Environmental Economics*, eds. J. List and A. de Zeeuw. Cheltenham, UK: Edward Elgar.

National Research Council (NRC). 2004. *Valuing Ecosystem Service: Towards Better Environmental Decision-Making*. Washington, DC: National Academies Press.

NRC. 2007. *Progress toward Restoring the Everglades: The First Biennial Review, 2006*. Washington, DC: National Academies Press.

Simberloff, D., D. C. Schmitz, and T. C. Brown, eds. 1997. *Strangers in Paradise: Impact and Management of Nonindigenous Species in Florida*. Washington, DC: Island Press.

U.S. Census Bureau (USCB). 2007a. "50 Fastest-Growing Metro Areas Concentrated in West and South." At www.census.gov/Press-Release/www/releases/archives/population/009865.html.

USCB. 2007b. "Interim Projections: Ranking of Census 2000 and Projected 2030 State Population and Change." At www.census.gov/population/projections/PressTab1.xls.

U.S. Government Accountability Office (GAO). 2007. *South Florida Ecosystem: Restoration Is Moving Forward but Is Facing Significant Delays, Implementation Challenges, and Rising Costs*. Report to the Committee on Transportation and Infrastructure. U.S. House of Representatives. Washington, DC: GAO.

Weisskoff, R. 2004. *The Economics of Everglades Restoration: Missing Pieces in the Future of South Florida*. Cheltenham, UK: Edward Elgar.

PART II

The Platte River

The main stem of the Platte River flows through the length of Nebraska after having been fed by snowmelt from the Rocky Mountains and delivered via the North and South Platte rivers. Fifteen major dams and reservoirs, supplemented by many smaller diversions and storage projects that serve agricultural and municipal needs, heavily impact Platte River flows. The U.S. Bureau of Reclamation (USBR) operates water facilities that account for 40 percent of the total storage capacity on the river. In chapter 4, after analyzing changes to the river imposed by dams, reservoirs, and water projects, David M. Freeman documents major themes in the negotiations that produced a Platte River Basin habitat recovery program.

The three affected states (Nebraska, Colorado, and Wyoming) and water providers along the river, under the leadership of USBR, facing serious curtailment of water operations by and endless consultation with the U.S. Fish and Wildlife Service (USFWS) in its capacity as enforcer of the Endangered Species Act (ESA), agreed in 1994 to negotiate a basinwide agreement. The aims of this collaborative effort were to develop and implement a habitat recovery program for the three endangered species (two birds, the whooping crane and interior least tern, and one fish, the pallid sturgeon) and one threatened species (the piping plover) and to enable water users in the basin to proceed with existing and some future uses. The three states and their constituent water providers and customers sought certainty in ESA enforcement from the USFWS in exchange for their investments in habitat recovery.

Located in a watershed where overall flow is insufficient, the Platte River project has at its core the conflict between agricultural and urban users, whose needs require withdrawal of water from the basin, and those advocating preservation of instream flows for habitat and other environmental purposes. In chapter 5, Thomas L. Crisman elaborates on the dynamics of the Platte River system and the ecological causes and effects of habitat loss. Stephen Polasky's chapter 6 explores from an economics perspective the competition among these types of uses in conditions of scarcity and uncertainty.

The conflict between consumptive water uses and the water needs of the natural system is a common theme in this book, lying at the heart of the Everglades and California Delta projects, for example. The difficult and lengthy negotiations conducted on the Platte River among federal government agencies, the states, and nongovernmental stakeholders are much like those experienced in the Everglades and Upper Mississippi River, though the mix of federal agency participants varies. As in all the ecosystem restoration projects, the integration of sound science into environmental policy making is of crucial importance to the Platte River Basin. Particularly instructive is Freeman's analysis of the dynamics of negotiations as project participants changed course in the face of new scientific understanding, confronting and putting into practice the theory of adaptive management on a large scale.

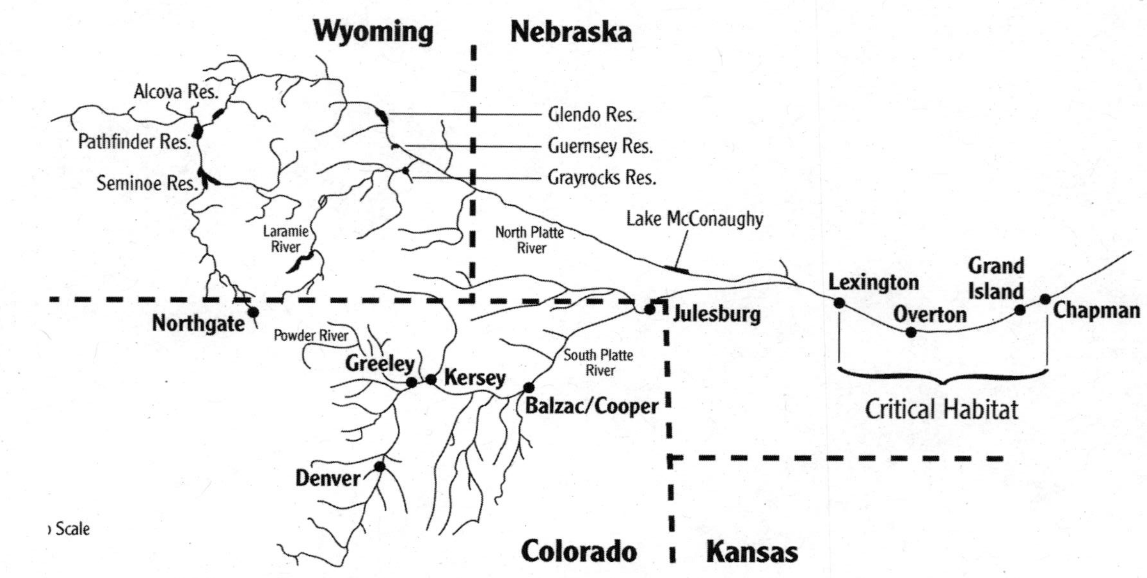

Figure 4.1. Platte River Basin (Created by David M. Freeman 2007).

Chapter 4

Negotiating for Endangered and Threatened Species Habitat in the Platte River Basin

DAVID M. FREEMAN

Brought to the table by the requirements of the Endangered Species Act (ESA), representatives of the U.S. Department of the Interior (DOI)—from the U.S Fish and Wildlife Service (USFWS) and the U.S. Bureau of Reclamation (USBR)—and the states of Colorado, Nebraska, and Wyoming successfully negotiated the terms and conditions under which they will collaboratively reregulate about 11 percent of the average annual surface flow of the Platte River, as measured near Grand Island, Nebraska. Negotiations took place in conjunction with restoring 10,000 acres of critical habitat for three species of birds—endangered whooping cranes (*Grus americana*), endangered interior least terns (*Sterna antillarum*), and threatened piping plovers (*Charadrius melodus*)—during the first thirteen-year increment of a cooperative program.[1]

The new basinwide recovery program will also test the hypothesis that the proposed program measures on behalf of bird habitat recovery in the Central Platte River Basin will enhance habitat for an endangered species of fish downstream, the pallid sturgeon (*Scaphirhynchus albus*), near the mouth of the Platte as it empties into the Missouri River. This chapter analyzes changes to the river imposed by dams and other water diversion and storage projects, documents major themes in the negotiations, and focuses on the human dynamics that produced the Platte River habitat recovery program.

Problem

The waters of the Platte River Basin are some of the most intensively exploited on the planet. By the time the South Platte River meets the North Platte River to form the main stem in western Nebraska, both tributaries have been harnessed repeatedly to serve the utilitarian needs of industrial agriculture, urban life, and recreation—a pattern sustained across Nebraska. Waters are used and repeatedly reused by agriculture, municipalities, and industry. Human engineering of Platte Basin waters for these multiple uses has exacted a high toll on the river and associated riparian ecosystems.

Near the Continental Divide, mountain snowpack in Colorado and Wyoming thaws into rivulets, gathering into plunging streams that flow through rough canyons and then abruptly run out on flat prairie, where water settles into wide beds. Located well before the Nebraska borders, these wide beds drop at an average rate of only 7 feet per mile. Traditionally, plains channels were broad, braided, and sandy, with low banks, sparse woody vegetation, and high sediment loads (Eschner et al. 1981). The Platte's original braided channel networks were multiple channels of the river separated by small islands and sandbars (see chapter 5, this volume). Prior to European settlement, the natural flow pattern consisted of a spring rise beginning in March, extending to a peak in late May or early June, and then a sharp decline in summer, fall, and winter months. Spring and early summer floods cleared vegetation from sandbars, islands, and riverbanks and distributed sediment across a wide path. A mile wide in some places, the Platte River was described as a burlesque of rivers, braided with islands and studded with sandbars. Early travelers complained that the Platte could not be ferried for lack of water and could not be bridged for lack of timber (Matter 1969, 239).

Over the course of the late nineteenth and twentieth centuries, Platte River Basin flows have been affected by fifteen major dams and reservoirs, supplemented by many smaller water diversion and storage projects. On the South Platte alone, 106 storage facilities hold an average of 2.8 million acre-feet of water (Eisel and Aiken 1997). Upstream from Lake McConaughy on the North Platte River, there are eighty-four storage works with a capacity of 4.3 million acre-feet. The total basin storage capacity is about six times the average annual flow of the Platte at Grand Island. Dams and basin reservoirs provide a total storage capacity of over 7.1 million acre-feet, with USBR projects accounting for 2.8 million acre-feet (Keyes 2002, 2). The volume of a body of water is calculated by multiplying its surface area in acres by its average depth in feet. One acre-foot is equal to a sheet of water 1 acre (0.405 hectare) in area, 1 foot (30.48 centimeters) in depth, and 43,560 cubic feet (1,233.5 cubic meters) in volume, or approximately 325,851 gallons.

In general, the traditional flow regime has been changed by these dams, reservoirs, and projects to one characterized by lower and less frequent spring flood pulses; clearer water flows, as sediment has been trapped behind dams; more incised, straighter channels; and higher summer, fall, and winter flows (McDonald and Sidle 1992, 149–58). All of these changes have caused a net loss of wide, sandbar, braided, shallow, water-filled channels and an associated increase in vegetated riverbanks and islands. These results, in turn, have meant loss of habitat essential to native species.

Habitat Requirements

The Big Bend stretch of the Platte River in central Nebraska presents an extremely favorable combination of habitat types, hosting bald eagles, peregrine falcons, over 10 million ducks and geese, Eskimo curlew, and, for a brief period each spring, over a half million sandhill cranes along with their rare cousins, a few whooping cranes. As human impact has destroyed the wide, shallow, braided Platte in most reaches, whooping cranes, least terns, and piping plovers are being crowded into ever smaller reaches of viable habitat, along with millions of other migrating birds that press into the same area.

The Central Platte is strategic because (1) its channel offers the most dependable listed bird habitat, especially as other wetland areas (for example, the Rainwater Basin to the southeast) are likely to be dried up or frozen over in early spring months; (2) wetlands are more conducive to disease transmission; and (3) wetland vegetation provides cover for predators. A National Academy of Sciences review team concluded that no apparently suitable alternatives exist to replace the Central Platte River habitat for migrating whooping cranes (NRC 2005, 149).

Whooping Crane

Whooping cranes are among the largest birds in the world. Standing over 5 feet tall, with a wingspan of 7.5 feet, they weigh on average about 14 pounds and frequently fly from 200 to 500 miles per day during migration. They lay two eggs per year in the Arctic north and live as long as 40 years. Brilliant white birds, with black wingtips and bare red head tops, they migrate typically from March through April and from September through October, moving from wintering grounds on the Texas gulf coast to breeding areas in northern Canada and then making their autumn return south. Accurate drawings and descriptions of whooping cranes are readily available (Matthiessen 2001; Walkinshaw 1973).

The whooping crane population, estimated in 1941 to be as low as sixteen (with only six to eight breeding birds), has rebounded a bit because of habitat acquisition, federal protection, and intense management of breeding and wintering areas. Over the decades, the wild population has fluctuated around a slowly increasing trend line until, in 1998, about two hundred whooping cranes made up the North American midcontinent flock, out of four hundred worldwide, many of which are in captivity (Austin and Richert 2001). Whooping cranes are the rarest of the world's fifteen crane species. Although whooping cranes do not breed on the Platte, nor do they remain on the Platte beyond a few days, they, along with over three hundred other species of migrating birds, use the Platte. Weeks later, the several migrating species spread out over wide-ranging, sparsely populated northern breeding grounds. However, in the narrow stretch of the Central Platte, the many migrating species, including the few whoopers, become highly concentrated in the late February through April period.

The habitat of interior least terns and piping plovers has also been degraded by intensification of human uses of the Central Platte. Increasingly, least terns and piping plovers are confined to ever more limited, fragmented, scattered habitats, due to faster, deeper summer flows moving through incised channels that lie between banks and islands, which support dense, woody vegetation.

Interior Least Tern

The interior least tern is the smallest of the tern species, approximately 9 inches in body and 20 inches in wingspan. Adults are recognized by a white patch on the forehead, contrasting sharply with a black crown, a bright yellow bill with a black tip, gray back, white underbody, and orange-yellow feet (Forbush and May 1955, 235–36). In recent decades, this species has been found on only a fraction of its former habitat, which stretched from Texas to Montana and from the front range of eastern Colorado and New Mexico to Indiana early in the twentieth century. The species was listed as endangered in 1985, and recent estimates place its population at about 4,800 (USFWS 1997). In Nebraska, interior least terns are found on sandbars of the Missouri, Loup, Niobrara, and Platte rivers; on the beaches of Lake McConaughy; and on shores of sandpits created by human extraction of gravel.

Piping Plover

Piping plovers are similar to least terns in that they require much the same habitat and compete for the same nesting sites. They differ slightly from least terns in that they are somewhat more tolerant of woody vegetation encroachment and do not eat small fish, as terns do. This species was listed under ESA as threatened in 1985 (NRC 2005). A 1991 census estimated the piping plover population in both Canada and the United States to be about 2,440 breeding pairs. The population is distributed from southeastern Alberta to northwestern Minnesota and along prairie rivers and reservoirs to southeastern Colorado. Piping plovers are a stocky, robin-sized shorebird about 6 to 7 inches long with a wingspan of about 14 to 15 inches. The head, back, and wings are pale brown to gray with black and white highlights.

Traditionally, least terns and piping plovers would await the decline of spring peak flows and then scrape out their shallow nests on beaches or sandbars. With the coming of human manipulation of river flows all summer long for purposes of irrigation, power production, and municipal use, these birds have become vulnerable to being flooded out and to high water, which continues long past the time they can wait to nest (Currier, Lingle, and Van Der Walker 1985, 38–39). One obvious adaptation to sustained higher flows is to nest on higher exposed sandbars, but today large parts of main channel areas are clogged with trees and shrubs. The two species often retreat to gravel pits, but mortality of the young is high due to inferior food sources and predators that can find them too easily.

Pallid Sturgeon

The Lower Platte, below the mouth of the Elkhorn River and continuing to the Missouri River, is a free-flowing stretch without the barriers that impede pallid sturgeon in their several life stages—larval, juvenile, and adult. The Missouri and Mississippi rivers once constituted a large-scale, connected primary habitat for the pallid sturgeon. However, suitable habitats in these rivers have been so reduced and fragmented by dams, diversions, levies, and rock-ribbed embankments that they provide, at best, few of the characteristics required by the pallid sturgeon: slow-moving waters in broad, braided channels with sandbars and islands to create refuge just beyond the downstream tips and warm backwater sloughs (Lutey 2002; NRC 2005). The extent of use and relative importance of the Lower Platte to pallid sturgeon (listed under ESA in 1990 as endangered) are now unknown (NRC 2005).

Therefore, at least two critical questions arise (1) what is the importance of the Platte River to recovery of pallid sturgeon? and (2) how will the basinwide efforts on behalf of the three birds on the Central Platte contribute to pallid sturgeon recovery? Without accurate answers to these questions, recovery program actions cannot be evaluated properly. Two steps, however, can be taken (1) to preserve and restore the hydrogeomorphic processes that provide desired habitat flows and (2) to document and assess the fishes' response to varying channel characteristics.

In sum, over the last three decades, Central Platte habitats for the listed species had deteriorated to the point that a series of USFWS biological opinions on water diversion and storage projects in the basin found that the degraded habitats were likely to jeopardize the listed three bird and one fish species under the terms of the 1973 federal ESA (Zallen 2005). Section 7 of ESA requires that all federal agencies

> confer with the Secretary [of DOI or Commerce][2] on any agency action which is likely to jeopardize the continued existence of any species proposed to be listed under section 4 ["Determination of Endangered Species and Threatened Species"] or result in the destruction or adverse modification of critical habitat proposed to be designated for such species (subsequently, "likely to jeopardize"; it is understood that the former phrase includes the possibility: "or not likely to jeopardize depending on context"). (USDOI, USFWS 1973, 15)

Federal ESA Requires Implementation

The ESA thus forced a confrontation between activities of water providers in the basin and the needs of the species listed as endangered or threatened under that law. The ESA has compelled a sustained, three-decade conversation about how to reconcile human water needs with improved and preserved critical habitats for the listed species. When water users are dependent upon federal government projects or when nonfederal water facilities need federal approvals, water users who plan to undertake actions that are likely to jeopardize a listed species must find ways to achieve ESA compliance (usually, to find a "reasonable and prudent alternative" to the original proposed action that is not likely to jeopardize a listed species) in order to gain essential permit(s). Since the 1970s, ESA has been an unwelcome guest at virtually every Platte Basin water provider dinner party.

Water providers, singly or in collaboration, thus seek to use ESA's section 7 consultation process to create a "reasonable and prudent alternative" that mitigates jeopardy to a listed species. To accomplish this, they must have the alternative reviewed by an environmental impact statement (EIS), as mandated by the 1969 National Environmental Policy Act (NEPA), and have it judged to be sufficient to offset harm to the species in a USFWS biological opinion. By fulfilling these requirements, they may receive permission to continue particular water operations with the promise of regulatory certainty.

Future without a Collaborative Recovery Program

Since the late 1970s, USFWS has issued "jeopardy" biological opinions for many water projects in Wyoming, Colorado, and Nebraska that deplete flows in the Central Platte River (that is, biological opinions that determine that a proposed action is likely to jeopardize an endangered or threatened species). More detailed chronologies of federal actions on behalf of the Central Platte listed species habitat are found elsewhere (Echeverria 2001; Freeman 2003; Zallen 2005). Brief mention of a few episodes will provide some flavor of federal actions that mobilized water providers to come, however reluctantly, to the negotiating table.

In 1978, USFWS issued a biological opinion on the impacts of a power plant under construction in Wheatland, Wyoming. Cooling water for this large, coal-fired facility was to be captured at Grayrocks Dam and Reservoir in eastern Wyoming on the nearby Laramie River, a tributary to the North Platte, thereby imposing a depletion on the Platte's system of flows. Earlier, the U.S. Army Corps of Engineers (USACE or the Corps)—without consulting USFWS—had issued a Section 404 permit, required by the Clean Water Act before dredged or fill material may be deposited in wetlands. The National Wildlife Federation and the state of Nebraska sued project promoters and the Corps for violating ESA section 7 by not consulting with USFWS before issuing this permit. The USFWS's eventual biological opinion offered alternatives that would permit the power plant project to proceed, given successful completion of mitigation actions, which included establishment of a $7 million Whooping Crane Maintenance Trust, granting advocates for the birds a voice and active player status.

On Colorado's South Platte, also in 1978, USFWS informed the Corps that a proposed Wildcat Reservoir, to be filled by South Platte waters, would impose depletions damaging to whooping crane habitat in central Nebraska. Subsequently, Wildcat Reservoir's developers learned that the Corps had to consult with the USFWS under ESA section 7. The promoters, facing the prospect of having to address listed species' needs far downstream in another state, terminated their attempts to build the reservoir. Water users, witness to these two shots across the bow of the "good ship water storage and diversion" were awakening to a new world.

During the years from 1979 to 1985, the organizational sponsors of Nebraska's Kingsley Dam–Lake McConaughy project for irrigation and electrical power production struggled to stay beyond the grasp of ESA review, knowing that their licenses, granted by the Federal Energy Regulatory Commission (FERC), were due for renewal in 1987. The USFWS wrestled with FERC and the two local districts that were operating the Kingsley–McConaughy complex: the Central Nebraska Public Power and Irrigation District (CNPPID), producer of electricity and irrigation water, and the Nebraska Public Power District (NPPD), distributor of electric. The story has been told elsewhere (for more details, see Echeverria 2001). However, by 1985, all parties could see that FERC would need to consult with USFWS under ESA section 7 and that the two districts, to gain their operating licenses from FERC, would have to undertake action on behalf of listed species habitat downstream on the Central Platte.

Meanwhile, in 1980, USFWS informed USBR that it would have to consult under ESA section 7 on its Platte Basin projects—the Colorado–Big Thompson project on northeastern Colorado's South Platte and USBR dams on the Wyoming North Platte. About 80 percent of the water in Nebraska's Lake McConaughy comes from USBR Wyoming projects; therefore, loss of sustainable water project yields for these Pathfinder and Kendrick enterprises would diminish flows downstream and damage McConaughy's yields. In 1983, USFWS issued a biological opinion on a large proposed project on Colorado's South Platte—the Narrows Dam and Reservoir, which had been promoted for decades (USFWS 1983). When USFWS's opinion concluded that the Narrows project would capture important sediment-carrying peak flows and would be likely to jeopardize the whooping crane, the repercussions were again felt throughout the basin.

In this highly charged context in 1983, viewing the wreckage of the post-Grayrocks political and administrative landscape, the regional directors in the USBR and USFWS initiated a Platte River Management Joint Study, which was completed ten years later (1993). Soon, in 1984, water interests in Colorado, Wyoming, and Nebraska petitioned the secretary of DOI to be included. As federal authorities planned to discuss the future of water in their basin, state authorities and water users also wished to participate. During the remainder of the 1980s and early 1990s, study participants cajoled, wrangled, and battled over how to design a solution that could serve the requirements of ESA, state appropriation doctrines, apportionments among the states made by the U.S. Supreme Court decrees on the North Platte, and the 1922 Colorado–Nebraska compact on the South Platte. They struggled to frame a vision for addressing species' needs while trying to figure out how to serve a wide range of urban, agricultural, and industrial consumers—all of whom demanded cheap, reliable, high-quality water supplies. By 1993, the Joint Management Study Committee had set forth a sketchy preliminary vision. The statement succeeded in anticipating how necessary habitat could be acquired and how water assets could be made available to that land. It presented an outline of a research and monitoring program, a budget, possible sources of funding, and a system of representative governance. The collaborative effort's preliminary vision had been hammered out.

Meanwhile, USFWS was conducting a series of workshops that, in 1994, produced instream flow recommendations (USFWS 1994) for the Central Platte River. Most striking was the agency's determination that there was an average annual 417,000 acre-foot shortage of water flow (measured at Grand Island, Nebraska), as compared with what the agency calculated to be necessary to fulfill habitat needs of the listed species. The number was, from a state and water provider perspective, shockingly high. Given the history of diversion and depletion, the contemporary Central Platte River produced an average annual flow of only about 1 million acre-feet per year. Water users faced the daunting task of shifting more than 40 percent of that flow to serve USFWS demands.

If, however, water managers failed to arrive at a satisfactory basinwide solution, they would have to undergo individual consultations, during which USFWS would evaluate each project against the severe target flow shortage numbers. Even though state authorities and water users were not asked for input and never agreed to any part of the agency's target flow analysis, they were bound to the numbers by the federal regulatory process. If a cooperative approach failed to produce an acceptable, "reasonable and prudent alternative," the USFWS would devise its own solution on an individual, case-by-case basis as federal permit renewals came up. The USFWS would do so within a frame that included not only the 417,000 acre-foot figure, but also a USFWS determination that there needed to be a total of 29,000 acres of high-quality land habitat in the Central Platte Basin. Reluctantly, states and water providers—realizing that a prolonged legal struggle would put them at risk of highly uncertain outcomes, fearing what individual ESA section 7 consultations would portend, and wishing to come up with some kind of collaborative deal—agreed to sit and talk. This led to an agreement to negotiate a Central Platte basinwide collaborative program, signed on June 10, 1994, by the secretary of DOI and the governors of Wyoming, Nebraska, and Colorado.

The agreement had two major objectives: (1) to develop and implement a habitat recovery program and (2) to enable water users in the Central Platte Basin to proceed with existing and new activities (pre- and post-July 1, 1997). The agreement also allowed the two Nebraska power and irrigation districts to comply with FERC interim relicensing requirements by undertaking specified mitigation actions on behalf of listed species habitat—for example, establishing an environmental water account at Lake McConaughy and acquiring and enhancing habitat lands. All parties agreed to work within existing federal and state laws, court decrees, and interstate compact arrangements. If the effort to construct a viable collaborative program were to fail, it was agreed that all relevant biological opinions would be reopened (a considerable "hammer").

Working Together

During the years from 1994 through 1997, a general plan emerged, strategies were sketched out, and principles were adopted. The task was to convert the general intent of the 1994 Memorandum of Understanding into a workable 1997

Cooperative Agreement that would point to construction of a defensible, basin-wide Central Platte habitat recovery program.

Getting to the 1997 Cooperative Agreement

The essence of the federal approach was to offer the promise of long-term regulatory certainty for water users in return for states' and water providers' willingness to fulfill species habitat needs and to blend in the concept of "milestones fulfillment," which would provide temporary relief from a USFWS finding of jeopardy during the negotiations. The FERC and the USFWS would also demonstrate federal willingness to collaborate in a mutual learning process called "adaptive management"—a disciplined effort to gather data and information scientifically to test program assumptions and habitat manipulations for purposes of better decision making (Walters 1986; chapters 5 and 6, this volume).

After July 1994, a series of meetings took place in motel conference rooms and public buildings along the I-80 and I-25 corridors of the Central Platte River Basin. As the process dragged on, negotiations varied from the tough and confrontational to the tedious (Zallen 1997). As time went by, representatives of all parties felt pressures to "just make a deal." However, representatives of the several interests believed that the price of acquiescing to unacceptable terms for their respective constituencies would be "to have their heads handed to them." For all parties, achieving regulatory certainty was paramount, but everyone wanted that certainty on terms unacceptable to some other party.

Negotiations centered on specifying program milestones intended to embody the concept of "sufficient progress," as defined and required by USFWS. Milestones represent specific benchmarks against which progress of the discussion and of eventual program implementation could be measured. They would allow adjustments for unforeseen circumstances via adaptive management and create incremental goals to bring the whole negotiation process closer to the objective: construction of a viable, "reasonable and prudent alternative." Milestones would be assessed year by year, state by state, and organization by organization. If sufficient progress were not in evidence, USFWS could threaten to withdraw relief from jeopardy determinations and reopen any biological opinions that had been issued. Milestones would be negotiated and employed in the domains of water, land, research and monitoring, as well as program governance and administration.

The USFWS promoted constructive discussion by calling for and collaboratively arranging workshops on troublesome issues—everything from organizing a land habitat entity to the needs of pallid sturgeon—every few months. Small groups and committees representing the several interests met to define problems, do homework and envision solutions, and feed proposals into the workshops.

A basic dynamic was repeated across the years. Members within each network or party (states, federal agencies, and the environmental community) would consult, sort out issues, and build their positions. Each party would then talk with individuals in other networks and strive to work with other parties in program building and critiquing—usually bouncing topics back and forth between interest group committees and the Governance Committee, which consisted of representatives of USFWS and USBR, water users, and environmentalists. As new issues and opportunities emerged, members of each network would meet and regroup to reconfigure their positions. Virtually all participants were busy with other responsibilities in their organizations; thus the work of building networks and developing and modifying positions required considerable sacrifices of time and energy.

Negotiations evolved from many conversations within and among the subnetworks. Leaders cannot simply sign documents and expect successful outcomes. Collaborative efforts, by their very nature, are slow because leaders must build coalitions to retain players' support. Constructing strong coalitions capable of implementing costly and difficult habitat restoration is a process that cannot be rushed.

Target Flow Challenge

Between 1994 and 1997, negotiators realized that their biggest substantive challenge was finding a way to deal with USFWS's judgment that the history of water diversions in the Central Platte River Basin had shorted the critical habitat of an annual water flow average of 417,000 acre-feet. The USFWS distinguished among three kinds of flows:

1. Species flows: minimum flows to keep habitats connected during summer months when heavy irrigation demands were made upon the stream
2. Peak flows: largely uncontrollable, short-duration flow spikes
3. Pulse flows: flows that occur during about 75 percent of the historical record, have a duration from 5 to 30 days, and are in the range of 2,000 to 9,000 cubic feet per second

Target flows were calculated as consisting of species flows plus pulse flows. The USFWS identified springtime pulse flows as a high priority because they were viewed as essential for the following reasons:

- To maintain and enhance the physical structure of wide, barren, and braided channels
- To supply soil moisture and pooled water during the growing season for plants and animals lower in the food chain in meadow grasslands

- To rehabilitate and sustain side channels as nursery habitats for fish, shellfish, and other aquatic organisms
- To facilitate nutrient cycling in floodplains

Pulse flows also raise groundwater levels in wetlands adjacent to rivers and bring organisms close to the soil surface for predation by migratory birds and other species. Pulse flows contribute to the breakup of winter ice, thereby inducing the scouring of vegetation from sandbars, which is especially important in years of low flow. Except for the driest years, USFWS desired that at least 50 percent of the pulse flows occur from May 20 to June 20 and wanted them to emulate traditional, pre-dam flow patterns of 10 days ascending, 5 days cresting, and 12 days descending. Target flow volumes were categorized by dry years, normal years, and wet years. Decisions for what kind of year managers would be facing were to be based on estimated gross water supply, plus estimates of groundwater, precipitation, and snowpack across the basin.

Water users did not agree with the numbers presented by USFWS's biological opinion. Water users believed that the USFWS number—417,000 acre-feet of water—for habitat recovery was not based on logic, history, or even river capacity. In the often heated discussions of target flows, even basic facts about the physical structure of the river could not be agreed upon, especially the most crucial element, which was how much water would actually be required to restore and sustain habitat for the listed species. The USFWS was fixed on its target flow analysis and absolutely would not negotiate away the very basis of its jeopardy opinions. Yet water users vehemently rejected that analysis. Target flows had become nonnegotiable, and this single issue threatened to bring prospects for a negotiated river basin solution to a halt.

Adaptive Management: Avoiding an Impasse

A path had to be found to keep the talks from floundering. The way out was found in the concept of adaptive management. A compromise agreement was reached that allowed water users and states to reject the federal "target flows" analysis, while the federal government held to them, thereby protecting its biological opinions. Adaptive management allowed all parties to make peace—not around any agreed target flow number—but around the idea that collaborative work, research, and monitoring over the duration of the first thirteen-year program increment would allow a more accurate determination of actual species needs. This mutual commitment to adaptive management allowed all parties to get past impossible target flow discussions by agreeing to disagree. In the end, DOI accepted a 1997 Cooperative Agreement that called for the first program increment's average annual reduction of shortage to target flows to be from 130,000 to 150,000 acre-feet. Using this calculation, USFWS could preserve the integrity of its biological opinions while advancing the discussion with water users. If the first increment's water objectives were fulfilled and volumes were found to be inadequate, DOI could always ask for more water under the terms of the original biological opinions that had launched the search for a program. Meanwhile, the talks were back on track.

1997 Cooperative Agreement

On July 1, 1997, the secretary of the Interior and the governors of Colorado, Wyoming, and Nebraska entered into the *Cooperative Agreement for Platte River Research and Other Efforts Relating to Endangered Species Habitats along the Central Platte River, Nebraska* (often called the Platte River Cooperative Agreement or CA). The parties agreed to negotiate a program to conserve and protect the habitat of four species listed as endangered and threatened under ESA. The CA was almost one hundred fifty pages long and included the following documents:

- A list of specific milestones describing obligations of the parties during the Cooperative Agreement period (originally 1997 to 2000, later extended to December 31, 2006)
- A water conservation/water supply document describing studies needed to acquire an annual average of from 130,000 to 150,000 acre-feet of program water to reduce shortages to target flows
- The proposed program, describing water supply projects, means to acquire and manage 10,000 acres of terrestrial habitat during the first thirteen-year program increment, and money to be provided (50 percent by the three states and 50 percent by the federal government)
- A governance document

The Cooperative Agreement period would be employed to (1) complete negotiations of unresolved issues pertaining to land, water, research, and monitoring; (2) provide time for a USBR/EIS team to assess the proposed program as required by NEPA; and (3) provide an opportunity for USFWS to produce a biological opinion that would evaluate the proposed program for sufficiency. A viable program would be launched within the first thirteen-year increment. Water users, meanwhile, would enjoy regulatory certainty by fulfilling milestones.

By 1997, water users had sketched out three sources of program water that would cumulate to an estimated average total of about 80,000 acre-feet per year—a solid start toward the target flow shortage reduction goal of 130,000 to

150,000 acre-feet per year. Nebraska's initial contribution would come from an environmental account established at Lake McConaughy. That account would be served by water inflows equal to 10 percent of the storable natural inflows to the lake in the months of October through April, up to a maximum of 100,000 acre-feet (CA 1997). Wyoming's proposed Pathfinder Dam and Reservoir modification project promised to increase storage by 54,000 acre-feet. About 34,000 acre-feet of this increased storage would serve the program's environmental account at Lake McConaughy. The remaining 20,000 acre-feet would be held for Wyoming's municipal uses. Colorado constructed what came to be known as the Tamarack Plan (after a defunct ranch by that name), designed to produce an annual average yield at the Colorado–Nebraska state line of 10,000 acre-feet via groundwater recharge (replenishment) that would generate return flows to the South Platte, at times needed for recovery of species habitat. In addition, each state and DOI pledged to construct a future depletions plan by the end of the CA period. Any new depletions imposed by water users upon the river would be replaced—whether or not users were in the nexus of ESA.

In the end, everyone could see the advantages of pursuing a collaborative recovery program. The federal government secured a path toward improved flows and restored habitat and had created the prospect of effective working relationships with water interests in the three states. It had also dodged the costly, time-consuming, and highly uncertain outcomes that would be associated with what would otherwise be over one hundred individual ESA section 7 consultations. Water providers and the states would gain regulatory certainty, and they could promise their nervous constituents that land and water acquisitions would be forthcoming on a willing seller/buyer basis. Nebraska obtained relicensing of Kingsley Dam. Wyoming would get a 40 percent portion of additional storage at Pathfinder Reservoir. Colorado would receive coverage for its high mountain dam, reservoir, and conveyance facilities. Environmentalists would see real habitat improvement in their lifetimes. All parties benefited from avoiding a permit-by-permit quagmire.

The July 1, 1997, Cooperative Agreement was far from perfect in the view of any negotiator, but it framed a potentially livable future. In the next three years, from 1997 to 2000, CA proponents converted a general vision of the agreement into a viable, "reasonable and prudent alternative" that could win the support of all the players. The several strategic pieces of a proposed solution appeared to be coming into alignment.

Moving toward a Program: 1997 to 2000

For the next three and one-half years (July 1, 1997–December 31, 2000), negotiators created a plan to supplement the 80,000 acre-feet per year already identified at Wyoming's Pathfinder Reservoir, Nebraska's Lake McConaughy, and Colorado's Tamarack with another 60,000 to 80,000 acre-feet. They hammered out ways and means to acquire habitat land, grappled seriously with the fundamentals of measuring progress toward defined program goals, and began to struggle with the ramifications of adaptive management.

During these three years, negotiations had established the outlines of a deal that took water providers and state authorities well beyond their comfort zones. States and water providers complained about the intrusion of a federal environmental agenda into their state appropriation systems. The USFWS fretted that the process was not advancing at full speed, given the opportunities that states had to delay coming to grips with essential issues. Environmentalists worried that federal negotiators would cut environmentally unprincipled deals to make coalitions with users that—by making their lives easier in dealing with Congress—would gut habitat improvements on the high plains.

The Federal Case

Federal authorities contended that they had demonstrated good partnership values in the negotiations leading up to the 1997 Cooperative Agreement and thereafter. They were flexible in handling target flows (agreeing to set aside the 417,000 acre-feet figure to accept a reduction in shortage of from 130,000 to 150,000 acre-feet during the first increment) and in agreeing to reduce the first increment land requirement from 29,000 acres of land to 10,000. The federal government had also committed itself to an adaptive management strategy, founded on a learning model designed to respond to emerging problems. Mutual learning, USFWS contended, would employ the best available science and management practices and, therefore, should not be too frightening to states because the process would be open to contributions and critiques from states and water users.

The dual role that USFWS must play in any such negotiation became a fundamental source of tension. According to federal law, the agency must retain final authority to implement ESA and to determine program sufficiency. Congress has required that only federal authorities can determine what is or is not acceptable as a "reasonable and prudent alternative." However, to keep negotiations moving forward, USFWS leaders also had to make deals in good faith when they found mutually acceptable opportunities to advance toward serving species habitat needs. Yet, at any given point in the negotiations, none of the participants could foresee the future well enough to prejudge any proposed program's sufficiency. While making deals at the table was one thing, making a subsequent final finding that any proposed program was sufficient was another.

Therefore, on behalf of their agency's managers and scientists, USFWS's representatives accepted responsibility for devising a regulatory dynamic to conduct the best possible EISs and biological opinions not beholden to the politics of deal making in the negotiating rooms. The USFWS could—and would—only make a finding of program sufficiency or insufficiency upon completion of its own internal evaluation processes. If the tension between the two roles of USFWS created uncertainty among resource users, they had little alternative but to press on and live with it.

The States' Case

One fundamental problem centered on the status of USFWS in the negotiations. When USFWS participants took the lead in putting together a document, were they helpfully doing staff work for others at the table? Would their document's contents be open for full discussion among equals, or were the agency's people laying down policy? State water representatives and, at times, environmentalists felt that USFWS negotiators were handing down a series of fiats not open to serious questioning by others. To the extent that water users and environmentalists sensed they were looking at USFWS's decrees, they seriously questioned the legitimacy of the collaborative agreement process. All participants knew that USFWS had to play a dual role as partner and judge and had to keep in mind that any agency representative's statement was likely to contain elements driven by both roles. Water users' and environmentalists' confusion as to what they were looking at and USFWS negotiators' reluctance to make themselves absolutely clear—because they were attempting to be inclusive, nonautocratic, and yet firm stewards of ESA—added to confusion and distrust.

Many of the states' representatives thought that federal representatives, acting in their "partner roles," would signal a willingness to arrive at a specific set of trade-offs in making an agreement. Then, at later points while negotiating the same issues, federal representatives, switching into their "judge roles," insisted on employing "best science," thus reversing or changing their original positions. Negotiators working on behalf of water users found their distrust fueled when USFWS negotiators would bring up new considerations after deals had been hammered out and written down. The water users' negotiators were repeatedly alarmed when USFWS negotiators would apparently see an opening, then keep exerting pressure to advance the perceived federal water, land, and research/monitoring agenda.

Many states' representatives could understand the USFWS's reluctance to renegotiate the biological opinion upon which the rationale for the habitat recovery program was based. Nevertheless, when some USFWS representatives zealously defended their agency's narrowly constructed biological analysis with seemingly minimal appreciation of its optional implications, states' representatives had difficulty maintaining their faith in federal representatives' capacity for constructive discourse. Repeatedly, USFWS negotiators refused to acknowledge the possibility of any flaw in the agency's biological opinions or in its judgments in applying the findings of that document. The water users then wondered what kind of openness and collaboration they could expect from USFWS representatives in the future. Would higher federal authorities be willing to reach out and discuss issues in ways that were not purely peremptory? Busy local people, often operating on inadequate information about the federal government's intent, constructed their own interpretations that fit their various and sometimes hostile preconceptions. Many viewed USFWS and its representatives as unpredictable and undependable. For these state negotiators and citizens, the future of the state–federal partnership became deeply problematic.

The Environmentalists' Case

While the states attacked USFWS for being out of control in its demands for water and land, environmentalists expressed concerns that the agency had been altogether too willing to compromise species' needs. During the summer and fall of 1996, some environmentalists walked away from the table for a brief period, primarily because they viewed USFWS as being too soft on issues such as target water flows and land habitat acquisition goals for the first program increment. Some in the environmental community concluded that their best chance for effective action could be the threat of or initiation of litigation. At that point in the negotiations, these environmentalists believed that further collaboration would simply assist water users in getting around the hard-core intent of ESA (Echeverria 2001). They called upon environmental community members to confront their major strategic choices squarely: (1) to work within the cooperative framework and make what were seen by the more radicalized to be unjustifiable compromises or (2) to stay "pure" outside the negotiations, be prepared to challenge any given negotiated outcome in court, and reject whatever modest gains for the species that might be forthcoming in the name of greater returns over a longer term.

The more radical environmental voices abandoned the negotiations at that time; others returned to the table. Those who remained—representatives of Audubon, the National Wildlife Federation, the Whooping Crane Maintenance Trust, and, for several years, Environmental Defense—chose to accept the compromises inherent in moving from a demand for 417,000 acre-feet of reorganized water to something between 130,000 and 150,000 acre-feet and from 29,000 acres of terrestrial habitat to 10,000 in the first thirteen-year program increment. These more moderate voices saw a desperate situation on the river that mandated action on behalf of the listed species sooner rather than later. They had worked hard to build good relationships with their neighbors on the river and foresaw huge setbacks for the local environmental agenda if national organizations pursued court actions. They also appreciated that the environmental community had won a major battle—a victory not to be abandoned—in that the discussion was no longer about preservation of the old water regime. Instead, by then the negotiations were about installing a new environmental agenda on the Central Platte; the only questions were how much and how soon.

Meanwhile, USFWS leadership sat in the middle of the contending forces, attempting to steer a path that would serve the species and keep opposition from getting out of hand—all this without sufficient personnel, time, or money. The U.S. Congress, then divided over the wisdom of amending ESA, had not devised any new legislative solutions. Congress had clearly determined one thing, however; it would hold USFWS on a short leash by seriously constraining its budgets. Whatever USFWS did by way of implementing ESA via large, landscape-scale programs, the agency would do with inadequate funding and staff.

In the end, what drove the negotiations ahead was the water users' quest for reestablishment of regulatory certainty, as all parties feared becoming bogged down for years with individual water project consultations under ESA section 7. Working something out collaboratively was far more palatable than the daunting prospect of expensive, time-consuming, individual consultations with no promise of any predictable result.

Negotiations Extended: 2000 to 2006

When negotiators gathered on August 3, 2000, they felt good about progress in several directions. During the initial three-year CA period (1997–2000), discussions had advanced on a number of crucial matters:

- A conceptual plan had been devised for providing an additional 50,000 to 70,000 acre-feet of water per year for habitat improvement beyond the 80,000 agreed to by June 1997.
- A rough version of a land habitat acquisition and management plan had been made.
- A preliminary monitoring and research plan had been developed.

Additionally, over the three-year period, USFWS/USBR researchers in charge of Platte River modeling had moved forward, especially with regard to the impacts of peak and pulse flows and the movement of sediment for sandbar building. During this time, the DOI modeling team had applied its methodologies to the hydrology and geomorphology of the river and expected that a viable programmatic EIS would soon be forthcoming. That accomplishment, in turn, was expected to be followed by a "nonjeopardy" ESA opinion. The forthcoming November 2000 national elections would put a new administration in place the following January. The negotiators and their support staffs held guarded hopes that the relevant authorities would endorse the proposed program, either in the last days of the Clinton administration or the in earliest days of a new one.

The August 3, 2000, meeting began with a bombshell announcement: The USBR/EIS team had completed a preliminary analysis of the proposed water action plan, the centerpiece of the 1997 Cooperative Agreement, and found that it could not serve as an ESA-required "reasonable and prudent alternative." The analytical team had found that the planned clear-water releases from the environmental account at Lake McConaughy would likely scrub out channel sediment and further incise it. Program flows would likely carve out deeper incisions in the river channel than already existed and thereby undercut prospects for a wide, shallow stream braided by shifting sandbars. The negotiated program, had it been acted upon, would have exacerbated the very problems it was supposed to solve. Years of work lay in disarray.

This announcement had the potential to unravel everything. Best science had undercut a key program premise—that more water was the cornerstone of a solution. More water was clearly a necessary but not a sufficient condition of improved habitat. Scientists would have to revise the conditions of water delivery carefully. Clearly, a major extension of the 1997 CA would be necessary. By the end of 2000, all parties agreed to extend the deliberations by thirty months, to June 30, 2003. Later, when that extension would prove insufficient, the CA negotiating period would be extended to December 2003, then to June 30, 2005, to December 31, 2005, and finally to December 31, 2006. The revised CA had to be completed by the end of 2006 for several compelling reasons: (1) DOI could not continue to dangle its carrot of a collaborative program forever; (2) budgeted funds for the 1997–2000 CA period were drying up, and no party wanted to go back to Congress or state legislatures to plead for more dollars after nine years of negotiations that had yet to yield a viable program; and (3) in 2006 a national election would occur. State and federal incumbents and challengers would be staking out their claims for votes in 15-second radio and television commercials. None of the negotiators wanted to risk some aspect of their intricate program getting caught up in electioneering. Everybody at the table understood the looming dangers.

Getting to a Viable Program

The years from 2001 through 2006 would provide time for representatives of the various interests to grapple with several highly charged issues:

- Ascertaining sedimentation–vegetation patterns
- Determining river capacity to deliver pulse water to the top of the critical habitat, given highly vegetated conditions upstream
- Tracking water provider depletions up and down the river
- Replacing future depletions that water providers could be expected to impose
- Putting together a more viable research and monitoring program
- Organizing a defensible, practical adaptive management plan

Many of the issues that had emerged since 1994 and had come into further focus by 2000 were now reexamined with attention to working out details that, before 2000, would have been left to the first thirteen-year program increment.

The Governance Committee—consisting of representatives of DOI (from USFWS and USBR), water users, and environmentalists—was committed to guide program efforts by employing an incremental approach, informed by systematic adaptive management principles (Holling 1978; Walters 1986). All parties agreed to treat important program manipulations as sets of experiments designed to test outcomes of thoughtfully prepared management actions that would permit appropriate comparisons and hypothesis testing.

Yet, what was the actual meaning of "adaptive management"? By 2004–2005, the issue of how to organize adaptive management took center stage in the deliberations. All participants could agree that the term meant funding an integrated research and monitoring program that would test assumptions and alternative hypotheses regarding program manipulations of the Central Platte critical habitat. Adaptive management also clearly meant evaluating results scientifically to improve program efforts. Nonetheless, an open-ended examination of a program's effects can be a frightening proposition for those political leaders and administrators who need to know in advance exactly what to expect in terms of financial commitments that need to be made. These administrators can see the dangers of a program's agenda getting out of control, making endless demands on treasuries.

Thus water users and state authorities feared that "adaptive management" could become little more than a catch phrase for tapping into their treasuries to fund half-baked environmental schemes. Environmentalists worried that adaptive management would become the state water users' excuse to tie up action proposals in endless peer reviews. Leaders of USFWS were concerned that adaptive management could be manipulated by the other parties to avoid doing what the agency's scientists deemed essential to habitat restoration and improvement. What was the vision for the river that adaptive management would ostensibly serve?

Conflicting Visions

For more than thirty years, USFWS had consistently advanced its vision, centered on restoration of some semblance of natural water flows to restore habitats for listed species. The natural flow vision, usually characterized by USFWS as a "river processes model," has been advanced in a rich, neatly synthesized scientific literature (Poff et al. 1997; Poff et al. 2003). This vision centers on the importance of highly variable river flows to ecosystem health and native biodiversity. Five critical components of any flow regime are (1) magnitude of discharge, (2) frequency of pulses passing a given point, (3) duration of high water discharges, (4) timing of the peaks, and (5) their "flashiness" (how quickly the flows change from one magnitude to another). Diversity in flow patterns at appropriate times is seen to generate a diverse mosaic of habitat types and to promote ecosystem integrity.

The path to sustaining and improving listed species habitat on the Central Platte River Basin in USFWS's view, therefore, was to reintroduce not only minimum flows but also variable flows in the forms of controllable flood pulses and preservation of peak flows that can scour out vegetation, create and re-create barren sandbars, and maintain wide, braided, shallow channel characteristics essential to the listed species. Analysts from USFWS explicitly recognized that, given a seriously damaged riverine ecosystem, any feasible regime of peaks and pulses could not carry the full burden of channel restoration by itself. To provide quality habitat, vegetation would have to be cleared, and densely vegetated sandy islands that had risen too far above typical summer flow levels would have to be leveled out. Such major work would require intervention with chainsaws, bulldozers, and other mechanical equipment. Yet USFWS wanted to insure that adaptive management would restore some modicum of traditional natural flows. The mantra of USFWS and the environmentalists was clear the channel, level larger higher islands, and rely as much as humanly possible on pulse flows to sustain needed habitat qualities. Negotiators representing USFWS would push hard for program language with "natural flow" requirements.

In contrast, over the years the states and water providers consistently had advanced another vision of how the program should apply adaptive management. Their concerns had always centered on protecting their own water projects' yields for their demanding customers, who expected the water they paid for to be cheap, reliable, and sufficient. The law mandates that water providers serve their customers; providers have no authorization to question the economic growth games that their clients play. Providers are also wary of getting into any kind of program with economic demands that cannot be anticipated in advance.

For water providers, adaptive management had to preserve project yields in an arid environment and operate within agreed-upon "defined contributions" of water, land, and money. To formulate a "reasonable and prudent alternative" that could deliver regulatory certainty, water interests countered the natural flow model with their own mantra—clear (mechanically), level (mechanically and as cheaply as possible), and plow (to whatever extent necessary to move the sand out of vegetated islands into barren sandbars). Water providers voiced deep skepticism that any reasonable amount of water organized into pulses would ever generate the needed habitat characteristics. They maintained that much heavier reliance upon mechanical intervention than USFWS was prepared to endorse would be cheaper and would keep waters in reservoirs—where water customers wanted it.

Water providers view themselves as stewards of water projects that have made our form of civilization possible in the Platte River Basin, projects assembled out of bold vision, engineering genius, and commitment to serve a fundamental need of society—paid for by generations of tax- and fee-paying citizens. To install a collaborative program on the Central Platte meant that their customers would pay not only a negligible amount more for an acre-foot of water (not a problem) and pay a fraction more for a kilowatt hour of electricity (maybe a slight problem), but also would trade away important increments of drought protection on behalf of environmental accounts and other manipulations (a much larger problem). To water providers, the real threat to the basin was the federal agenda—as it opened a crack in the states' doors that, over time, would expand to accommodate a federal intruder with designs on state appropriations that would reduce their projects' water yields (a frightening problem).

In an arid, high plains environment, each water manager must capture more water than needed in average to wet years to serve demands in dry years. Water tends to come in a few (frequently from one to four) short, intensely stormy precipitation spikes each year. To watch a significant amount of these precious waters surge by in the name of improved habitat for three bird and one fish species far downstream was not easy for water providers and users to accept. Attempting to convince state legislators and most of their clients, who advocate economic growth, to accept the environmentalists' vision seemed to be politically impossible. Anything that would reduce project yields would fuel vehement protests from social organizations that require water to fuel their economic and recreational aspirations. Whenever water is not stored, one or more water rights holders on a state's priority list must be shorted that amount in a basin where all rivers are already overappropriated (overallocated) except in the few wettest years.

Authorities responsible for high-quality catchment, diversion, storage, and delivery of water to the public must ensure that water projects' yields available to their users are safeguarded, regardless of whether as individual water users they may have sympathies with environmental needs for water throughout the Platte Basin. Before signing any type of agreement, water providers need to be assured that resources are not going to be allocated in an open-ended process that implicitly says, "Let's wait to find out how much more water, land, and money will be needed next." Instead, water providers spent years skeptically examining USFWS's variable flow river vision and then insisted on protecting their water projects' yields. They pressed hard for an initial adaptive management program commitment that rejected the federal natural flow model. The water providers finally agreed to coexist with that federal vision only within a clear set of defined state and water user contributions—10,000 acres of land, 130,000 to 150,000 acres of target flow shortage reduction, and a specified dollar budget.

The states' representatives feared losing water project yields. They believed that USFWS, relying upon its variable program flow data, had little realistic chance—in the first program increment of thirteen years—of actually obtaining much by way of improved habitat. Water providers did not wish to contemplate what a disappointed USFWS might do in the future when it sent its representatives back with demands for greater inputs of water, money, and land. Therefore, the states' negotiators fiercely objected to any attempts to "hardwire" elements of the "natural flow" model into the adaptive management program. Essentially, they pled, "Judge us on our willingness to provide defined contributions, not upon the success or failure of any particular adaptive management effort guided by a 'natural flow' vision."

These two visions—natural flow and defined contribution—would frequently be pitched against each other around the negotiating table and would drive the discussion on many program particulars. The states wanted a USFWS "non-jeopardy" biological opinion and regulatory certainty, but not at the price of any explicit acknowledgment of the validity of USFWS's philosophy and science of river restoration.

For the states and water users, a critical question came to be: What will happen if a given river manipulation to improve habitat, advanced within the adaptive management plan, fails? What if USFWS becomes wedded to that particular action and simply calls for additional, larger-scale, more expensive efforts to advance futile efforts? If states were to withdraw support for such manipulations, would USFWS then find jeopardy to the species and withdraw the "carrot" of regulatory certainty? If so, how much more water, time, and money would be required to obtain a given habitat's enhancement objective?

The solution was to make a clear distinction between accomplishing program milestones and defining the degree of success associated with any particular adaptive management activity. Milestones specified organizational, land, water, research, and monitoring activities to be undertaken by specific dates. They identified essential program inputs that must be delivered by the parties at specified times. Fulfilling milestones has been deemed essential to defining program success. Failure to do so can justifiably lead to withdrawal of regulatory certainty. Negotiators then adopted meticulously worded language to the effect that specific adaptive management actions will be viewed as explorative steps in the quest for habitat recovery. Failure of any particular set of adaptive management activities will not raise issues of water program sufficiency. Negotiators for USFWS had advanced the distinction between milestones and adaptive management actions for years, but states and water users previously had seen large potential problems in the gray areas between the two concepts. Working out specific language proved to be a major challenge and, at times, almost brought the proceedings to a halt.

By April 2005, however, negotiators crafted language with the necessary balance of mutual flexibility and constraint. Participants could look forward to an adaptive management plan that would be systematic in its approach to habitat restoration. Although the USFWS retained its vision of natural flow restoration, at no point in the program's language did the states and water providers embrace any aspect of that vision. For them, the adaptive management plan accepted their "defined contributions." Somehow, adaptive management itself will have to adapt to the coexistence of two sharply different visions of how to proceed within the management process. The contesting parties may never agree on an overarching vision for the river, but they have agreed that listed species' habitats need to be enhanced and sustained.

The Deal

When negotiators worked out a way for all parties to live with their conflicting visions in the adaptive management plan, they removed a major obstacle to proceeding with the recovery program. The program package is intended to accomplish the following:

- To provide ESA compliance for existing and new water providers' activities
- To help prevent the need to list more endangered and threatened species
- To mitigate negative impacts of new depletions on USFWS target flows

- To establish an organizational structure that will ensure stakeholder involvement in program implementation

Mitigating Water User Depletions through June 30, 1997

By 1997, the parties had agreed to reduce the shortage to USFWS's target instream flows by an average of 130,000 to 150,000 acre-feet of water per year, as measured at Grand Island, Nebraska, during the first thirteen-year program increment. By then about 80,000 acre-feet per year had been identified for the program. During the three and one-half-year period from July 1, 1997, through December 31, 2000, the states pieced together an additional set of projects that analysts, performing reconnaissance-level analysis, established could fill the gap between the 80,000 acre-feet already identified and the objective of 130,000 to 150,000 acre-feet.

This second set of water supply projects centered heavily on water stored in Wyoming and Nebraska being leased from farmers who traditionally have irrigated their farmland. These water supplies will be freed for instream flows, retiming water via manipulation of reservoirs and irrigation ditches in Nebraska. Farmers with irrigation water stored in a Wyoming or Nebraska reservoir will be paid by the program to reduce their own agricultural demands, freeing up water for the birds and the sturgeon.

Electrical power interference arrangements will be exploited in Nebraska; these entail making cash payments to electrical power producers to change the patterns of water releases through their turbines on behalf of species habitat. In Colorado, the plan calls for mounting a groundwater recharge and retiming project that will capture flows of the South Platte just upstream of the Nebraska border and release them into sandy pits to recharge groundwater near the river channel. The location of the pits is calculated to generate return flows to the surface channel at times of shortage at the Grand Island gauge. Diversions in the neighborhood of 30,000 acre-feet per year in times of excess river flow are expected to produce, on average, about 17,000 acre-feet of additional water returning to the river at times of shortage at the Grand Island gauge and thereby earn program credit. This returned water will enhance base flows, upon which reservoir releases from Lake McConaughy (in turn, also supplied by the environmental account at Wyoming's Pathfinder Reservoir) can supply springtime pulse flows and/or more steady-state summer flows. An environmental account manager employed by USFWS will be based at the agency's Grand Island field office to oversee all such projects. These water supplies will offset the historic depletions imposed by all basin water projects in place before July 1, 1997.

Mitigating Water User Depletions on or after July 1, 1997

What about future—post July 1, 1997—water project depletions? The three states and DOI devoted considerable energy to designing new depletions plans to provide offset water for everything from consumptive uses associated with new U.S. Forest Service campgrounds and other federal facilities to increases of the states' agricultural, municipal, and industrial consumptive uses in the Platte River Basin. By 2000, negotiators had established general concepts, but their discussions of ways to provide offset water became more focused and detailed in the years from 2001 through 2006.

The core of Wyoming's depletions plan is to establish baseline benchmarks that will define water use in agriculture, municipalities, and industries as of June 1997. Wyoming has a small agricultural economy on the North Platte; thus, during the course of many years, its water providers are likely to operate below the program caps. If and when a given benchmark is exceeded, Wyoming authorities will provide offset water to the North Platte channel at the Wyoming–Nebraska state line, with the expectation that the program will see that the water arrives at Lake McConaughy. From there it can serve habitat recovery.

Nebraska's future depletions plan centers heavily upon offsetting depletions caused by continued agricultural well installations after mid-1997. Well owners are a powerful political force in Nebraska. After years of difficult discussions, negotiators agreed that the state treasury would finance provision of offset water for all "new" Nebraska depletions from July 1, 1997, to December 31, 2005. After that date, the party proposing a project that would impose a new depletion will generally provide offset water. The state of Nebraska will share some of the costs of providing offset water, depending on the nature of its use and on the extent that a particular new project threatens state-protected flows and USFWS target flows.

In Colorado, the South Platte River has increased its flow over the past 125 years, due to water imports from the western slope's Colorado River Basin to serve the needs of growing east-slope Colorado cities and the state's remaining agriculture. The South Platte also has accommodated the shift of water usage from agriculture to densely populated urban and suburban areas. Cities and suburban areas consume less water to sustain businesses, factories, and human beings than farmers use to grow corn and dry beans. Therefore, the foundation of Colorado's future depletions plan is to reprogram these increased river flows from times of excess (late fall and winter) to times of need (late spring and summer) for listed species' habitats. Colorado employed this same logic to address its pre-1997 depletions. Federal planners have estimated that during the first thirteen-year program, the new nonforest, vegetation-related federal activities will deplete less than an average of 1,050 acre-feet of water per year around the basin. The federal government owns no water rights within Colorado, Nebraska, and Wyoming. Thus USFWS has negotiated an arrangement whereby up to 350 acre-feet per year may be leased from each state within state priority systems in return for appropriate cash payments.

Land

A central strategic objective was to work out procedures for acquiring high-quality terrestrial habitat. The USFWS has sought to maximize acquisition of habitat complexes, composed of barren, sand-island, shallow, river channel acreage where fields of vision for whooping cranes would have a radius from 400 to over 600 feet, with adjacent wet meadows and buffer zones to protect the birds from human disturbance. The first program increment objective is to secure 10,000 acres of suitable habitat on a willing seller/lessor/buyer basis. The program will incorporate habitat lands acquired prior to July 1, 1997, by the Platte River Whooping Crane Maintenance Trust, Audubon, the Nature Conservancy, and lands acquired for river restoration by the CNPPID and NPPD as part of their FERC relicensing requirements.

A Successful Outcome

Negotiators representing DOI, the three states, and the environmental community produced a mutually satisfactory program in December 2005, secured a nonjeopardy biological opinion by mid-2006, and—after much haggling over details—were witnesses to state and federal endorsements in November 2006 (USDOI 2006; USFWS 2006a; USFWS 2006b). Implementation of the thirteen-year first program increment began on January 1, 2007 (Governance Committee, PRESP 2006a and 2006b).

Conclusion

In the end, DOI's promise of regulatory certainty for water users and relief from ESA "jeopardy" determinations, to be realized via plodding fulfillment of milestones, has produced a program that can potentially serve the needs of the listed species. Representatives of three states, water providers, environmental interests, and federal agencies—none of whom had a history of mutually supportive, warm relationships—succeeded in putting together a major program of land, water, and empirically informed adaptive management strategies that represents, whatever its inadequacies, a major departure and achievement in basin water planning and management.

Placing an environmental agenda on the dockets of Platte Basin utilitarian water use is no small accomplishment. The process had its tedious and antagonistic moments. It was time consuming and expensive, if measured against costs of many other environmental programs of lesser scale and challenge. Yet this process turned out to be comparatively expeditious and inexpensive, considering the nation's higher priorities during the late 1990s and the first six years of the twenty-first century.

This process worked because all parties acknowledged that ESA required water providers to accept accountability—either through a high-cost, uncertain game of individual project consultations or by participating in a cheaper, quicker, more systematic, voluntary collaborative program. The issue is not, Shall we have water community accountability for impacts on listed species habitat? The issue, instead, is, How and at what cost will water providers, environmentalists, and state and federal agencies in the basin conduct the essential policy consultations and implement solutions?

The effort to assemble a Platte River habitat recovery program reflects a fundamental shift in this nation's water history. Even though the states and water users have avoided explicitly embracing USFWS's natural flow conceptual model, the mandates of ESA have helped create something new and important under the high plains' sun. Amid the cacophony of basin voices, whenever water and electricity consumers want to increase their diversion of basin water flows, their representatives must first address the needs of three birds and one fish and the rich strands of the biotic web that ride with their fate.

Acknowledgments

The author extends warm thanks to Annie Epperson and Troy Lepper, sociology graduate students at Colorado State University, for their assistance with the first edition of a larger manuscript. This chapter is drawn from the author's larger, book-length study of the Platte River's endangered and threatened species program, entitled *Something New Under the Platte River Basin Sun: Mobilizing Water Users to Implement the Endangered Species Act by Producing a Large, Landscape-Scale Collective Good.*

Notes

1. Endangered species are those in danger of extinction; threatened species are those considered likely to become endangered. The ESA protects all species "listed" as either endangered or threatened by USFWS, which—along with the National Marine Fisheries Service (NMFS)—administers ESA.

2. "The FWS of the DOI and the NMFS of the U.S. Department of Commerce share responsibility for administering ESA. . . . Essentially, the Secretary of the Interior (through

FWS) is responsible for terrestrial species, freshwater species, and some marine species (marine birds and sea otter). . . . The Secretary of Commerce (through NMFS in the National Oceanic and Atmospheric Administration [NOAA]) is responsible for most marine species and most anadromous fish" (Drew 2007, note 81, 10490).

References

Austin, J. E., and A. L. Richert. 2001. *A Comprehensive Review of Observational and Site Evaluation Data of Migrant Whooping Cranes in the United States, 1943–1999*. Jamestown, ND: U.S. Geological Survey (USGS), Northern Prairie Wildlife Research Center.

Cooperative Agreement for Platte River Research and Other Efforts Relating to Endangered Species Habitats along the Central Platte River, Nebraska. 1997. (Platte River Cooperative Agreement). At www.platteriver.org/library/CooperativeAgreement/CooperativeAgreement.htm.

Currier, P. J., G. R. Lingle, and J. G. Van Der Walker. 1985. *Migratory Bird Habitat on the Platte and North Platte Rivers in Nebraska*. Grand Island, NE: Platte River Whooping Crane Critical Habitat Maintenance Trust.

Drew, Cynthia. 2007. "Beyond Delegated Authority: The Endangered Species Act [ESA] Counterpart Consultation Act Regulations." *Environmental Law Reporter* 37 (June) note 81: 10490.

Echeverria, J. 2001. "No Success Like Failure: The Platte River Collaborative Watershed Planning Process." *William and Mary Environmental Law and Policy Review* 25 (3): 559–604.

Eisel, L., and J. D. Aiken. 1997. *Platte River Basin Study: Report to the Western Water Policy Review Advisory Commission*. Springfield, VA: U.S. Department of Commerce, National Technical Information Service.

Eschner, T., R. Hadley, and K. Crowley. 1981. *Hydrologic and Morphologic Changes in Channels of the Platte River Basin: A Historical Perspective*. Denver: USGS.

Forbush, E. H., and J. B. May. 1955. *A Natural History of American Birds of Eastern and Central North America*. New York: Bramhall House.

Freeman, D. M. 2003. *Organizing for Endangered and Threatened Species Habitat in the Platte River Basin*. Special Report 12. Fort Collins, CO: Colorado Water Resources Research Institute, Colorado State University.

Governance Committee, Platte River Endangered Species Partnership (PRESP). 2006a. *Platte River Recovery Implementation Program Cooperative Agreement*. October. At www.platteriver.org/library/Program%20Agreement/program_agreement_final.pdf.

Governance Committee, PRESP. 2006b. *Platte River Recovery Implementation Program (Attachment 3) Adaptive Management Plan*. October 24. At www.platteriver.org/library/Program%20Document/adaptive_management_plan.pdf.

Holling, C. S. 1978. *Adaptive Environmental Assessment and Management*. New York: John Wiley.

Keyes, J. W., III. 2002. "Statement on The Platte River Cooperative Agreement." Presentation by the commissioner, U.S. Bureau of Reclamation. Oversight Hearing on the Endangered Species Act: Platte River Cooperative Agreement and Critical Habitats. Field hearing at College Park, Grand Island, Nebraska, February 16.

Lutey, J. M. 2002. *Species Recovery Objectives for Four Target Species in the Central and Lower Platte River (Whooping Crane, Interior Least Tern, Piping Plover, Pallid Sturgeon)*. Report prepared for the U.S. Fish and Wildlife Service. June 26. Denver: URS Greiner Woodward Clyde.

McDonald, P. M., and J. G. Sidle. 1992. "Habitat Changes Above and Below Water Projects on the North Platte and South Platte Rivers in Nebraska." *Prairie Naturalist* 24 (3) (September): 149–58.

Matter, M. J. 1969. *The Great Platte River Road.* Lincoln: University of Nebraska Press.

Matthiessen, P. 2001. *The Birds of Heaven: Travels with Cranes.* Paintings and drawings by R. Bateman. New York: North Point Press.

National Research Council (NRC) of the National Academies. 2005. *Endangered and Threatened Species of the Platte River.* Final Report. Washington, DC: The National Academies Press.

Platte River Management Joint Study. 1993. "Platte River Habitat Conservation Program." Unpublished document. May. Denver: USBR and USFWS.

Poff, N. L., J. D. Allan, M. B. Bain, J. R. Karr, K. L. Prestegaard, B. D. Richter, R. E. Sparks, and J. C. Stromberg. 1997. "The Natural Flow Regime: A Paradigm for River Conservation and Restoration." *Bioscience* 47 (11) (December): 769–84.

Poff, N. L., J. D. Allan, M. A. Palmer, D. D. Hart, B. D. Richter, A. H. Arthington, K. H. Rogers, J. L. Meyer, and J. A. Stanford. 2003. "River Flows and Water Wars: Emerging Science for Environmental Decision Making." *Frontiers in Ecology and Environment* 1 (6): 298–306.

U.S. Department of the Interior (USDOI or DOI). 2006. *Record of Decision Platte River Recovery Implementation Program Final Environmental Impact Statement.* September. At www.platteriver.org/platte_record_of_decision_092706.pdf.

USDOI, U.S. Fish and Wildlife Service (USFWS). 1973. Endangered Species Act of 1973 as Amended through the 108th Congress. "Interagency Cooperation." Section 7: 15–23. (ESA sec. 7 16 U.S.C sec. 1536). At www.fws.gov/endangered/pdfs/ESAall .pdf.

U.S. Fish and Wildlife Service (USFWS). 1983. *Biological Opinion on the Proposed Narrows Dam and Reservoir on the South Platte River in Colorado.* Denver: USFWS.

USFWS. 1994. *In-stream Flow Recommendations for the Central Platte River.* Report prepared by David Bowman. May 23. Fort Collins, CO: National Ecological Research Center of the National Biological Survey. Unpublished.

USFWS. 1997. *Biological Opinion on the Federal Energy Regulatory Commission's Preferred Alternative for the Kingsley Dam Project (Project No. 1417) and North Platte/Keystone Dam Project (Project No. 1835).* Grand Island, NE: USFWS.

USFWS. 2006a. *Biological Opinion on the Platte River Recovery Implementation Program.* June 16. At www.fws.gov/filedownloads/ftp%5Fregion6%5Fupload/Platte%20River%20Final%20Biological%20Opinion/Platte_River_FBO(June16).pdf.

USFWS. 2006b. *Platte River Recovery Implementation Program Final Environmental Impact Statement,* Volume 2. April. At www.platteriver.org/library/FEIS/Volume2/Volume2.pdf.

Walkinshaw, L. 1973. *Cranes of the World.* New York: Winchester Press.

Walters, C. J. 1986. *Adaptive Management of Renewable Resources.* New York: Macmillan.

Zallen, M. 1997. "Integrating New Values with Old Uses in the Relicensing of Kingsley Dam and Related Facilities." Paper presented to the Natural Resources Law Center, 18th Summer Conference. June. Boulder, CO: University of Colorado School of Law.

Zallen, M. 2005. "Proposed Platte River Recovery Implementation Program: How and Why Did We Get Here and Where Is Here?" Paper presented to Colorado Water Law Conference. Denver: Continuing Legal Education International. March 8. Unpublished.

Chapter 5

Platte River Basin Ecology

A Three-Dimensional Approach to Adaptive Management

THOMAS L. CRISMAN

The Platte River Basin comprises two major sub-basins, the North Platte River, with a length of 1,070 kilometers (km) and a drainage basin of 90,390 square kilometers (km^2), and the South Platte River (length 724 km, drainage basin 62,936 km^2). The two rivers join at North Platte, Nebraska, to form the Platte River, which flows for 499 km to its confluence with the Missouri River and drains an additional 77,181 km^2 (USGS 2006). Before European settlement (preimpact) of this part of the West, both the North and South Platte rivers flowed from their headwaters in the Rocky Mountains down through deeply cut canyons until reaching the plains and spreading into a network of braided streams, characterized by sparse vegetation and high loads of sediment (Eschner et al. 1981). At the time of European contact, spring floods, resulting from high mountain snowmelt, kept woody vegetation in check in the channel network, which often exceeded 1.5 km in width (Matter 1969). River flow during the remainder of the year was low.

In the twenty-first century, the entire three-river basin can be divided into three distinct divisions: upper, middle, and lower. The upper river segments of both the North and South Platte rivers drain high-mountain, relatively pristine landscapes and represent the best water quality of the basin, with the exceptions of interbasin water transfers, impoundments (reservoirs created by dams), leaching of heavy metals from abandoned mines (Heiny and Tate 1997), and nonnative species of fish (Vincent and Miller 1969). Scientists and others are concerned that climate changes may reduce habitat for cold-water fish species in the upper segments of both rivers (Rahel et al. 1996).

The middle segments of both rivers begin at the base of the mountains and extend eastward to their confluence to become the Platte River. Both segments begin with large reservoirs. The South Platte passes through the only significant metropolitan area in the entire basin, Denver, where it receives urban runoff that can contain elevated concentrations of nutrients (some of which are chemicals that algae and other aquatic plants respond to by excessive growth), heavy metals, and pesticides (Kimbrough and Litke 1996; Sprague 2005). The middle river segments are experiencing the greatest ecosystem alteration through reservoir construction, channel modifications, altered riparian zones (areas adjacent to the river), groundwater extraction and irrigation, and habitat loss for several endangered bird species. Flow in the lower segment of the Platte River is the least regulated in the basin, with pronounced loss of fish habitat, especially for sturgeon, as this segment receives and combines environmental impacts from the entire upper and middle basins.

While recognizing the important management issues of both the upper and lower river segments, this evaluation focuses on structural and functional alterations of the middle segments of the North Platte and Platte rivers. Particular emphasis is placed on in-channel and riparian responses to human impacts within a broad ecohydrological context, especially those associated with altered channel flow and irrigation.

Reservoirs

Surface water flow in the Platte River Basin has been altered significantly by construction of fifteen major dams and numerous storage works. The South Platte segment of the basin has 160 storage facilities, holding on average a total of 2.8 million acre-feet of water, while the North Platte has 84 works with a total capacity of 4.3 million acre-feet (Eisel and Aiken 1997). Total water storage capacity of the Platte River Basin, including 7.1 million acre-feet in reservoirs, is over six times the average annual flow at Grand Island, Nebraska (see chapter 4, this volume).

Hynes (1975) noted that rivers are dynamic ecosystems intricately linked with their valleys laterally and longitudinally. A river's structure and function evolve downstream in direct response to changing interactions with its increasingly complex valley. Vannote et al. (1980) expanded on the ideas of Hynes and developed the river continuum concept to explain the progressive, predictable changes in structure and function that rivers experience from their headwaters to their terminuses. More recently, Ward and Stanford (1995) developed the serial discontinuity concept to explain how dams and other channel alterations can disrupt the progression of evolving patterns in downstream structure and function, creating disconnects within the system. Dams disrupt seasonal hydrological patterns and associated sediment grain size, transport, and deposition. Environmental alterations in ecohydrological relationships within the middle segments of the Platte River Basin have the potential to remove the heart of the entire basin, break system continuity, and irrevocably alter the structure and function of both upper and lower river segments.

Dams and associated reservoirs have created major impacts upstream and downstream within the Platte River Basin. Upstream impacts have severely affected the oldest major reservoir in the system, Lake C. W. McConaughy, in Nebraska, built in 1941 (Joeckel and Diffendal 2004). Erosion of the river channel immediately above this reservoir has cut more than tens of meters of soil off the riverbank, and a progressively growing delta, formed from built-up sediment, extends approximately 4 km into the lake. When a lake is drained or exposed, the soil at its bottom is eroded by the wind (eolian erosion) and by the movement of sediment from one place to another (redeposition); sodium chloride (salt) can also be carried into the lake via evaporation (salinization). When a lake is formed, the water table of the surrounding watershed rises.

Water discharges from reservoirs of the Platte River Basin are now clearer than preimpoundment (before being dammed), and, as a result of sediment deposition behind dams, spring floods have been reduced in frequency and magnitude, and water discharges in summer, fall, and winter are generally higher (McDonald and Sidle 1992). Lowered water levels in the river downstream from dams often result in elevated summer temperatures that can be detrimental to wildlife and aquatic biota (plant and animal life) (Gu et al. 1998).

Channel Configuration

By far the most pronounced changes downstream have been caused by alterations in the configuration of the river channel. Dam construction has resulted in downstream river flow being confined within a single or a few incised, relatively straight channels created by erosion action from sediment-starved, clear water discharge from dams. The original braided channel network (characterized by shallow, multiple streams separated by small islands and sandbars over a wide area), which existed before the dams were built, was changed dramatically by surges of water let out from the dams that scoured out the small streams and flooded the streams' banks. All of these actions reduced the width of the floodplain (Johnson 1997). Such changes have been associated with lower magnitude and less frequent spring flood pulses and generally higher summer, fall, and winter flow regimes (McDonald and Sidle 1992).

Scientists' understanding of longitudinal changes in the historical structure and function of aquatic biota within the Platte River Basin is incomplete, complicating their ability to reconstruct the preimpoundment interrelationships between the river and the biota in and around it. Few studies have been published on benthic macroinvertebrates (creatures without backbones that live on the river bottom and in its sediment), and data on in-channel macrophyte (visible aquatic plants) and algal communities are practically nonexistent. Communities of benthic macroinvertebrates in mountainous segments of the South Platte River are statistically more species-rich than comparable communities of the middle river, but significant differences within the latter are associated with channel width and configuration (braiding) and nutrient levels (Tate and Heiny 1995).

Yu and Peters (2003) suggested that observed seasonal differences in fish communities of the Platte River in Nebraska were likely related to habitat characteristics. In spite of profound alteration of the river channel associated with dams, Chadwick et al. (1997) found that the species composition of the fish that live in the central Platte River is somewhat more diverse than that of the lower river and has not changed appreciably between 1939–1941 and 1987–1995. With a critical loss of habitat throughout the Missouri and Mississippi river basins, the status of the pallid sturgeon in the lower Platte River is poorly known but is a critical component of basin management (see chapter 4, this volume).

Water quality varies longitudinally in the Platte River system as an integration of local geology, land use, and hydrologic conditions. Both urban and agricultural areas are major exporters of pesticides and herbicides to the river, especially during summer (Kimbrough and Litke 1996), with higher concentrations of both pollutants in fish tissues in areas of mixed urban/agricultural land use than in areas of pasture lands (Tate and Heiny 1996). While often clearly reflecting local geology, heavy metal concentrations in water and fish were elevated in mining and urban areas (Heiny and Tate 1997). Many water chemistry parameters are elevated in the Platte system during droughts, with clear differences noted among land use categories (Sprague 2005). In central Nebraska, fish community structure clearly reflects differences in the water's specific conductivity, that is, its electrolytic strength (Frenzel and Swanson 1996). Distinct differences were noted between rangeland and cropland, with the latter having the lowest scores for the index of biotic integrity of any land use examined, due to alterations in water quality.

Habitat heterogeneity in a river basin, preferably a broad, braided river channel as the presettlement Platte River Basin was, is considered essential for wildlife. Migratory waterfowl display clear species-specific preferences for open water habitat areas that offer differences in feeding strategies (Johnson et al. 1996). Beginning about 1900, woodlands dominated by cottonwood, poplar, and aspen trees (*Populus spp.*) and willows (*Salix spp.*) began expanding along the river course and, by the 1930s, had occupied most of the former channel areas of the North and South Platte rivers and were moving into the Platte River (Johnson 1994). Loss of open channel areas in the Platte River has exceeded 10 percent of total area per annum during droughts. No significant declines in channel area have been reported since 1969, and channel area in some river reaches has actually increased. Since 1986, channel to woodland proportions have remained relatively stable because a steady state has been achieved through flow and channel configuration, coupled with erosion of some established woodlands (Johnson 2000).

Expansion of woodlands into former channels reduces potential roosting habitat of sandhill and whooping cranes (Johnson 2000). The Big Bend stretch of the Platte River alone hosts over 10 million ducks, geese, and Eskimo curlews and approximately 500,000 sandhill cranes (see chapter 4, this volume). The loss of formerly extensive braided river channels to woodlands has pushed migratory birds into ever smaller areas of open water and poses a major conservation problem of national and international importance. Of equal concern is the habitat fragmentation and overall loss of sandbars for endangered least terns and threatened piping plovers due to altered flow, erosion, and expansion of woodlands. In addition to habitat loss, bird communities of the Platte River Basin are susceptible to chemical contamination from industrial and agricultural sources (Custer et al. 2001). Proposed rehabilitation of bird habitats in the central Platte River Basin through a reregulation of 11 percent of the average annual surface flow of the river and restoration of 4,046 hectares (ha) of sandbar and channel habitats will not only aid conservation of cranes, plovers, and terns, but also likely enhance habitat for endangered species downstream, including the pallid sturgeon, near the confluence of the Platte and Missouri rivers (see chapter 4, this volume).

Riparian Zones

The Platte River is different from most major river systems in North America because, rather than expanding broadly across a normally dry floodplain, the river's seasonal flooding traditionally filled both the channel network and intervening sandbars and uplands of the wide, braided river channel. Johnson (1998) developed a two-phase model to explain the response of the Platte River and its riparian vegetation to dam construction. Phase 1 is relatively rapid, lasting a few decades and characterized by channel narrowing and deepening with expansion of pioneer woodland (*Populus–Salix*), whereas phase 2 follows geomorphic (natural landscape features) stabilization and represents the successional process—that is, the series of changes taking place in an ecosystem after cessation of a disturbance—leading to old-growth forests that can take longer than a century to attain.

Expansion of riparian woodland, especially cottonwood (*Populus deltoides*), was most rapid in the 1930s and 1950s, and by the 1960s, total forested area stabilized (Johnson 1997). Katz et al. (2005) noted that the structure of riparian woodlands in some regulated versus unregulated segments of the Platte River system were extremely similar because most forests are located on terraces that rarely flood. This result was considered a temporal artifact, reflecting the lack of major floods since construction of upstream dams. Even taxa (plant or animal species) living along the riverbank, such as box elder (*Acer negundo*), are thriving below dams due to less flooding and the dams' physical removal of sediments (Friedman and Auble 1999).

Attempts to remove riparian vegetation to improve wildlife habitat have met with mixed results. Removal of riparian vegetation in one section of the Platte River to improve habitat for migrating sandhill cranes and waterfowl led to disequilibrium of the river channel downstream, with a 10-percent loss of channel area due to sediment mobilization and agradation (physical degradation) of the channel bed (Johnson 1997). The latter stimulated increased tree and shrub growth.

The impact of altered river hydrology extends far from the channel and riparian zone. Dam construction on the North Platte River resulted in a 75-percent decline in wetted area of the riparian zone from 1937 to 1990 (Miller et al. 1995). One consequence of this was a reduction in woodland areas with greater than 70 percent canopy closure (the degree to which forests block sunlight), so that cottonwood populations shifted from dense, young stands to more open stands dominated by mature trees. Sedgwick and Knopf (1991) suggested that seasonal, prescribed cattle grazing in the riparian zone, especially during the dormant season, had little impact on riparian vegetation compared with altered hydrology. Additionally, Henszey et al. (2004) found that riparian grassland species distributions are most affected by high rather than low, mean, or median water levels. Thus the dams, built to increase available water from reservoirs for growing populations, caused many unintended consequences that affected the river's hydrology and taxa.

Wetlands

While not a major landscape contributor to the Platte River Basin, wetlands provide valuable habitat and resources for wildlife in general and migratory birds in particular. Wetlands represent a diversity of types, from fully fresh to the saline-sodic (high sodium content) wetlands of Nebraska (Joeckel and Clement 2005). Macroinvertebrates, an integral component of fish and migratory bird diets, display greatest community diversity in intermittent wetlands, but overall, populations are greatest in permanently inundated wetlands, suggesting the need for multiple wetland types for wildlife conservation (Whiles and Goldowitz 2005). Wetlands of the central Platte River that are flooded from nine to eleven months each year appear to maximize development of macroinvertebrate communities (Whiles and Goldowitz 2001). In addition to natural ecosystems, many ephemeral wetlands, that is, those that are flooded for only a few months a year, have been created as a consequence of irrigation operations beyond the riparian zone (Peck et al. 2004). While these systems enhance habitat for wildlife conservation, they are likely to be destroyed if water transfer plans, proposed as part of the Platte River Basin management plan, are not properly implemented.

Hyporheic/Groundwater

Hydrologic alterations in the Platte River system have had far-reaching, landscape-level impacts via interactions of the channel proper with groundwater via the hyporheic zone (area beneath the stream channel that interacts vertically and laterally with groundwater). One third of the area of the South Platte River Basin has been converted to croplands; 8 percent of the total basin area is irrigated agriculture with groundwater (Baron et al. 1998). This has resulted in a 37-percent increase in basin atmospheric water loss via evapotranspiration.

The Platte River in central Nebraska loses water to the aquifer in the western part of the region and gains it from the aquifer in the east, with overextraction of groundwater likely to result in pronounced depletion of water in the river channel (Shu and Chen 2002). Similarly, Sjodin et al. (2001) noted that direction and volume of water moving between the hyporheic zone and the aquifer in the South Platte River Basin is a reflection of season and intensity of groundwater pumping for irrigation.

Overextraction of groundwater for irrigation deprives trees and plants of sufficient water. For example, abrupt drops in the alluvial (stream-influenced) water table in proportion to the amount of water extracted for irrigation causes cottonwood trees' leaves to die (Cooper et al. 2003). Nutrients can be elevated in groundwater of the Platte River Basin (McMahon et al. 1994), and where nutrient-rich groundwater enters the river channel proper via the hyporheic zone, extremely high denitrification rates have been reported (Sjodin et al. 1997). Understanding the three-dimensional nature of landscape-level interactions is critical for management of the Platte River Basin.

Conclusions

As discussed in chapter 4 of this volume, beginning with the Platte River Management Joint Study initiated in 1983 by the U.S. Bureau of Reclamation

(USBR) and the U.S. Fish and Wildlife Service (USFWS), and the subsequent 1997 Platte River Cooperative Agreement, restoration of the Platte River Basin has embraced adaptive management to provide a flexible program developed within the reality framework of competing needs for agriculture, domestic needs, and conservation purposes. An integrated research and monitoring program was proposed that would test assumptions and hypotheses considered key elements of an overall basin management plan. Sound science was considered the basis for policy development. Key elements of the adaptive management plan included (1) providing compliance for existing and new activities of water providers, (2) preventing the need to list new species, (3) mitigating the negative impacts of new depletions on target flows, and (4) establishing an organizational structure to ensure stakeholder involvement in program implementation (see chapter 4, this volume). The program stressed cooperation among the various interest groups that would lead to restoration of the basin in ways that would not affect landscape use by any party.

Embracing an adaptive management approach for the Platte River Basin is the most appropriate strategy to address long-term variability in land uses and climate. However, like the Upper Mississippi River project (see chapters 13, 14, and 15, this volume), management of the basin should be viewed within the context of rehabilitation rather than restoration to a preimpact condition. Dams and associated reservoirs disrupt longitudinal patterns in the evolution of stream structure and function, resulting in serial discontinuity (Ward and Stanford 1983, 1995). Without removal of dams, an unlikely scenario, the downstream evolution of river and riparian structure, function, and interactions envisioned in the stream continuum concept (Vannote et al. 1980) are difficult to attain. While complete hydrologic restoration in a human-dominated basin such as the Platte is generally not an option, it is better to consider flow-related impacts on biota (Strange et al. 1999). This is a reality-based approach to management that begs the questions of how little water an ecosystem needs and when it needs it. Finally, the Platte River is likely to be very sensitive to climate change, especially changes in evapotranspiration. One critical link in the water balance for the basin is winter snow precipitation and the timing of spring snowmelt (Baron et al. 1998).

References

Baron, J. S., M. D. Hartman, T. G. F. Kittel, L. E. Band, D. S. Ojima, and R. B. Lammers. 1998. "Effects of Land Cover, Water Redistribution, and Temperature on Ecosystem Processes in the South Platte Basin." *Ecological Applications* 8:1037–51.

Chadwick, J. W., S. P. Canton, D. J. Conklin, and P. L. Winkle. 1997. "Fish Species Composition in the Central Platte River, Nebraska." *Southwestern Naturalist* 42:279–89.

Cooper, D. J., D. R. D'Amico, and M. L. Scott. 2003. "Physiological and Morphological Response Patterns of *Populus deltoides* to Alluvial Groundwater Pumping." *Environmental Management* 31:215–26.

Custer, T. W., C. M. Custer, K. Dickerson, K. Allen, M. J. Melancon, and L. J. Schmidt. 2001. "Polycyclic Aromatic Hydrocarbons, Aliphatic Hydrocarbons, Trace Elements, and Monooxygenase Activity in Birds Nesting on the North Platte River, Casper, Wyoming, USA." *Environmental Toxicology and Chemistry* 20:624–31.

Eisel, L., and J. D. Aiken. 1997. *Platte River Basin Study: Report to the Western Water Policy Review Advisory Commission*. Springfield, VA: U.S. Department of Commerce, National Technical Information Service.

Eschner, T. R., R. F. Hadley, and K. D. Crowley. 1981. Hydrologic and Morphologic Changes in Channels of the Platte River Basin: A Historical Perspective. Open-file Report 81-1125. Washington, DC: USGS.

Frenzel, S. A., and R. B. Swanson. 1996. "Relations of Fish Community Composition to Environmental Variables in Streams of Central Nebraska, USA." *Environmental Management* 20:689–705.

Friedman, J. M., and G. T. Auble. 1999. "Mortality of Riparian Box Elder from Sediment Mobilization and Extended Inundation." *Regulated Rivers Research and Development* 15:463–76.

Gu, R. C., S. Montgomery, and T. Austin. 1998. "Quantifying the Effects of Stream Discharge on Summer River Temperature." *Hydrological Sciences Journal* 43:885–904.

Heiny, J. S., and C. M. Tate. 1997. "Concentration, Distribution, and Comparison of Selected Trace Elements in Bed Sediment and Fish Tissue in the South Platte River Basin, USA, 1992-1993." *Archives of Environmental Contamination and Toxicology* 32:246–59.

Henszey, R. J., K. Pfeiffer, and J. R. Keough. 2004. "Linking Surface and Ground Water Levels to Riparian Grassland Species along the Platte River in Central Nebraska, USA." *Wetlands* 24:665–87.

Hynes, H. B. N. 1975. "The Stream and Its Valley." *Verhandlungen der Internationale Vereinigung fur Limnologie* 19:1–15.

Joeckel, R. M., and B. J. A. Clement. 2005. "Soils, Surficial Geology, and Geomicrobiology of Saline-sodic Wetlands, North Platte River Valley, Nebraska, USA." *Catena* 61:63–101.

Joeckel, R. M., and R. F. Diffendal. 2004. "Geomorphic and Environmental Change around a Large, Aging Reservoir: Lake C. W. McConaughy, Western Nebraska." *Environmental and Engineering Geoscience* 10:69–90.

Johnson, G. D., D. P. Young, and W. P. Erickson. 1996. "Assessing River Habitat Selection by Waterfowl Wintering in the South Platte River, Colorado." *Wetlands* 16:542–47.

Johnson, W. C. 1994. "Woodland Expansion in the Platte River, Nebraska: Patterns and Causes." *Ecological Monographs* 64:45–84.

Johnson, W. C. 1997. "Equilibrium Response of Riparian Vegetation to Flow Regulation in the Platte River, Nebraska." *Regulated Rivers Research and Management* 13:403–15.

Johnson, W. C. 1998. "Adjustment of Riparian Vegetation to River Regulation in the Great Plains, USA." *Wetlands* 18:608–18.

Johnson, W. C. 2000. "Tree Recruitment and Survival in Rivers: Influence of Hydrological Processes." *Hydrological Processes* 14:3051–55.

Katz, G. L., J. M. Friedman, and S. W. Beatty. 2005. "Delayed Effects of Flood Control on a Flood-dependent Riparian Forest." *Ecological Applications* 15:1019–35.

Kimbrough, R. A., and D. W. Litke. 1996. "Pesticides in Streams Draining Agricultural and Urban Areas in Colorado." *Environmental Science and Technology* 30:908–16.

Matter, M. J. 1969. *The Great Platte River Road*. Lincoln: University of Nebraska Press.

McDonald, P. M., and J. G. Sidle. 1992. "Habitat Changes above and below Water Projects on the North Platte and South Platte Rivers in Nebraska." *Prairie Naturalist* 24:149–58.

McMahon, P. B, D. W. Litke, J. E. Paschal, and K. F. Dennehy. 1994. "Ground-water as a Source of Nutrients and Atrazine to Streams in the South Platte River Basin." *Water Resources Bulletin* 30:521–30.

Miller, J. R., T. T. Schulz, N. T. Hobbs, K. R. Wilson, D. L. Schrupp, and W. L. Baker. 1995. "Changes in the Landscape Structure of a Southeastern Wyoming Riparian Zone following Shifts in Stream Dynamics." *Biological Conservation* 72:371–79.

Peck, D. E., D. M. McLeod, J. P. Hewlett, and J. R. Lovvorn. 2004. "Irrigation-dependent Wetlands versus Instream Flow Enhancement: Economics of Water Transfers from Agriculture to Wildlife Uses." *Environmental Management* 34:842–55.

Rahel, F. J., C. J. Keleher, and J. L. Anderson. 1996. "Potential Habitat Loss and Population Fragmentation for Cold Water Fish in the North Platte River Drainage of the Rocky Mountains: Response to Climate Warming." *Limnology and Oceanography* 41:1116–23.

Sedgwick, J. A., and F. L. Knopf. 1991. "Prescribed Grazing as a Secondary Impact in a Western Riparian Floodplain." *Journal of Range Management* 44:369–74.

Shu, L. C., and X. H. Chen. 2002. "Simulation of Water Quality Exchange between Groundwater and the Platte River, Central Nebraska." *Journal of Central South University of Technology* 9:212–15.

Sjodin, A., W. M. Lewis, Jr., and J. F. Saunders. 1997. "Denitrification as a Component of the Nitrogen Budget for a Large Plains River." *Biogeochemistry* 39:327–42.

Sjodin, A., W. M. Lewis, Jr., and J. F. Saunders. 2001. "Analysis of Groundwater Exchange for a Large Plains River in Colorado (USA)." *Hydrological Processes* 15:609–20.

Sprague, L. A. 2005. "Drought Effects on Water Quality in the South Platte River Basin, Colorado." *Journal of the American Water Resources Association* 41:11–24.

Strange, E. M., K. D. Fausch, and A. P. Covich. 1999. "Sustaining Ecosystem Services in Human-Dominated Watersheds: Biohydrology and Ecosystem Processes in the South Platte River Basin." *Environmental Management* 24:39–54.

Tate, C. M., and J. S. Heiny. 1995. "The Ordination of Benthic Invertebrate Communities in the South Platte River Basin in Relation to Environmental Factors." *Freshwater Biology* 33:439–54.

Tate, C. M., and J. S. Heiny. 1996. "Organochlorine Compounds in Bed Sediment and Fish Tissue in the South Platte River Basin, USA, 1992–1993." *Archives of Environmental Contamination and Toxicology* 30:62–78.

U.S. Geological Survey (USGS). 2006. At www.usgs.gov.

Vannote, R. L, G. W. Minshall, K. W. Cummins, J. R. Sedell, and C. E. Cushing. 1980. "The River Continuum Concept." *Canadian Journal of Fisheries and Aquatic Sciences* 37:130–37.

Vincent, R. E., and W. H. Miller. 1969. "Altitudinal Distribution of Brown Trout and Other Fishes in a Headwater Tributary of South Platte River, Colorado." *Ecology* 50:464–66.

Ward, J. V., and J. A. Stanford. 1983. "Serial Discontinuity Concept of Lotic Ecosystems." In *Dynamics of Lotic Systems*, ed. T. D. Fontaine and S. M. Bartell. Ann Arbor, MI: Ann Arbor Science.

Ward, J. V., and J. A. Stanford. 1995. "The Serial Discontinuity Concept: Extending the Model to Floodplain Rivers." *Regulated Rivers: Research and Management* 10:159–68.

Waller, A., D. McLeod, and D. Taylor. 2004. "Conservation Opportunities for Securing In-stream Flows in the Platte River Basin: A Case Study Drawing on Casper, Wyoming's Municipal Water Strategy." *Environmental Management* 34:620–33.

Whiles, M. R., and B. S. Goldowitz. 2001. "Hydrologic Influences on Insect Emergence Production from Central Platte River Wetlands." *Ecological Applications* 11:1829–42.

Whiles, M. R., and B. S. Goldowitz. 2005. "Macroinvertebrate Communities in Central Platte River Wetlands: Patterns across a Hydrologic Gradient." *Wetlands* 25:462–72.

Yu, S. L., and E. J. Peters. 2003. "Diel and Seasonal Abundance of Fishes in the Platte River, Nebraska, USA." *Fisheries Science* 69:154–60.

Chapter 6

Navigating the Shoals

Costs and Benefits of Platte River Ecosystem Management

STEPHEN POLASKY

The presettlement Platte River was a wide, shallow, muddy, slow-moving river with complex, braided channels. Early attempts to use the river to transport furs and other goods met with frustration and failure, as boats frequently ran aground and had to be dragged over numerous sandbars. Nineteenth-century humorist Artemus Ward described the Platte as "a mile wide and an inch deep" and said that it would be a considerable river if turned on edge (Willoughby 2007).

Attempting to navigate a successful plan for ecosystem management on the Platte is even more difficult than attempting to navigate a fully loaded boat on the river itself. Like the Platte's physical description, negotiations to reach agreement on how to manage water flows in the Platte River are wide-ranging; often slow moving; involve complex, interconnected sets of interest groups; and have an unclear (muddy) resolution.

While the Platte River has not proved to be important for transportation, it is vitally important in other ways. The Platte flows through one of the most arid regions of the country, from central and eastern Colorado and Wyoming through western and central Nebraska, before emptying into the Missouri River at the relatively wetter, eastern end of Nebraska. In dry years, the Platte is the only significant source of surface water in much of its drainage basin: "The Platte River is a consistent source of relatively well-watered habitat . . . with its water source in distant mountains' watersheds that are not subject to drought cycles as severe as those of the Northern Plains" (NRC 2004, 8). As such, water from the Platte River and the habitat along the river is in high demand by people and other species.

What the river is good for and how it should be managed depend on one's point of view. There are numerous complex interest groups with a stake in watershed management along the Platte; it is useful to categorize them into three main groups: (1) urban water users, (2) agricultural water users, and (3) environmental interests. The first two groups have primary interests in water uses that require withdrawal of water from the river, while the environmental interests seek to maintain natural flow regimes and habitat along the river necessary to support species, especially three federally listed endangered species: whooping crane, interior least tern, and pallid sturgeon and one threatened species: piping plover.

Urban and Agricultural Water Use

Within the Platte River Basin along the Front Range of the Rocky Mountains are rapidly growing metropolitan areas from Denver to Cheyenne. The area is attractive because of its sunny weather and the nearby mountains' beauty and recreational opportunities. In Colorado, the population living in the South Platte Basin increased by 34 percent between 1990 and 2003, from 2.3 million to over 3 million (Thorvaldson and Pritchett 2005). Rapid population growth is expected to continue. Population within the South Platte Basin in Colorado is expected to grow by almost 2 million people (a 65-percent increase) between 2000 and 2030 (DiNatale et al. 2005). More people will mean more water demands for residential, commercial, and industrial uses. Total water demand in the South Platte Basin in Colorado is expected to increase by 630,000 acre-feet per year, roughly a 50-percent increase from 2000 to 2030 (DiNatale et al. 2005).

While urban water uses are growing rapidly, agriculture remains the dominant water user in the Platte River Basin. Water is the limiting factor for agricultural production in much of the western United States, certainly the case in the Platte River Basin. Rainfall in much of the basin averages between 10 and 20 inches per year, too little to support agricultural production without irrigation. ("Dryland" agriculture, which does not require irrigation, is practiced but requires fallow years between crop production years.) Only near the mouth of the Platte in eastern Nebraska is rainfall sufficient to support annual crop production without significant irrigation.

In the western states as a whole, agriculture accounted for over 75 percent of all surface water diversions in 1990 (CBO 1997). During the same year, municipal and industrial uses accounted for roughly 10 percent of total surface diversions (CBO 1997). In the South Platte Basin in Colorado, agricultural diversions are currently 3.4 times those of municipal and industrial uses (DiNatale et al. 2005). Approximately 2 million acres of land are irrigated in the Platte River Basin (NRC 2004). While agriculture uses the lion's share of the basin's water, the economic contribution of the agricultural sector to the overall economy is small. Agriculture contributed less than 1 percent of the total value of annual revenues from agricultural products in the South Platte Basin in Colorado (Thorvaldson and Pritchett 2005). Agriculture constitutes a higher percentage of the economy in Nebraska, 4.6 percent of total gross state product in 2006 (BEA 2007), because agricultural production is larger and the rest of the economy is smaller.

Water for agricultural use in the Platte River Basin is likely to decline over time because agriculture cannot compete economically or politically with water demand for urban uses, and overall water diversions will be limited by environmental concerns. In the South Platte Basin in Colorado, irrigated acres are expected to fall by between 133,000 to 226,000 acres from 2000 to 2030 (DiNatale et al. 2005).

As well noted by David Freeman (see chapter 4, this volume), though diverting water from agriculture to other higher valued uses is supported by economic logic, the drying of agriculture will have negative consequences for small towns and rural areas dependent upon agriculture. Many small communities in the Great Plains have suffered from declining populations and stagnant economies for decades. Loss of water for irrigation will hasten the decline and literally drain much of the remaining vitality from these communities.

The recent surge of interest in the production of renewable energy from biomass crops (biofuels) might keep agriculture afloat in the region. The increased demand for corn from ethanol production helped push corn prices from around $2 per bushel in early 2006 to nearly $4 per bushel in early 2007 before prices fell back slightly (ERS 2007). The increased demand and higher price for corn make its production more profitable and water use for agriculture more valuable. Calls for even higher production of biofuels in the future could result in further increases in agricultural prices. In the State of the Union Address in 2007, President George Bush called for increasing renewable and alternative fuel production to 35 billion gallons by 2017. By way of contrast, ethanol production in 2006 was 4.85 billion gallons (RFA 2007) . Ethanol production in the Platte River Basin, however, will likely be limited by water availability. In addition to water used for irrigation of corn and other crops, water for ethanol production amounted to 4.2 gallons of water per gallon of ethanol produced in 2005 (IATP 2006).

Environmental Interests and the Endangered Species Act

Engineering the Platte River to deliver water when needed for agricultural, industrial, and municipal uses has resulted in fundamentally altered ecosystems along the river. Water has been diverted, thereby lowering the amount of overall flow through the river. Storage dams change the timing of flows, water temperature, and the amount of sediment in the water. Changes in timing of flows have altered flood regimes, which have had the further consequence of altering vegetation along the river. Reduced flooding has allowed encroachment of trees and permanent vegetation on previously sandy or grassy terrain. These altered conditions have had detrimental effects on several species that evolved under previous conditions. For example, whooping cranes "have specific requirements for roosting areas that include open grassy or sandy areas with few trees, separation from predators by water, and proximity to foraging areas such as wetlands or agricultural areas" (NRC 2004, 8). For whooping cranes, a river a mile wide and an inch deep with spring floods that sweep away woody vegetation provides prime habitat. Piping plover, a federally listed threatened species, have also been adversely affected by the encroachment of trees, reducing the number of open and sandy or grassland areas they use for nesting.

For most of the time since European settlement, environmental interests have been relegated to a back seat. Agricultural water users, and, increasingly after World War II, industrial and municipal water users, have been in the driver's seat. This changed with the passage of the Endangered Species Act (ESA) in 1973.

The ESA gave environmental interests a powerful place at the negotiating table to make decisions regarding water use and watershed management in the Platte River Basin. As the National Research Council Committee on Endangered and Threatened Species in the Platte River Basin stated, "The federal ESA . . . is a major regulatory force that limits land and water use in the basin. . . . [It] ha(s) been a primary motivating force behind development of the Platte River Cooperative Agreement" (NRC 2004, 74). Section 7 of the ESA forbids actions by any federal agency that would cause "jeopardy" to the continued existence or recovery of a listed species. Several proposed water projects in the Platte River Basin have been challenged under ESA section 7, including the Grayrocks Dam in Wyoming, The Narrows Project in Colorado, and the Two Forks Dam and Reservoir in Colorado. The Grayrocks Dam was subsequently built after the original plans were changed to reduce storage capacity of the dam and all parties agreed to create and fund the Platte River Whooping Crane Critical Habitat Maintenance Trust. Plans for the Narrows Project and the Two Forks Dam and Reservoir were shelved, and the projects were not built.

The decision on whether to build a project involves complex negotiations among a variety of parties; thus it is too simplistic to say that the ESA caused these projects not to be built. Still, the ESA fundamentally changed the structure of these negotiations, giving opponents of water projects a powerful bargaining position and requiring proponents to prove that the proposed project would not cause jeopardy to listed species before proceeding. In the face of more powerful opposition and facing much higher costs for getting projects approved, many proponents have simply given up fighting for controversial water projects, including the Narrows Project and the Two Forks Dam and Reservoir.

Allocation among Competing Uses

Recent years have seen a number of bitter fights in the western United States over water management among a variety of water users and environmental interests: "The problem of reconciling the management of water for species and for other beneficial uses is typical of many rivers, and the Platte is not an unusual case in this respect. Similar debates occur regarding the Rio Grande, Snake, Klamath, Trinity, Truckee, Sacramento, Missouri, and Colorado Rivers" (NRC 2004, xxii). Many of these fights have occurred recently because, as a society, we have not agreed upon a new set of management rules and expectations that match the new era, in which the environment figures much more prominently vis-à-vis water users' desires. Endangered species issues have changed the balance of power between extractive water users and environmental interests. Gradually, new rules and expectations will emerge, and the frequency and bitterness of western water fights will surely subside. However, peace or even détente in the western water wars is likely to take many more years of patient negotiations, clearer regulations, and more moderate expectations on all sides.

Water allocation issues are difficult to resolve in part because they have strong elements of being a "zero-sum game": more water for one use means less water for another use. Facing an overall cap on water withdrawals, increasing the amount of water going to urban uses means decreasing the amount of water going to agriculture. However, just because urban versus agricultural water use is a zero-sum does not mean that the value generated by water is a zero-sum in and of itself. Water has different values within and among different uses. Within agriculture, irrigating lands with rich soils will give greater crop yields than applying the same water to land with poorer soils. Water is often much more highly valued per unit in municipal and industrial uses than in agriculture. By allocating water to high-value uses, the total benefits generated from water use will be greater; that is, water allocation is not a zero-sum game when examined through the value lens.

Economists have argued for establishing water markets in a number of water basins, primarily in the western United States, which would allow trade among water users as a way to increase the value realized from a fixed allocation of water (see Easter et al. 1998). In principle, trading allows those users with higher valued uses to buy water from those with lower valued uses. Actual water markets have emerged in a number of locations, including in the South Platte Basin in Colorado (Howe and Goemans 2003). In addition to creating potential efficiency gains, water markets can be useful as a way to discover the value of water in different uses and as a means to resolve thorny water allocation problems. In practice, there are often institutional details and special-interest politics that must be overcome for water markets to realize their potential. Difficulties can arise over ill-defined property rights; equity concerns; the small number of potential traders that raise transactions and bargaining costs (coincidentally called an "illiquid" market); and the existence of third-party effects from return flows when trades occur across basins (Colby 2000). Nevertheless, given the increasing demand for water, water markets may be an effective partial solution for water management because they allow for the orderly transition of water allocation from traditional uses to new, higher valued uses. However, before any water trades in the Platte River Basin are approved, they should be shown not to cause harm to endangered and threatened species.

Water management disputes between extractive users and advocates for endangered and threatened species, similar to disputes among water users, may at first appear to be zero-sum games. For example, water diverted from a river for use means less water instream for species. However, stakeholders can negotiate outcomes that go a long way toward meeting the objectives of both interests. For example, it may be possible to meet the needs of endangered species while not imposing large costs on extractive users, by modifying when and where water is diverted and is returned to instream flow. Many of the major threats to species, such as loss of habitat, may not be closely tied to the total amount of water diverted; instead, they may have more to do with disturbance regimes, such as flooding or fire. Along the Platte in central Nebraska, to help generate whooping cranes' and piping plovers' requirement of sandy or grassy habitat, a pulse of high water could be allowed down the river to scour banks and sandbars, or excess woody vegetation could be cleared away. Ensuring high-quality habitat is not necessarily antithetical to the needs of water users. In addition, there may be a number of alternative ways to meet the requirements of endangered and threatened species. These alternatives will likely differ in terms of the costs they impose on water users. If a flexible enough process is undertaken, negotiations between water users and trustees for endangered species' interests might be able to reach a bargain that protects species while imposing minimum costs on water users.

Dealing with Uncertainty

Reaching an agreement between water users and environmental interests or knowing what outcomes will meet the needs of endangered and threatened species while imposing low costs on water users is often complicated by a lack of ecological understanding of the effects of alternative water management on these species. This is certainly the case in the Platte River. Clearly, current conditions, with large-scale modifications of historical flow regimes and habitat, have been detrimental to listed species. Less clear are what changes need to be made to ensure survival and recovery of these species.

In its review of the state of the science, the National Research Council Committee on Endangered and Threatened Species in the Platte River Basin pointed to numerous gaps in understanding that limit our current ability to predict how alternative management options would affect the survival and recovery of listed species. Among important knowledge gaps were lack of understanding of the following factors:

- The effects of variations in flows in one part of the basin on other parts of the basin
- Connections between surface flows and groundwater
- Effects of changes in flows on vegetation
- Effects of changes in water quality
- Detailed information about population sizes and dynamics
- Interactions among species
- Connection between habitat and populations in the Platte and the larger region
- The ability to predict species' responses to changes in land-management or water use
- The cost-effectiveness of conservation actions (NRC 2004).

The perennial refrain from scientists is the need to do more research to reduce the gaps in our understanding. While this is clearly necessary, it is also true that decisions will be made based on the understanding that exists at the time. A large body of literature addresses decision making under uncertain conditions and offers guidance for such situations (French 1986; Raiffa 1997). The basic approach for decision making under uncertainty involves specifying the range of potential outcomes, how desirable each outcome is (its "utility" score), and the probability of each outcome occurring under a management strategy. The optimal management strategy is the one that yields the highest expected utility, derived by summing the product of the utility score and its probability of occurrence over all potential outcomes. For example, in comparing the current situation with one that restored periodic flooding to the Platte River through central Nebraska, we would assess the probability of various levels of recovery for the whooping crane and piping plover, as well as the cost to water users, and determine whether the increased probability of recovery were worth the cost.

Two important additions to the basic approach for decision making under uncertainty have special relevance for water management in the Platte. First, when a series of management decisions are made throughout a fairly long period of time, these decisions should be designed partly as experiments with opportunities to learn, so that learning occurs as mistakes are corrected, parameters are adjusted accordingly, and uncertainties are reduced or defined in more specific terms for future decision making. This is the essence of "adaptive management" approaches (Lee 1993; Walters 1986).

Second, when important, irreversible consequences, such as species extinction, are at stake, an additional value, called "option value," comes into play to ensure that options are kept open, so that irreversible consequences are avoided until uncertainty is reduced or resolved. In a classic example, Arrow and Fisher (1974) showed how preserving a wilderness from development—in an area where it would be impossible to re-create wilderness after development occurred—would generate an option value for preservation. Such thinking could be applied usefully to development of dams or other infrastructure along the Platte whose consequences are uncertain and whose effects would be difficult to reverse.

Difficulties arise when decision making under uncertainty is put into an explicitly political process, like that involved in reaching accords over water management in the Platte River. What complicates the political process is the combination of large potential costs and benefits under the cloud of unavoidable uncertainty. In such cases, those who favor the status quo over potential changes can play up uncertainties and make legitimate arguments for delaying actions until more information is available. What makes the political process particularly difficult in the case of the Platte is that no one can tell water users who give up water or bear other costs exactly what will be accomplished for endangered species conservation as a result of their sacrifices. Having identifiable, concentrated groups who will bear costs with certainty tied to environmental benefits that are uncertain and incalculable is a fairly toxic combination for a political process.

These complexities are largely what make gaining agreement on ecosystem management of river systems like the Platte difficult; as we learn more about them, we understand why we cannot expect immediate success. However, the prospect of not reaching agreement, which may entail higher costs or lowered probability of species recovery, should provide more than enough incentives to keep negotiating despite the sandbars and other obstacles in the path.

References

Arrow, K., and A. Fisher. 1974. "Environmental Preservation, Uncertainty, and Irreversibility." *Quarterly Journal of Economics* 88: 312–19.

Bureau of Economic Analysis (BEA). U.S. Department of Commerce. 2007. *Gross Domestic Product by State*. www.bea.gov/regional/gsp. Accessed June 22, 2007.

Colby, B. G., 2000. "Cap-and-Trade Policy Challenges: A Tale of Three Markets." *Land Economics* 76 (4): 638–58.

Congressional Budget Office (CBO). 1997. *Water Use Conflicts in the West: Implications of Reforming the Bureau of Reclamation's Water Supply Policies*. Washington, DC: CBO.

DiNatale, K., W. Davis, R. Brown, S. Morea, J. Rehring, and N. Brown. *Building a Unified Strategy: The Colorado Statewide Water Supply Initiative*. 2005. American Water Works Association Annual Conference. The Statewide Water Supply Initiative (SWSI) is sponsored by the Colorado Water Conservation Board, Denver.

Easter, K. W., M. W. Rosegrant, and A. Dinar, eds. 1998. *Markets for Water: Potential and Performance*. Boston: Kluwer Academic Publishers.

Economic Research Service (ERS). U.S. Department of Agriculture. 2007. *Feed Grains Database*. At www.ers.usda.gov/Data/FeedGrains/StandardReports/YBtable12.htm.

French, S. 1986. *Decision Theory*. New York: Halsted Press (www.directtextbook.com/publisher), a division of John Wiley and Sons.

Howe, C. W., and C. Goemans. 2003. "Water Transfers and Their Impacts: Lessons from Three Colorado Water Markets." *Journal of the American Water Resources Association* 39 (5): 1055–65.

Institute for Agriculture and Trade Policy (IATP). 2006. *Water Use by Ethanol Plants: Potential Challenges*. Minneapolis: Institute for Agriculture and Trade Policy.

Lee, K. N. 1993. *Compass and Gyroscope: Integrating Science and Politics for the Environment*. Washington, DC: Island Press.

National Research Council (NRC). 2004. *Endangered and Threatened Species in the Platte River Basin*. Washington, DC: National Academies Press.

Raiffa, H. 1997. *Decision Analysis: Introductory Readings on Choices under Uncertainty*. New York: McGraw-Hill.

Renewable Fuels Association (RFA). 2007. *Industry Statistics*. At www.ethanolrfa.org/industry/statistics/#A.

Thorvaldson, J., and J. Pritchett. 2005. *Profile of the South Platte River Basin*. Economic Development Report EDR 05-03. Fort Collins, CO: Department of Agricultural and Resource Economics, Colorado State University.

Walters, C. 1986. *Adaptive Management of Renewable Resources*. New York: Macmillan.

Willoughby, S. 2007. Ward, A. Quoted in "Wonder Waters: The South Platte." *The Denver Post*, July 8. At www.denverpost.com/extremes/ci_6323409.

PART III

The California Bay-Delta

Located at the confluence of the Sacramento and San Joaquin rivers as they flow into San Francisco Bay, the San Francisco Bay–Sacramento–San Joaquin Delta is the largest wetland habitat in the western United States and the most extensive, diverse wetland ecosystem on the west coast of the Americas. The Delta serves the water needs of 22 million people in California and is crucial to the viability of the state's water supply. Water occurring in relative abundance from rainfall and snowmelt in northern California is captured in the state's two main storage projects, the federal Central Valley Project, managed by the U.S. Bureau of Reclamation, and the State Water Project. These facilities convey water through and from the Delta to high consuming agricultural lands and big cities in the south. Urban users and farmers to the north also draw water flowing toward the Delta. In addition to its crucial role in meeting human demands, the Delta supports hundreds of species of fish and wildlife, including two fish species listed as threatened under the Endangered Species Act. Home to one of the state's most important commercial and recreational fisheries, the Delta serves as a crucial migratory corridor for salmon and waterfowl. Water diversions and withdrawals have had adverse effects on Delta water quality and on threatened species, causing bitter conflicts among stakeholders and profound challenges for those responsible for California water policy at the federal and state levels. Thomas L. Crisman's chapter 8 looks closely at the tidal–freshwater dynamic in the Delta and how water extractions and diversions have altered its ecology.

In 1994, after years of litigation and negotiations on water quality and wildlife protection issues, the parties created CALFED, a joint federal–state program designed to develop and implement watershed-wide, science-based environmental policy to improve conditions in the Delta. CALFED seeks to balance competing objectives and interests while addressing four interrelated, at times competing goals: water supply reliability, ecosystem restoration, water quality improvement, and levee integrity.

Like the Platte River and Everglades projects, CALFED is concerned with water allocation in a situation where demands exceed supplies and the needs of the natural system to maintain instream flows compete with consumptive use demands of urban and agricultural users. Stephen Polasky's chapter 9 explores these water allocation conflicts from an economics perspective. As in the Everglades, water quality issues in the Delta provided the catalyst for the first collaborative efforts among federal agencies and then between the federal government and the state, which led to the formation of CALFED. The federal–state partnership at the heart of CALFED has not developed in the robust way envisioned in the 1990s; yet, as in the Everglades and Platte River, collaboration continues to move forward, if haltingly. Federal and state funding for CALFED has fallen below anticipated levels, and stakeholders have turned to litigation and political maneuvering in the past few years. On the positive side, the CALFED science program has gained wide respect as a model of independence and utility, which has helped policy makers agree on the best available methods that provide sound bases for the adaptive management approach. Like the Platte River experience, a major lesson of CALFED is that collaboration in environmental decision making, once structured and practiced, will endure even under harsh political climates. Many stakeholders and government agencies participating in CALFED, similar to their Chesapeake Bay, Everglades, and Platte River counterparts, have come to see collaboration as the best and only way forward.

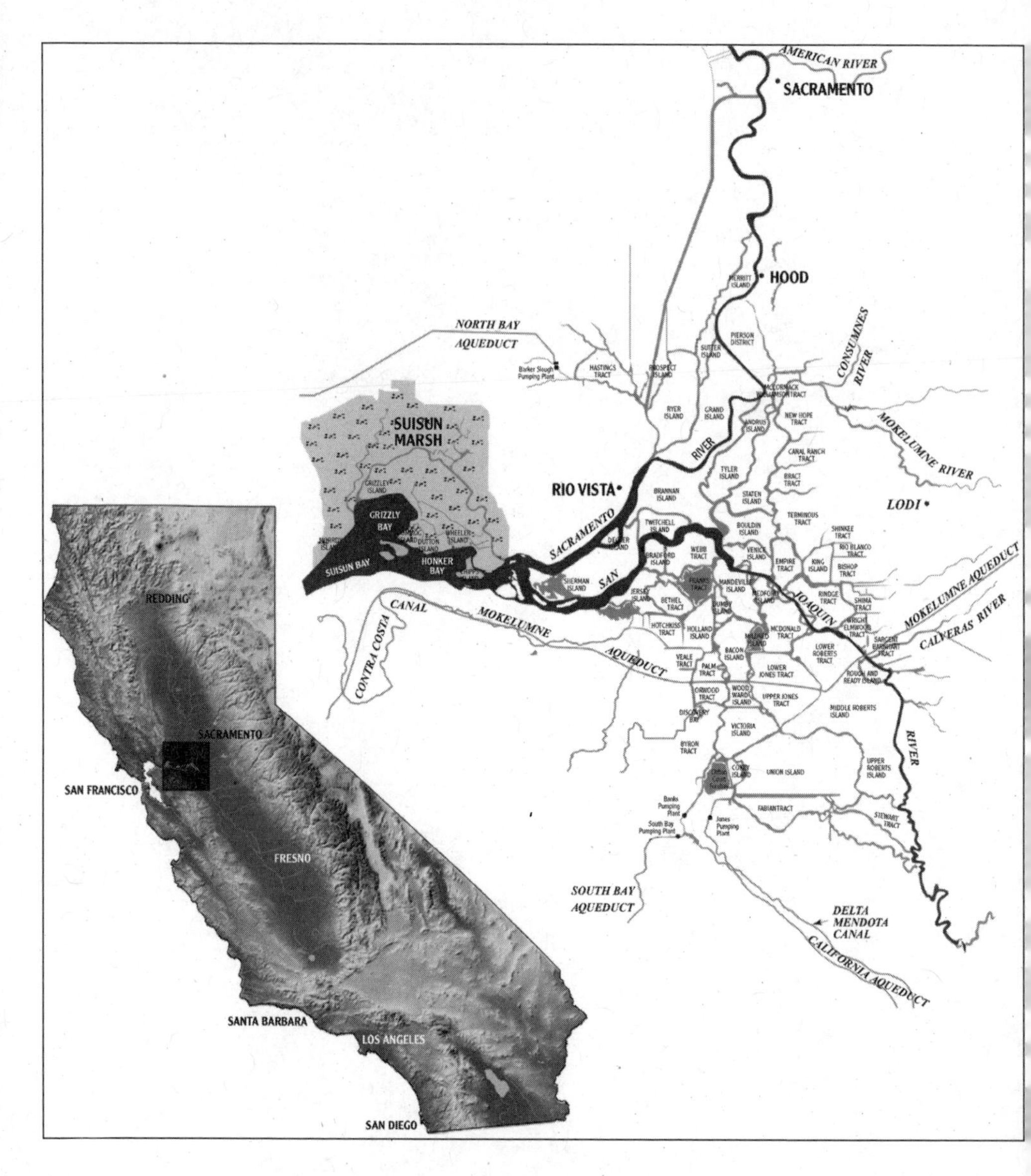

FIGURE 7.1. The California Delta (State of California, Department of Water Resources 2007).

Chapter 7

The California Bay-Delta

The Challenges of Collaboration

David Nawi and Alf W. Brandt

The San Francisco Bay–Sacramento–San Joaquin Delta ecosystem is the most valuable estuary ecosystem on the west coast of North or South America, a natural resource of hemispheric importance. Created by the confluence of the Sacramento and San Joaquin rivers as they flow into San Francisco Bay from the north and south, respectively, the estuary is a maze of tributaries, sloughs, and islands. It contains the largest brackish estuarine marsh on the West Coast. The Delta ecosystem, the largest wetland habitat in the western United States, supports more than 750 wildlife species and more than 120 species of fish, as well as one of the state's largest commercial and recreational fisheries. The Delta estuary also provides migration corridors for two-thirds of the state's salmon and nearly half of the waterfowl and shorebirds along the Pacific flyway.

The Delta also serves as the heart and a critical crossroads of California's water supply and delivery structure. California's precipitation falls predominantly north and upstream of the Delta, whereas much of the state's urban and agricultural water uses occur south of the Delta. The state's two major water projects, the federal Central Valley Project (CVP) and California's State Water Project (SWP), store water in major reservoirs upstream of the Delta, convey water through the Delta, and export the Delta's water south from project pumps in the south Delta. As the water flows from the Sierra toward the Delta, cities and farmers draw water from the system.

The Delta's value as an ecological resource and its role in meeting California's water supply needs have resulted in inherent conflict. The disparate functions and values of the Delta and the competing demands for its resources have long been sources of bitter conflicts and profound challenges for stakeholders and policy makers. Between the state and federal governments, at least twenty agencies share and sometimes contest responsibility for Delta issues. Local entities within the Delta's watershed multiply that number severalfold. Affected stakeholders number in the hundreds. These interests have engaged in conflict for decades.

From a long history of conflict and after much hard work, the joint federal–state CALFED Bay-Delta Program and its successor, the California Bay-Delta Authority (both referred to in this chapter as CALFED or the Program), emerged to forge creative and collaborative policies and a structure to carry them out.[1] CALFED was designed to address four interrelated objectives simultaneously: water supply reliability, ecosystem restoration, water quality, and levee integrity. Any success CALFED achieves depends on maintaining balance among competing objectives, interests, and regions.

In its struggle to maintain balance, CALFED and its participants have learned lessons that may assist others who face similar challenges in ecosystems across the country and around the world. The federal and state agencies have learned to share in governing the Program while engaging countless stakeholders and elected officials. The "CALFED process" includes transparent governance, a comprehensive science program that supports an adaptive management approach, and local projects and decisions to support regionwide objectives. These characteristics are essential for a complex, ecosystem-wide program to address the challenges of collaboration and continue to progress toward multiple objectives.

Despite a positive beginning and a strong, well-supported collaborative structure, the CALFED Program has been unable to maintain sustained progress toward achieving its objectives. The condition of the Delta, as evidenced by the perilous condition of the federal- and state-listed delta smelt, has become increasingly critical. Support for CALFED has largely evaporated in favor of a new effort, Delta Vision, created by Governor Arnold Schwarzenegger and the state legislature, and litigation and resulting judicial orders have come to dictate actions affecting the Delta. Thus, while CALFED can provide a model for a functioning, collaborative ecosystem restoration program, it also demonstrates the difficulty of any effort to achieve success in resolving long-standing conflicts that involve both the protection and the utilization of a scarce and unique natural resource.

Background

To understand the complexity and lessons of CALFED, one must have some knowledge of the Delta's hydrology, interests in conflict, and history. By addressing these three areas in a coordinated fashion, CALFED has succeeded in continuing to engage stakeholders and a multiplicity of federal and state agencies in an ongoing process of collaboration necessary for progress toward CALFED's objectives.

Hydrological Setting

The Delta serves, literally and figuratively, as the heart of the state's north–south water conveyance system and a vital estuarine ecosystem. Water from upstream reservoirs on the Sacramento River and its tributaries flows into the Delta, mixes with flows from several other river systems, and then flows out to the San Francisco Bay. At the Delta's south end, large federal and state pumping facilities move an average of more than 5 million acre-feet of water south to San Joaquin Valley farms and Southern California cities. At the same time, the Delta's unique mix of fresh- and saltwaters creates a rich estuary ecosystem for fish and wildlife.

Waters from the Delta's watershed provide drinking water for two-thirds of Californians and irrigation water for over 7 million acres of the most highly productive agricultural land in the world.[2] Agricultural water users in the Delta use approximately 1 million acre-feet (net) each year, producing $500 million of gross agricultural value. Agricultural crops with even greater value are produced with water exported from the Delta south to the San Joaquin Valley.

The Sacramento–San Joaquin Delta, as defined by California statute, includes 738,000 acres (1,153 square miles). Its watershed covers more than 61,000 square miles, 37 percent of the state. The Delta's waterways stretch for more than 700 miles and include more than 1,100 miles of levees. Some Delta islands are more than 20 feet below the adjacent water level. Historical Delta inflow ranges from 6 to 69 million acre-feet per year, with an average annual inflow of 24 million acre-feet. More than 7,000 diverters draw water from the system, including 1,800 in the Delta itself.

Interests in Conflict

The importance of Delta water for the environment and for disparate human uses has led to long-standing and deep conflicts among three major interest groups: agricultural water users, urban water users, and environmentalists (including their closely allied fishery interests), as well as among federal and state agencies with divergent missions. The history of the settlement of California and the ensuing development of its water resources is replete with political and legal battles among these interests, which have had ample political, legal, and financial resources for battle.[3]

Agriculture

Agriculture is one of California's leading industries and exports products worth $20.8 million per day, with a substantial amount coming from areas south of the Delta that are dependent on water provided by the state and federal projects.

Urban Water Users

The Delta conveys drinking water to more than 24 million people in Southern California and the San Francisco Bay Area, including Silicon Valley. In addition, other urban areas, notably San Francisco, take water upstream of the Delta, thereby reducing Delta inflow.

Environmental Community

For decades, environmental groups seeking to protect the Delta's resources have participated in continuous debates and conflicts over the Delta. These groups have brought to the conflicts both scientific expertise and the concerns of the sport and commercial fishery communities.

State and Federal Agencies

Responsibility for managing the Delta's resources lies with a multiplicity of state and federal agencies with widely divergent interests, management structures, cultures, and missions. As collaboration has replaced the historic Delta conflicts among these agencies, the magnitude of the challenge of "coordinating" their actions has become increasingly apparent.

The notable set of stakeholders that, until recently, remained mostly absent from the Delta debates were the five counties and numerous cities with land-use authority over the Delta. As the pressures of urban development and the need for an adequate level of flood protection markedly increased in and around the Delta, local governments have begun to participate in the larger discussions about the future of the Delta.

Geographic Interests

In addition to interest groups, geography has dictated the politics of water in California. A long-standing fixture of the state's water politics has been a north–south conflict over moving water that falls in northern California watersheds through the Delta to farms and cities to the south. Geography often resulted in internal divisions among agricultural and urban users because both types of uses occur on both sides of the Delta divide. Geography also determines the nature of each user's water supply and rights. In contrast with in-Delta and north-of-Delta users, whose rights date back earlier than the creation of the CVP and SWP, many south-of-Delta users depend on project water with more recent rights.

The long-standing conflicts over the Delta provided the basic institutional challenge facing CALFED. According to Patrick Wright, former director of the California Bay-Delta Authority, "For the two decades prior to the formation of CALFED, California water planning and politics [were] characterized by conflict rather than cooperation" (1995, 27). Prior to CALFED's beginnings, attempts to address Delta issues constructively resulted in gridlock time after time.

History of Delta Conflict

The roots of Delta conflict can be traced back to the earliest water diversions in the Sacramento Valley and Sierra foothills in the nineteenth century. Hydraulic mining that took place in the 1870s and early 1880s deposited vast amounts of sediment and debris in the Sacramento River system. By the time the dumping from hydraulic mining waste was banned in 1884, Sacramento Valley farmers were diverting large amounts of water that otherwise would have flowed to the Delta. In 1922, the California Supreme Court determined that a town in the Delta could not require upstream farmers to reduce diversions to prevent salt-water intrusion into the Delta.[4]

In the 1930s, the state adopted a water plan for a CVP but lacked the financial resources to build it. As a result, the federal government authorized its construction by the U.S. Bureau of Reclamation (Reclamation or USBR), an agency within the U.S. Department of the Interior (Interior). In 1960, California voters approved bonds to construct the State Water Project (SWP), a separate system that, like the CVP, conveyed northern California water to farms in the southern part of the Central Valley and to cities in Southern California. The CVP and the SWP were substantially complete by the 1980s, making the Delta the crucial pivot in California's water supply system. In 1982, California voters rejected plans to build a "peripheral canal," a facility initially envisioned as part of the SWP to convey water around the Delta to CVP/SWP export facilities in the southern Delta. In the 1980s, extensive conflicts arose regarding the obligations of the CVP and the SWP to meet water quality standards. Litigation resulted in a landmark California appellate court decision, rejecting the state's 1978 Delta water quality control plan, in large part based on the court's conclusion that the State Water Resources Control Board (SWRCB or State Board) had not adequately considered permitted diversions and fishery needs throughout the watershed.[5]

Several events in the early 1990s changed the nature of water and ecosystem management in the Delta. First, winter-run chinook salmon and delta smelt, species that live in or pass through the Delta, were listed as threatened under the federal Endangered Species Act (ESA). As a result, the federal CVP was required to consult, pursuant to section 7 of the ESA, with the U.S. Fish and Wildlife Service (USFWS) and the National Marine Fisheries Service (NMFS) on the effects of project operations on the listed species. Second, in 1991, the Environmental Protection Agency (EPA) disapproved the state's 1978 water quality standards for the Delta as not meeting the requirements of the Federal Water Pollution Control Act (1972), amended in 1977 as the Clean Water Act (CWA).

Third, in 1992, despite strong pressure to veto the legislation from water users and Governor Pete Wilson, President George H. W. Bush signed legislation enacting the Central Valley Project Improvement Act (CVPIA).[6] This major reform legislation added fish and wildlife protection as one of the CVP's authorized purposes, facilitated transfers of CVP water, and imposed several environmental requirements, including the dedication of a specified amount of CVP yield primarily for fishery restoration purposes.

The long-standing federal–state conflict that grew in intensity during the early 1990s could not be sustained if either government wished to satisfy the needs of vital constituencies and achieve progress in addressing Delta environmental and water conveyance problems. A lawsuit filed by an environmental group against EPA provided the lever to force resolution of the immediate conflict over Delta water quality standards for fishery needs. A resulting consent decree required EPA to promulgate federal water quality standards for the Delta by December 15, 1994.

In the summer of 1993, the federal agencies, NMFS, EPA, USFWS, and USBR, began to communicate effectively and collaborate on Delta issues. They formed the Federal Ecosystem Directorate, familiarly known as Club Fed, and eventually signed an "Agreement for Coordination on California Bay/Delta Issues," committing them to collaborate and undertake an ecosystem-based approach. Club Fed's first major accomplishment was to issue an integrated set of regulatory proposals to protect the Delta, published on December 15, 1993, in a single Federal Register notice (Federal Ecosystem Directorate 1993).

In addition to cooperation among the federal agencies, federal–state collaboration was critical to alleviating tensions created by federal decisions and actions under the CWA and ESA, which state officials perceived as unilateral and ignoring the state's interests. In June 1994, the federal and state administrations signed an agreement to coordinate activities in the Delta, particularly the development of Delta water quality standards.[7] Notably, the fact that the federal administration was Democratic and the state administration was Republican did not prevent progress toward a coordinated approach.

Faced with a looming judicial deadline of December 15, 1994, for EPA to adopt water quality standards, federal and state agencies and interest groups worked together to develop a scientific basis for agreement on the standards. A science-based proposal, developed by urban and agricultural water users, achieved wide acceptance and allowed the federal and state agencies, with the encouragement of the environmental community, to create a broad-based agreement on water quality standards. The agreement was entered into on the day of the deadline and was entitled Principles for Agreement on Delta Standards between the State of California and the Federal Government (dated December 15, 1994), subsequently referred to as the Bay-Delta Accord or the Accord.[8]

The Accord represented a fundamental shift in the world of California water. Its importance at the time it was executed is difficult to overstate. The Accord was a critical milestone, marking a visible and meaningful change from the basic model of conflict to the collaborative approach embodied by CALFED. In addition to setting forth the substance of the new water quality standards, the Accord provided for close CVP–SWP coordination in complying with the new standards and meeting Delta fishery needs for an interim period of three years. The Accord contained measures to protect anadromous fish species (those that migrate from the sea up rivers to spawn, like salmon) and export limits, and it addressed water users' desire for certainty regarding the amount of project water that would be used to meet regulatory requirements.

Development of the CALFED Bay-Delta Program

With the achievement of the Bay-Delta Accord in 1994, the agencies and stakeholders initiated planning for a long-term program to improve Delta conditions, which became the CALFED Bay-Delta Program. For the next five years, the agencies collaborated with stakeholders to prepare an environmental analysis of options for long-term improvements to the Delta. In the summer of 1999, the agencies completed a draft of a comprehensive, programmatic Environmental Impact Statement/Environmental Impact Report (EIS/EIR) pursuant to the National Environmental Policy Act (NEPA) and the California Environmental Quality Act (CEQA). The range of alternatives reflected in the draft document continued to engender substantial stakeholder debate.

As debate on alternatives continued with no end in sight, state and federal government leaders moved into private negotiations—without the stakeholders. These high-level negotiations focused on a few key issues: creating an environmental water account, setting well-defined milestones for carrying out specific actions and measuring progress toward their implementation, selecting storage options to pursue with site-specific environmental analysis, and committing to program funding.

Based on these negotiations, Secretary of the Interior Bruce Babbitt and Governor Gray Davis released *California's Water Future: A Framework for Action* on June 9, 2000. In the next twelve weeks, the state and federal governments completed a final CALFED EIS/EIR and worked nonstop to develop the federally required record of decision (ROD), as well as other agreements and action documents to support the ROD and create the structure necessary to begin implementing the CALFED Program. Issuing the ROD on August 28, 2000, was the next long-awaited milestone in CALFED's progress following the 1994 Accord. It set forth, in a formally adopted, federal–state document, a roadmap for future development and implementation of the CALFED Bay-Delta Program.

Summary of the CALFED Program

The CALFED ROD of 2000 forms the basis of the long-term solutions and processes embodied in the CALFED Bay-Delta Program. The state legislature carried forward the basic structure of the Program through legislation (Act of 2002) that created the California Bay-Delta Authority (the Authority). The U.S. Congress authorized federal agency participation two years later (CALFED Bay-Delta Authorization Act 2004). Although the state legislature has adjusted some of the Program's structure in recent state budget legislation, the ROD has served as a cornerstone for Delta programs and decisions.

As structured in the ROD, the CALFED Program is a cooperative undertaking of twenty-one state and federal agencies with responsibilities in the Delta and its watershed. The Program provides for extensive participation by stakeholders, who accepted it largely because it includes elements specifically designed to address the basic needs of all the major stakeholder groups. Although the Program may contain elements that particular groups do not agree with, each group has supported CALFED to obtain the benefits that address its core interests.

The fundamental structure of the Program rests on an integrated approach to the entire ecosystem. The Program addresses four primary concerns or objectives: (1) ecosystem restoration, (2) water supply reliability (including storage and conveyance), (3) water quality, and (4) levee system integrity. The preferred alternative in the ROD addresses these four concerns in an Action Plan that includes the program elements described later in this section.

Two significant, substantive elements of the CALFED Program merit separate discussion. First, additional surface water storage capacity was one of the most difficult issues facing CALFED. Environmentalists opposed any additional surface water storage facilities because they believed that additional storage would inevitably result in increased diversions and a continuation of practices that led to the Delta's decline. Water users, in contrast, sought a commitment to construct additional surface storage facilities as the surest way to address their water supply concerns. Ultimately, faced with these conflicting positions, the designers of CALFED chose to include additional surface storage as an activity to be pursued through further study and subject to subsequent authorization.

A second, extremely contentious issue was how to convey water through the Delta, specifically, the volatile question of whether to build an "isolated facility," which many saw as a resurrected plan for a peripheral canal, rejected by voters in 1982. CALFED's leaders chose a middle course to address this issue, at least for the first seven years: Water from the Sacramento River would continue to be taken through, not around, the Delta. This approach was accompanied by and dependent upon improvements in water supply reliability, water quality, ecosystem health, and levee integrity. CALFED's leaders rejected the immediate development of a new facility to take water around the Delta to the water project pumps in the southern Delta. In order to prereserve options for long-term water quality, however, they decided to continue studying the effects of building a screened diversion facility on the Sacramento River in the northern Delta that would support conveyance separate from the Delta.

The elements of the Action Plan are set out in the following sections. Their scope, variety, and number clearly reflect the comprehensive nature and magnitude of the CALFED undertaking. As the first five years after the ROD progressed, state and federal legislators provided funding below the level of support envisioned in the ROD, primarily because both governments' fiscal conditions deteriorated rapidly. CALFED projects, therefore, have not proceeded on the timeline envisioned in the 2000 ROD, but the two governments have not abandoned the ROD's programmatic outline. They have proceeded with actions to implement CALFED, based on available funding and statutory authorizations for the various projects. Thus the Program has remained largely unchanged, but the timing and sequence of actions have been altered. In 2007, however, the state began examining new options to protect the Delta.

Water Supply Reliability

Integration and balance among the four (at times) competing program objectives form one of the ROD's cornerstones. Improving water supply reliability, particularly over the long term, depends upon progress on all the other objectives. The Delta's ecosystem conditions need to improve in order to reduce the conflict between fishery and water supply needs and to avoid water supply reliability problems. The entire CALFED Program rests on the interdependence of water supply reliability and the other program elements. Certain program elements, however, provide a more direct linkage to improving water supply reliability, including the following four key points, all of which can be found in the ROD under these same subheadings (CALFED ROD 2000).

Storage

The ROD sets ambitious goals for water storage expansion: 950,000 acre-feet in increased surface storage and from .5 million to 1 million acre-feet of groundwater storage. The ROD identifies five regions (including existing reservoirs) where surface storage investigations will proceed.

Conveyance

For the first seven years (2000–2007), the ROD requires that agencies use their best efforts to improve through-Delta conveyance, as opposed to an isolated facility to take water around the Delta. This option is intended to improve water supply reliability for in-Delta and export users, support continuous improvement in drinking water quality, and complement ecosystem restoration.

Water Use Efficiency and Conservation

The water use efficiency element is intended to "implement an aggressive water-use efficiency program to make the best use of existing water supplies." Much of this element relies on actions of local agencies, which receive CALFED grants and incentives to promote increased water-use efficiency. Increased efficiency and conservation contribute to reduced reliance on exports through the Delta.

Water Transfers

The goal of the water transfer element is to "promote development of an effective water transfer market that protects water rights, the environment and local economies." This smaller program element creates an agency-stakeholder working group to help streamline the water transfer process.

The participants anticipated that these four elements—(1) storage, (2) conveyance, (3) water use efficiency and conservation, and (4) water transfers—in combination with all the other Program activities, would achieve increased water supply reliability and avoid difficulties experienced in the Delta during the early 1990s.

Ecosystem Restoration Program/Watershed Management

Pre-ROD federal funding focused on environmental improvements in the Delta to address the conflicts that arose in the early 1990s out of fishery concerns regarding the effects of the Delta export pumps. Some debate had taken place about whether the most valuable assistance to commercial and recreational fishery interests would be achieved through reduced pumping in the Delta or habitat restoration upstream. To address this issue, the ROD's ecosystem restoration element activities extend throughout the Delta's watershed. The restoration element also applies to more than just ESA-listed species. Using California's natural community conservation planning (NCCP) structure (1991), the Ecosystem Restoration Program (ERP) is part of a Multi-Species Conservation Strategy (MSCS). The ecosystem restoration elements include the ERP, Watershed Management, and the Environmental Water Account (EWA).

Perhaps the most innovative element of the 2000 ROD is the EWA. The ROD established a four-year experiment involving a set of water assets derived from purchases and CVP/SWP operational efficiency adjustments. To avoid adverse impacts to water project contractors, these assets are used when the fishery agencies choose to reduce pumping beyond the regulatory baseline or otherwise use water to promote fishery restoration. Closely connected to the EWA is a commitment by the state and federal agencies to ensure that no additional delivery reductions will result from endangered species requirements for the first four years of the Program. In fall 2004, the agencies extended that commitment for an additional three years.[9]

Water Quality

While Delta water quality is important to several different beneficial uses of water, the ROD concentrates on drinking water quality issues. Improvements for drinking water would improve quality for all other uses—from Delta farming to fishery needs. Drinking water quality advocates debated the comparable values of treatment technology and source control. The CALFED ROD does not choose between these two methods of drinking water quality improvement; instead, it supports investments in technology and source control, as well as drainage management, which is a problem in portions of the San Joaquin Valley.

Levee System Integrity

In initially choosing through-Delta conveyance rather than an isolated facility, the ROD effectively highlights the importance of improving the integrity of the Delta's levee system. The levees protect valuable farmland and create the channels drawing water to the south Delta export pumps. Local reclamation districts are responsible for much of the Delta levee system, and the state oversees maintenance of certain levees important to conveyance. The ROD reflects a choice to invest in greater support for local levee maintenance to bring Delta levees up to a common baseline standard. At the same time, CALFED accepts the challenge of a comprehensive risk assessment of the levee system.

A major levee failure in June 2004 at Upper Jones Tract was the subject of statewide attention and caused the loss of thousands of acres of valuable farmland and farm workers' housing, as well as a shutdown of export pumping and railway traffic. The state spent approximately $45 million to restore this levee and the island it protected west of Stockton.

Integrated Agency Actions Incorporated in the Record of Decision

In recognition of the need to address regulatory and permitting issues in a comprehensive and integrated fashion, the agencies took several important actions when they issued the ROD and included the documentation of those actions as attachments to the ROD. Ordinarily, a ROD serves under NEPA to document a federal agency's decision on a proposed action. The CALFED ROD, however, is far broader in scope. Reflecting coordination among the multiple agencies involved, the CALFED ROD documents state actions as well as several separate federal actions.

The CALFED ROD incorporates the biological opinions issued contemporaneously by NMFS and USFWS pursuant to ESA section 7. As a joint federal–state document, the ROD incorporates findings and a certification under CEQA, as well as the California Department of Fish and Game (CDFG)'s approval of the Program's MSCS. The ROD also includes a programmatic consistency determination under the Coastal Zone Management Act, agreements addressing Sections 401 and 404 of the CWA, operation of the Environmental Water Account, overall Program implementation, and a conservation agreement regarding the MSCS. All these agency actions reflect the high degree of coordination the agencies were able to achieve, a remarkable accomplishment in light of the earlier fragmented nature of their actions and the disputes among them. The coordinated suite of actions taken by the agencies was consistent with the principles of "one-stop shopping" and a strong affirmation to stakeholders and others that the agencies could work together in a meaningful and effective manner.

The Principle of Balanced Implementation

The aspect of the CALFED Program that ties the preceding elements together and serves as the linchpin intended to foster continued collaboration and stakeholder and agency support is the principle of balanced implementation. This principle is intended to assure that progress toward implementation of each of the major program objectives (water quality, ecosystem quality, water supply reliability, and levee system integrity) occurs concurrently and that progress in a single area benefiting one stakeholder group does not occur while areas of interest to other stakeholders are neglected.

Balanced implementation carries forward the approach of including elements in the Program that address the needs of all stakeholder groups. Continued stakeholder support and involvement was seen to require assurance that all elements will, in fact, be implemented. To carry out the principle of balanced implementation, the ROD provides that either the governor of California or the secretary of the Interior may determine that CALFED is not achieving balanced progress in all Program areas. In the event of such a determination, state or federal funding would be suspended until the two senior officials agree on a new Program implementation schedule.

When, however, the Program appeared out of balance and the ecosystem was reported to be crashing, neither the governor nor the secretary of the Interior chose to exercise this option. Instead, Governor Schwarzenegger called for a comprehensive review, and the state legislature transferred all funding authority for CALFED to the Office of the Secretary of Resources (CSB 2006–2007).

Challenges of Collaboration

Over the last decade, CALFED agencies and stakeholders have struggled to develop and sustain a culture of collaboration in addressing difficult water supply and resource protection issues. This culture necessarily needs to survive changes in state and federal administrations, as well as changing priorities and new challenges for the Delta and its watershed. In building the CALFED collaboration, the agencies and stakeholders have addressed many issues and challenges that are common to watershed programs across the country. Some of the major challenges that CALFED has faced and its responses are discussed in the next sections.

Stakeholder Participation

Stakeholders have continued to play a critical role in developing the CALFED Program, even as litigation has resumed and attention has shifted to the courts, the governor, and legislators. Stakeholders have stood squarely in the middle of conflicts and have contributed to their resolution. The CALFED agencies realized that if Delta issues were to be resolved effectively, stakeholders needed to perceive that their interests were recognized. The agencies incorporated stakeholders into the collaborative conflict resolution process, rather than simply viewing them as part of the conflict.

When CALFED first began to develop in the early 1990s, various government agencies and stakeholders had been fighting for decades, leading to a stalemate in the Delta. The agencies—particularly USBR and the California Department of Water Resources (DWR)—had worked closely with their allied stakeholders in these battles. Typically, when one interest group would advocate some kind of Delta "improvement," other interests would line up in opposition. The groups' lawyers and "experts" would march off to court to challenge the legal and scientific bases for each other's positions.

The 1994 Accord did not end stakeholder conflicts. While continuing stakeholder conflicts, including litigation against agencies, did not stop CALFED's progress, they had the effect of limiting CALFED's ability to address certain issues. Tom Clark, the retired general manager of the Kern County Water Agency and a leading advocate for export agriculture, described this ongoing litigation as "trench warfare" (Clark interview 2002).

Despite the adversarial positions of stakeholders, they have played positive roles in resolving conflicts. As Tim Quinn, former vice president of the Metropolitan Water District (MWD) of Southern California and a leading urban advocate, has noted, stakeholders organized the first "three-way process" among agricultural, urban, and environmental interests from 1989 through 1991. Quinn calls CALFED "a stakeholder-created process," noting that "the agencies were content to keep fighting each other" (Quinn interview 2002). Stakeholders, particularly large agencies like Quinn's MWD, could offer expertise, funding, and independent leadership. When they have united, stakeholders have provided the political support to allow CALFED to continue to receive substantial government funding and authorization.

From the government agency perspective, engaging stakeholders in positive ways has remained a constant challenge. Ensuring participation by all the necessary stakeholder representatives created a cumbersome process because some stakeholders would attempt to dominate or derail discussions if the developing outcomes did not favor their positions. However, when all stakeholders were not included in the decision-making process, those who were excluded sometimes impeded progress. After the 1994 Accord, for example, water districts controlling reservoirs on tributaries to the San Joaquin River filed suit against the new "consensus" water quality standards because the stakeholders had not been invited to participate in crafting the standards. These districts settled the lawsuit after the State Board said it would not require them to contribute water to meet the new standards.

Finding a productive format for stakeholder involvement was not easy. For example, during high-level federal–state discussions in fall 1998, more than forty people crowded around conference room tables. This format was not conducive to effective progress. The public processes prescribed by the Federal Advisory Committee Act (FACA 1972) tended to discourage frank interchanges among agencies and stakeholders, in part because convening in informal settings to develop consensus was made difficult.

To facilitate uninhibited interagency discussions, the agencies' representatives frequently met separately from stakeholders. The agencies' normal processes allowed for public comments on draft documents, such as an EIS/EIR, but generally did not provide for effective nongovernmental participation in the actual decision-making process. When agency leaders made final decisions, they typically met behind closed doors. As Gary Bobker, senior policy manager for the San Francisco Bay Institute, observed, "There is a point at which it's really important to actually defer to stakeholders and encourage them to be leaders in the process, and there's a point at which that starts to be counterproductive" (Bobker interview 2002). The agencies recognized that point when they went behind closed doors to negotiate the 2000 ROD.

The stakeholders' role in CALFED has changed and grown because of broad stakeholder support for the ROD. Stakeholders now do much more than advise the state and federal agencies. Some local agencies have carried out their own projects that have been incorporated into the CALFED Program. As state and federal funding has dwindled, CALFED has relied increasingly on local projects—particularly those focused on water use efficiency. The local entities plan, fund, and implement beneficial CALFED-related projects, with or without incentive funding or cost sharing. In other cases, local stakeholders participate as members of a project's steering committee with federal and state agencies.

Even as the Delta's ecosystem health has declined and litigation has resumed in recent years, stakeholder participation in CALFED activities has remained a constant. By participating in stakeholder-agency working groups, such as the Delta Smelt Working Group, stakeholders can maintain their position of knowledge and influence as the agencies decide how to respond to the latest crisis. As stakeholders have shifted some of their resources from the meeting room to the courtroom, they have not completely abandoned collaboration. With water supply, levee protection, and ecosystem health at issue, stakeholders have found that their participation in CALFED helps to maintain their position of knowledge and influence, as the agencies decide on actions to take to respond to frequently critical situations.

The importance of stakeholder participation cannot be overstated. The long history of mistrust and conflict and the natural tendency to revert to those historic attitudes can be overcome only if stakeholders remain engaged and continue to talk to each other and the agencies' representatives and if stakeholders perceive that their voices are being heard and their interests considered throughout decision making and program implementation. The CALFED Program's director and state and federal agency representatives have developed effective mechanisms and attitudes that have encouraged stakeholders to participate as valued partners.

Governance

The role of stakeholders in CALFED, though essential, is just one of many issues that must be addressed for successful governance of such a comprehensive program. Although the first major collaborative success was the 1994 Bay-Delta Accord, the actual CALFED Bay-Delta Program began as an informal network of agencies and stakeholders talking about how to "fix the Delta." The agencies coordinated their efforts through shared and co-located state and federal staffs and regular meetings of agency leaders. However, until the 2002 passage of the California Bay-Delta Authority Act, CALFED progressed without a formal legislative charter and was essentially a creature of interagency agreements and budget allocations. In the absence of a formal structure, CALFED was able to proceed because of its unique ability to resolve conflicts and coordinate multiparty actions to address Delta issues.

From the agency perspective, a major challenge in creating a governance structure for CALFED was sharing sovereignty, or decision-making power, between the federal and state governments. The U.S. Constitution divides power and authority between the two levels of government. Within their respective realms, each is empowered to make its own decisions and not be called to account in the other's courts. In the absence of specific legislation, this basic structure constrains the range of options for collaboration between federal and state agencies. As the CALFED agencies and stakeholders sought to forge a strong federal–state partnership, they were unable to find any appropriate models for the two governments to share their sovereignty.

Further complicating the sovereignty issue was the question of how to incorporate participation of American Indian tribes in CALFED deliberations. Although no recognized tribes reside in the Delta itself, certain tribes within the watershed expressed interest in participating. Each of these tribes enjoys sovereignty separate from the state, adding an element that had to be addressed in forging a partnership among the sovereigns. While some tribal representatives accepted seats on the advisory committee, others would accept nothing less than an equal role in decision making.

The issue at the heart of discussions of CALFED governance was the scope of authority of a new entity. Numerous federal, state, and local agencies held either regulatory or programmatic responsibility for the Delta. In participating in CALFED, they had agreed to coordinate their individual decisions. In contrast, some stakeholders suggested giving a proposed commission authority over certain key regulatory programs, including those under the federal ESA. Consequently, the creation of a new entity raised the specter of a super-agency usurping individual agencies' authority.

Regulation under ESA had been a critical issue leading to CALFED's creation. Some stakeholders advocated creating a defined stakeholder role in ESA regulatory decisions that would reduce CVP and SWP diversions from the Delta. Kern County's Clark commented that the agencies make too many decisions behind closed doors: "One of the failings of CALFED is that there's way too much 'inside baseball.'" Clark explained that the original interagency agreement was supposed to provide for more decisions to be made "in the light of day" (Clark interview 2002).

The agencies uniformly opposed any transfer of regulatory authority, whether under ESA or CWA. The federal–state negotiators agreed, advocating the agencies' position, as they opposed the transfer or dilution of regulatory authority. The negotiators also discussed whether a new entity should simply coordinate agency activities or actually implement the entire program. The agencies and some stakeholders supported a role for the new governing entity as a coordinator. In discussing the debate over the 2002 state governance legislation, former Authority Director Wright stated that CALFED's strongest supporters "wanted a much stronger authority, precisely because they feared that the agencies would continue to be responsive primarily to their traditional constituencies and to resist many of the core values of CALFED (such as interagency coordination; competitive grants; and the commitments to balance science, planning, and tracking) unless compelled to coordinate under the new structure" (Wright interview 2002).

The CALFED ROD, issued in 2000, addressed the issue directly, introducing its Plan of Action with a section emphasizing the importance of a new Delta governance process. This Governance section, while acknowledging federal constitutional concerns, proposed a new "joint commission made up of high-level appointees [to] maintain visibility inside and outside the government, assure agency coordination, help secure funding, and provide policy leadership and accountability" (CALFED ROD 2000). In the interim, before the state and federal legislatures authorized a new entity, responsibility for each program element was held by a state agency and by a federal agency. The CALFED agencies agreed to continue coordinating their activities with the assistance of a new federal advisory committee of stakeholders.

Although the ROD proposed a new entity in 2000, the California State Legislature took two more years to create a state entity, and the U.S. Congress took another two years to authorize only limited federal participation in that state entity. The 2002 state legislation creating the state's California Bay-Delta

Authority represents a compromise between CALFED as a super-agency and CALFED as a coordination group.[10]

The state agencies continue to exercise what the Authority's former Director Wright called "ultimate decision-making authority" (Wright interview 2002) and to implement the CALFED Program in coordination with their sister federal agencies. The Authority, however, exercises oversight of the state agencies' CALFED activities through annual review and approval of each agency's work plans for those activities. The Authority's board includes six members representing state agencies, six members representing federal agencies (subject to congressional authorization), five members representing five regions that affect the Delta, and two at-large members who are appointed by state legislative leaders. In addition, the state legislature appoints four ex officio, nonvoting members from its own ranks. The statute creating the Authority did not address the role of sovereign tribal nations, but tribal representatives were appointed to the Bay-Delta Public Advisory Council.

State legislation provides for federal agency participation as voting members upon congressional approval. In 2004, however, Congress refused to allow the federal agencies to vote at the Authority and banned state participation in federal agency decisions. Thus, to this point, the federal agencies may coordinate with but may not join in decisions of the Bay-Delta Authority, and while they may participate in Authority discussions, they rarely do so in the public forum of an Authority board meeting.

Former California Secretary of Resources Mary Nichols predicted this outcome in 2002: "When we first started talking about long-term governance for CALFED, even before the record of decision was signed, many people had hoped for the creation of a new entity, which would be truly a joint state–federal entity. [They hoped for] something that would have a free-standing life of its own but be authorized at both levels and be equally balanced between the two. I think that's not likely to happen at this point. Meshing the budgets for these two entities is hard enough without also trying to decide whose personnel laws will govern" (Nichols interview 2002).

Legislative Support and Funding

The wide-ranging debate regarding CALFED governance was a sign of a serious challenge facing CALFED and its stakeholders because the debate engaged the legislative branch at the state and federal levels. From its inception, CALFED was developed through agency action. The debate among stakeholders occurred in an agency context and generally did not engage legislators. Only after consensus emerged did stakeholders go to legislators for funding or authorization. In the case of one bond election, stakeholders went around the legislature by putting a bond initiative on the ballot. While some leading legislators expressed concern that the agencies had not involved them, stakeholders offered legislators little political choice but to acquiesce in the face of unified support from farmers, water agencies, and environmentalists. Based on that unified support, agencies and stakeholders achieved substantial CALFED funding and general acceptance of the 2000 ROD.

Legislative support, however, has not been uniform. The state legislature took two years to create the California Bay-Delta Authority, with debate focused on membership of the Authority's board. The U.S. Congress took four years to authorize CALFED and its programs as it debated the scope of agency authorization, particularly related to construction of storage reservoirs. These debates reflected the challenge that broad stakeholder support for CALFED presents to legislators accustomed to deals that resolve conflicts among interest groups. As former Authority Director Wright explained, "To members of Congress and the legislature, CALFED represents a threat to the traditional way of doing business. They have been asked to authorize and fund the plan as a whole, in the interest of balance, rather than voting on the elements they support" (Wright interview 2002).

Legislative oversight over CALFED's current progress also presents political challenges. Without new authorizations, elements of the CALFED Program are still able to move forward. However, to acquire and retain continued legislative support, the Program's progress requires legislative oversight from members who are not necessarily focused on CALFED issues. According to Wright, "Members are focused on needs of their districts, rather than statewide issues. Water policy is extremely complex, and few members have the time or experienced staff. Outside of a handful of members on the relevant policy committees, few members have any knowledge of water issues" (Wright interview 2002).

Central Valley legislators, who have a more direct interest in CALFED, often lead legislative discussions. On the state level, however, term limits lead to regular shifts among individual legislative leaders, so that new Central Valley legislators have to learn quickly about the Delta if they wish to become leaders on CALFED issues. To some extent, these political challenges arise out of the nature of CALFED as a comprehensive program addressing different priorities throughout a 400-mile-long watershed. The Bay Institute's Bobker suggests that such problems arise not because of the nature of CALFED but because of the nature of the legislative branch: "We should design the solutions that, to the least extent possible, rely on legislative involvement. The legislature is a really good place for goals to be set and resources to meet those goals to be provided. It's a bad place to try to work out the complex details of an implementation package" (Bobker interview 2002).

Like all public agencies, CALFED needs adequate funding from the legislative branch. Apart from normal competition for public dollars, CALFED faces an additional difficulty because funds to implement the Program must be provided in the budgets of multiple agencies at each level of government: federal, state, and local. Each authorizing or appropriations committee traditionally controls the progress of its projects. When multiple committees are involved, they are forced to share control of the entire CALFED Program. The fundamental philosophy of CALFED—that beneficiaries of CALFED projects should pay their fair share of project costs, the so-called beneficiary pays principle—also presents a challenge for legislators because they usually determine who benefits and who pays.

Stakeholder disputes and lack of funding have created obstacles hampering the Program's progress. Some stakeholders have filed lawsuits,[11] and others have challenged certain projects by political means. Legislators have raised questions about the funding and continued viability of the program, often prompted by certain stakeholders. Nevertheless, CALFED has continued to make progress, albeit at an increasingly halting pace, because the Program has appeared to offer the only viable way to resolve issues in the Delta.

The key to the Program's united stakeholder support is a core group of what MWD's Quinn labels "dedicated centrists." Quinn explains, "There was a group of dedicated centrists that always protected the process from partisan politics, which is a problem. . . . The other problem has been protecting the process from the radical extremes on both the left and the right in California. It's Westlands [Water District] on the one hand and the hard-core environmentalists on the other. I've got bruises on my right and my left from going in and trying to protect something that I thought was sustainable in the middle from very vicious attacks from either the ag guys, who are never going to know a good deal when it's presented to them, and the environmentalists, who want to keep the growth war going instead of solving problems" (Quinn interview 2002).

Quinn believes the answer to CALFED legislative support is "to build a big, respectable centrist coalition," and he says that "incremental successes" at CALFED help the centrists convince legislators that addressing all interests with moderation can provide political success (Quinn interview 2002).

While unified stakeholder support has been critical to gaining legislative support, CALFED has also relied on strong congressional leadership from Democratic Senator Dianne Feinstein. David Guy, former executive director of the Northern California Water Association, emphasizes Senator Feinstein's "personal crusade" for CALFED as a key element in securing congressional support for CALFED (Guy interview 2002). CALFED has received bipartisan support through much of its development for a variety of reasons, including the fact that the federal and state administrations have commonly represented different political parties. California Republican members of Congress who led key authorizing and appropriating committees also succeeded in gaining broad support for CALFED from their fellow federal legislators.

At the state level, however, legislators reached a turning point in 2005. In the first five years after the 2000 ROD, CALFED enjoyed substantial state bond funds, leading to an imbalance between federal and state appropriations. As those state bond funds dwindled, there was increasing pressure to shift at least some of the financial burden to local agencies that benefited from CALFED projects, applying the "beneficiary pays" principle. Defining those benefits and the necessary payments, however, proved problematic for stakeholders, agencies, and legislators. In 2005, after state legislative committees cut CALFED's budget to extend its financial life on bond funding, Governor Schwarzenegger agreed that CALFED needed a long-term finance plan that would provide stable funding beyond the two years of remaining bond funds. In the years that followed, both the governor and the legislature reviewed the CALFED program and began the process of crafting "new" Delta policies, actions consistent with the ROD's call for a program review and decision in 2007.

Relations between CALFED and the Agencies

Critically important to CALFED's progress are effective coordination and collaboration between the California Bay-Delta Authority, responsible for oversight, and the federal and state agencies responsible for implementing the CALFED Program. As the Program has moved forward, the relationships among the federal and state agencies and the Authority have been fraught with difficulty. From former Authority Director Wright's perspective: "In many respects, it has been more difficult to obtain agency buy-in than stakeholder buy-in. As with the stakeholder groups, there is a core group of leaders within the agencies that support the program, but the vast majority of staff would largely prefer to simply continue running their own programs" (Wright interview 2002).

Agency staff members see the same issue, albeit from a different perspective. They express the concern that CALFED, under the rubric of coordination, may interfere with agency autonomy in such important areas as budgeting, allocation of resources, and program and policy decision making.

Tensions among the Authority's staff and agencies' staff exist in virtually every aspect of Program implementation. The roots of this tension are not difficult to discern. First, agencies in any area of government are characteristically resistant to sharing their authority and to participating in structures that constrain their independence or normal decision-making process. Closely related to this, former

CALFED Director Wright identifies the oversight role assigned to CALFED. He observes that tensions between any agency subject to oversight and the overseeing entity occur almost invariably simply because of the nature of the oversight relationship and natural agency resistance to outside oversight (Wright interview 2002). Similarly, in contrast with CALFED's structure, based on a collaborative multiagency model, agencies and their staffs are used to a model based on an individual agency's decisions and actions.

Specific aspects of the CALFED structure reinforce these typical sources of institutional tensions and disagreements. Most fundamentally, each state and federal agency responsible for CALFED's operations has retained its separate and independent statutory authority. This structure developed as the CALFED Program evolved in the 1990s, and, in the absence of specific statutory authority, it has been continued in the Bay-Delta Authority Act. The statute seeks to bring about cooperative and coordinated agency actions that are consistent with the Bay-Delta Program, but it also provides that nothing in the statute "shall be construed to override constitutional, statutory, regulatory, or adjudicatory authority . . . of any local, state or federal agency" (CWC 2002, 79403.5(c)).

As an offset to these tensions, the separation of the Authority from state and federal agencies responsible for its functioning brings benefits perceived by stakeholders. From the stakeholder perspective, the Authority staff is a natural ally because its ability to carry out its basic mission rests on stakeholder support and requires collaboration and coordination. The Bay Institute's Bobker explains, "Because people were staff for CALFED, they were looking at things in a more global light, and they were probably also more directly responsive to the interests of stakeholders. They tended to reflect more of where stakeholders were going with new approaches, while agencies were more resistant to that" (Bobker interview 2002). While acknowledging the difficulty of generalizing about agency staff and Authority staff, Bobker suggests that "it's the nature of where [Authority staff members] are positioned" that induces their inclination toward stakeholders (Bobker interview 2002).

Some stakeholders have supported Authority staff and encouraged the state and federal agencies to play a more active role in CALFED. Northern California's Guy comments, "The agencies have come kicking and screaming into CALFED in a lot of ways, particularly when it comes to the budget." Guy suggests that the agencies will find greater success in collaboration: "The agencies that are going to succeed are the ones that are going to have partners working alongside them, like the water districts, like the counties. Those partnerships are going to be the way that the funding is going to flow" (Guy interview 2002).

The division of responsibilities among the Authority and the state and federal agencies has ebbed and flowed. During the planning phase, various agency staff members served on many of the work groups that prepared and analyzed detailed plans for CALFED. During development of the CALFED ROD, agency leaders served as the key negotiators, and their staffs worked intensely to support the negotiations. When the ROD was signed, a state agency and a federal agency assumed responsibility for leading the implementation of each particular element of the CALFED Program. According to former Secretary of Resources Nichols, "Since the ROD was signed, however, there was a problem with keeping the federal agencies at the table with their state counterparts at the higher levels. It's not just because of the change of political leadership in Washington. It has to do with the natural tendency of things to fly apart. Everybody has their own mandates, their own problems. Working with and through the somewhat cumbersome process of CALFED is something that, institutionally, the agencies are always going to prefer to avoid, unless they are required to do it" (Nichols interview 2002).

While there has been the necessary degree of agency collaboration to move CALFED forward, basic tensions remain unresolved between state and federal agencies' staff members, who are inclined to operate with their traditional independence and not compromise their autonomy, and the Authority's director and staff, who desire to coordinate agency actions in the context of integrated Program implementation. Although the Bay-Delta Authority Act created a state agency, the Authority, to play the major role in CALFED, the state legislation affirms the principal role of individual state and federal agencies in Program implementation. The tensions between and among these twenty-odd agencies and the Authority have continued. The 2004 federal legislation went further, explicitly denying state agencies any role in federal decisions.

The nature and quality of each agency's leadership and the level of attention and support from the state and federal administrations determine how staff members' disagreements, inherent in the relation between the Authority's leadership and agency staff, are handled. To the extent that the top management of the Authority and the agencies set an example of effective collaboration and convey to their staff members their support for the collaborative model and an expectation that it will be followed, relations are likely to be smooth. Of equal importance is the attention CALFED receives from policy and political levels of the state and federal administrations. In the absence of effective leadership and strong direction from the various agencies' managers and direction from high levels in Washington and from the California governor's office, tensions between the agencies and CALFED are likely to continue as long as the Program exists in its present form.

Science

Integration of credible, sound science in the CALFED Program is a sine qua non to its effectiveness, in terms of addressing the substantive issues of the Delta and dealing with stakeholder concerns while maintaining their support. The manner in which CALFED has dealt with this core issue constitutes one of its most positive, laudable achievements.

In the early stages of the conflict that led to CALFED, science was used more for its combat potential than for its analytical qualities. Quinn, of Southern California's MWD, has labeled these enduring conflicts the battles of "advocacy scientists." As Quinn explains, when regulators presented science to support their agency's policies, stakeholders would hire their own scientists to combat the regulators' scientific conclusions. He describes science in the pre-Accord period: "Every action led to conflict. There were huge fights over science because there was no gain for us [water users] in objective science. I think the Fish and Wildlife Service and others would have pretty weak science out there sometimes. We would attack even the good science because there was nothing in it for us in promoting good science" (Quinn interview 2002).

Quinn adds this insight: "That's the way we did science in California before the mid-nineties. It was a lot of 'I'm going to beat you. My arguments are going to beat yours, and they're going to lead to the regulatory outcome I want,' instead of, 'What do the fish need, and how do we accomplish that in the least-cost way that we can think of?'" (Quinn interview 2002).

Kern County's Clark echoes Quinn's concern, commenting, "Our biologists and scientists that work for the agencies had become what I call advocates instead of scientists" (Clark interview 2002). Such scientific combat weakened scientific credibility on both sides. Few trusted the scientific conclusions of agencies or stakeholders.

The use of science as an advocacy tool to advance a particular interest is not uncommon in the arena of resource and regulatory conflict. As described by CALFED's founding Lead Scientist Sam Luoma, under the old model, the conflicting scientific positions are put forth, and then someone decides who is right. This process is conducive neither to a result based on sound science nor to acceptance by stakeholder groups of a decision based on divergent scientific views. Perhaps more important, as Luoma says, the advocacy process does not help decision makers address the difficult "gray areas," where the science is uncertain and the decision makers' need for objective information is the greatest.

As collaborative efforts to address Delta issues developed, the agencies invested time, effort, and funding to expand and improve the scientific basis for the CALFED Program. In particular, the CALFED agencies and stakeholders placed increased emphasis on science-based adaptive management. By 1998, the concept of adaptive management was an essential part of every CALFED Program element. The 2000 ROD formally established the CALFED Science Program "to provide a comprehensive framework and develop new information and scientific interpretations necessary to implement, monitor, and evaluate the success of the CALFED Program."

The CALFED science program that has emerged since issuance of the ROD incorporates several positive elements that have led to its achievements and widespread acceptance. First, according to Luoma, the program does not seek to provide definitive answers to the difficult and frequent issues where the science is not certain, but rather, focuses on areas of uncertainty and seeks to discuss and explain them. The areas of uncertainty, as well as areas of scientific agreement, form a common ground among scientists that can be described and presented to decision makers to provide an objective picture of the relevant science. With this knowledge, decision makers can reach scientifically based, credible decisions to effectuate preferred policy choices.

A second important element is the use of independent outside scientists who do not work for any of the agencies or stakeholders. Outside scientists from universities and other research institutions provide an increased level of expertise and objectivity. Their input and participation help to ensure that results are credible and widely accepted by agencies, stakeholders, and decision makers. The integration of standard peer review procedures adds to the objectivity and credibility of the scientific results.

The independence of the science program prevents it from being used or perceived as an advocate in regulatory conflicts. Luoma believes that agency scientists are increasingly accommodating to the open, participatory structure of the CALFED science program (Luoma interview 2002). However, Kern County's Clark suggests that it may be impossible to change the culture in which agency scientists appear to be advocates for their agencies' regulatory programs. He adds that the CALFED science program has done well because the CALFED "super scientists" are above the conflicts. He urges, "Elevate them above the muck and the mire of what all of us are doing, and, hopefully, give us all something that we can hang our hats on that is objective. They follow the scientific method that wherever the chips fall, they fall where they may" (Clark interview 2002).

A final benefit of CALFED's ability to use outside scientists is simply a matter of expanded resources. Because agencies generally lack adequate staff and funding to address the numerous complex scientific issues facing CALFED, the use of outside scientists fills an important need.

A third important element is the open nature of the CALFED science program. From its inception, the program has been conducted in a transparent manner, with stakeholders, agencies, and others afforded full opportunities to observe and contribute to the process. Like other elements of the program, this open process leads to the most credible, well-informed results and to their widespread support and acceptance.

Keen competition exists among proposals to CALFED to fund specific projects. This competition has allowed CALFED, following a thorough review process, to select the best developed, most valuable projects. Former CALFED

Director Wright points out that funding for the science program, some of which comes from state bonds, also provides the very important ability to support projects that focus on system-wide issues, rather than on specific projects only. The results of system-wide research will provide a basis for developing and evaluating specific projects, proposals, and other actions as CALFED moves forward (Wright interview 2002).

MWD's Quinn emphasizes the importance of science-based adaptive management, citing the fish screen program as one in which the scientists look at how best to support fish passage instead of how to build the best fish screen. He believes that the science program's autonomy has been critical and that the substantial funding needed to sustain the program's autonomy is well justified (Quinn interview 2002).

Former Secretary of Resources Nichols adds, "The science in CALFED has been one of the great success stories of the program" (Nichols interview 2002). On the same point, Clark agrees, "Clearly, science is the linchpin of the process" (Clark interview 2002). Nichols credits CALFED's Luoma for building the science program: "He looks for ways that the science program can be helpful. He understands how not to get drawn into debates where the science is being used as the surrogate for people's underlying disputes with each other. Adaptive management will happen. The most important thing for the ongoing success of the [science] program is not just that it produces science, but that the science, if it suggests that some changes are necessary, will actually be used" (Luoma interview 2002).

The 2002 California Bay-Delta Authority Act carries forth these elements of the science program. California Water Code (CWC) Section 79452 provides for the appointment of the lead scientist, who is responsible, in consultation with the implementing agencies, for the science program. The statute directs the lead scientist to ensure that (1) adaptive management is scientifically applied, (2) projects are monitored to reduce uncertainties, and (3) peer review is extensively applied. The functions of the science program include (1) providing authoritative and unbiased reviews of relevant scientific knowledge; (2) providing a comprehensive framework to integrate, monitor, and evaluate the use of adaptive management; and (3) reviewing the overall technical and scientific performance of the Bay-Delta Program. Separate provisions of the statute direct the Bay-Delta Authority to establish an Independent Science Board (CWC 2002, 79470) and to authorize the lead scientist to establish independent science panels to review and provide advice on individual program elements and other issues (CWC 2002, 79471).

Watershed-Wide Management versus Regional Management

Water and ecosystem issues throughout the Central Valley are closely interrelated. As with any watershed, upstream activities substantially affect the Delta's ecosystem and hydrology. The Delta's watershed is particularly challenging because of its size and complexity. In the 1930s, when the CVP was envisioned, only the federal government could offer the resources and infrastructure to address the Valley's water issues. By the 1990s, however, many regional and local agencies had developed the expertise and resources to tackle some of the Delta's challenges. During various CALFED meetings, it was not unusual for stakeholders to express concern regarding state or federal agency domination of local projects and activities.

As the Delta's watershed and the area served by Delta exports are so large and complex, participants in the CALFED process face the constant challenge of balancing the system-wide needs and the needs of individual regions that either contribute to or rely on water from the Delta. The Delta's health depends on contributions from upstream watersheds. Yet, when CALFED participants focused solely on the Delta's needs, they appeared to ignore the needs of other regions.

Over the years, local agencies feared that in the interest of addressing Delta issues, the state or federal government would impose a solution on entities throughout the Central Valley that elevated Delta needs over local needs and that local resources would be used for the benefit of the Delta only. As the CALFED Program has been implemented, disagreements have continued between those responsible for addressing the needs of the entire Delta watershed and those responsible for addressing the needs of each of its contributing watersheds, as suggested by Thomas L. Crisman's chapter 8 (this volume). A critical 1986 California court decision, which led in part to the negotiation of the 1994 Accord, identified this connection and required the state water board to incorporate upstream water supply and demand variables in setting Delta water quality standards (*United States v. State Water Resources Control Board* 1986, 100).

Upstream water users and water agencies that rely on exports from the Delta have also disagreed. The Sacramento Valley has the greatest water resources but fewer representatives in the state legislature and U.S. Congress. Southern California, which depends on water from the Sacramento Valley, previously sought to rely on its greater political power to assure that its needs were met. The tensions among the various regions made it difficult to forge a compromise solution that would address the needs of the entire Delta watershed and Southern California.

To respond to these interregion rivalries, leaders within the state and federal governments began considering how to deal with each region's needs without losing focus on improving the Delta. During the final CALFED negotiations, issues were divided among five regions: Sacramento Valley, San Joaquin Valley, Delta, San Francisco Bay, and southern California. The issues in each region received individual attention, and negotiators often met with stakeholders in each region to focus on that region's particular needs.

During the post-ROD implementation phase, when individual regions assumed some responsibility for implementation, the second significant shift toward addressing issues on a regional basis occurred. Wright, former director of the California Bay-Delta Authority, explained, "CALFED has facilitated the development of locally and regionally based plans to meet the program's goals. Under this vision, local and regional entities have primary responsibility for implementing the key actions in the plan, while the CALFED agencies focus on system-wide improvements, science, balance, and integration" (Wright interview 2002).

Former Secretary of Resources Nichols believes this shift to regional implementation has contributed significantly to CALFED's ability to move forward: "The shift on the part of the CALFED staff to the regional approach to the delivery of the CALFED Program has made an enormous difference in keeping a number of the more anxious players involved. It was a smart move from a political perspective to simply take the political position that CALFED was not going to impose projects or decisions from the state level but was going to use the groups that operate at the regional level to make key decisions about priorities and projects. . . . [It] has given a number of key players the confidence that this really is their program, not just something that is being done to them (Nichols interview 2002).

Also enthusiastic about local solutions, Northern California's Guy says, "Local partnerships are really going to be the key to driving the success of these programs and making sure they do get integrated with all of the other pieces in CALFED. In a lot of ways we [local agencies] are going to do the work for CALFED. If you empower the locals, provide the regulatory streamlining, provide the funding and resources, I think the locals can deliver the CALFED solution" (Guy interview 2002).

While the Delta's problems cannot be addressed in isolation from the entire watershed, local implementation offers the opportunity to integrate upstream and downstream needs. Guy explains why the focus on local projects works for the entire watershed: "If you can make it so that it's their project, I think [local agency leaders] take pride in the fact that it is contributing to a larger regional project" (Guy interview 2002).

The Bay Institute's Bobker adds a note of caution to the enthusiasm for regional solutions: "There are linkages, negative and positive, that you don't see if you look at the site-specific level. Seeing linkages is important. Take that next step, identifying the benefits and adverse impacts. Use the global planning process to push the benefits and avoid the impacts. That's extremely important. There are benefits you can't see if you're only looking at a small area. There are also benefits that may not be important to people in a local area but are important to the society as a whole. Conversely, there are adverse impacts that you don't see because they're outside your small area, or they're ones you don't particularly care about because you don't see them. But a large-scale planning process can take actions to avoid those even if they're not the motivation of the local planners" (Bobker interview 2002). Bobker's comments suggest that CALFED's integration of regional solutions and watershed-wide coordination combines the best of all worlds—local, regional, and watershed-wide improvements.

Delta Issues and Property Ownership

Despite CALFED's efforts to resolve conflicts and promote collaboration, various parties have resorted to litigation, raising important legal issues.[12] Perhaps the most well-known and controversial court decision related to CALFED, commonly called the "Tulare decision," involved Fifth Amendment "takings" claims filed in the Court of Federal Claims. The claims were filed by two agricultural contractors against the federal government, based on lost water they claimed had been taken from them when the SWP reduced Delta pumping in response to federal ESA regulation in 1992–1994, during the last major drought. Subsequent decisions from the same court have rejected the *Tulare* analysis.[13] From an economic perspective, Stephen Polasky, in chapter 9 (this volume), considers the *Tulare* decision and suggests that clarification of property interests in water rights would encourage resolution of Delta conflicts.

A clear definition of property interests in water rights and the extent to which any diminution of or interference with those rights requires compensation would, as Polasky suggests, facilitate market-based allocation. Water would presumably flow, in the words of many water practitioners, "uphill to money," or, in economic terms, toward its highest and best use. Establishing the circumstances under which compensation would be required could eliminate the uncertainty resulting from defined actions, such as reductions to protect endangered species or other resources. The uncertainty resulting from the variation in precipitation and other natural factors would, of course, remain, but legal and regulatory uncertainty would be reduced.

From a political and legal point of view, providing greater definition to water rights would not be an easy task. In California, water rights are subject to a variety of limitations that may affect water use and do not require compensation. Unlike land-based property rights, water rights are usufructuary only. That is, water rights are only for use, not the physical holding, of water (*U.S. v. SWRCB* 1986, 100). Water rights, by their nature, are "limited and uncertain" (*People v. Murrison* 2002, 359). California water rights remain subject to the "public trust" for such water in the public waterways, with regulation a "pervasive" limitation on California water rights (*Nat'l Audubon Society v. Superior Court* 1983, 419; *People v. Weaver* 1980, 30). In addition to the public trust doctrine, California's "reasonable use" doctrine requires ongoing analysis, reflecting changes in relevant conditions, and may result in changes to a water right as to its purpose, method, or amount (*U.S. v. SWRCB* 1986, 129–30). Limitations may also be imposed to comply with requirements of the state and federal endangered species acts and water quality laws.

If all the issues relating to providing compensation for limitations on water use arising from existing legal requirements could be resolved, the result would likely substantially facilitate economically based water transfers. However, other issues would remain. As described fully in this chapter, a fundamental aspect of the problems facing the Delta ecosystem arises from the conveyance of water through the Delta. Regardless of the transferability of water rights, as long as water is transferred from north to south of the Delta, the issues regarding conveyance and the consequent health of the Delta will confront Delta policy makers. Similarly, the impacts of water transfers on local communities and watersheds that would be affected by transfers between individual water users would also have to be addressed. Given the importance and complexity of these issues, it is likely that they would have to be resolved through the political process or in the courts.

CALFED's Future in Doubt

Despite the apparent early achievements of CALFED and the optimism engendered by its collaborative approach, its progress has not matched expectations, giving rise to the question of CALFED'S continued viability. In 2005, the governor's comprehensive program review was critical of CALFED. In 2006, the legislature moved all Delta funding from the California Bay-Delta Authority to the Office of the Secretary of Resources (CSB 2006–2007). In 2007, Assemblywoman Lois Wolk, chair of the assembly's water committee, told members of Congress that CALFED was "dysfunctional" (*Congressional Briefing Book*). Shortly thereafter, Governor Schwarzenegger appointed a Blue Ribbon Task Force to develop a long-term Delta Vision after studies concluded current Delta policies were not sustainable ("Delta Vision" 2007).

Despite these negative developments, CALFED has continued to function. The ROD provided for a review of CALFED progress and a decision on long-term water conveyance through or around the Delta in 2007. The California Bay-Delta Authority has begun that review, working toward the "End of Stage 1 Decisions" (www.calwater.ca.gov). The legislature has continued funding some CALFED programs, such as the Ecosystem Restoration Program (CSB 2007–2008). Stakeholders continue to support CALFED science projects to help determine what is best for the Delta as a whole. Both water project and fishery agencies continue to coordinate closely when crises in the Delta arise, as reflected in the June 2007 SWP export pump shutdown to protect delta smelt.

Offsetting these positive actions, and perhaps most tellingly, conditions in the Delta have worsened, and conflict and litigation have returned. Numbers of delta smelt, a key indicator species listed as threatened under ESA, have declined to critically low levels. This warning of ecological peril has resulted in a federal court assuming a dominant decision-making role in operations of the CVP and SWP.[14] In *Natural Resources Defense Council v. Kempthorne* (2007), a federal district judge invalidated a biological opinion on the operation of the CVP and SWP and issued an order prescribing conditions the projects must comply with until a new biological opinion is issued, expected in autumn 2008.

The central and active involvement by a court to carry out the mandates of ESA to protect a Delta species marks a significant departure from the usual after-the-fact review of project operations and reliance on agency discretion. As the legislature and interest groups struggle to develop long-term solutions that will take years to implement, critical decisions to assure that the projects comply with the basic mandate of ESA to preserve species may be left to the courts.

Conclusion

Essential to any efforts to address the Delta, whether through CALFED or a successor effort arising from the governor's Delta Vision Blue Ribbon Task Force, is the commitment of the leadership of interest groups to collaboration. Effective leaders must have a certain maturity, frequently gained from weathering extended conflicts. Guy, former executive director of the Northern California Water Association, described a key element leading to the creation of CALFED and its early effectiveness: "It really comes down to leadership. You need it at all levels. We have some strong leadership in the boardrooms across the Valley. Some of the agencies have it. You've got to have people who are willing to collaborate rather than fight. God knows it's a lot easier to fight. We all think we're pretty good at fighting. We're not always quite as good at collaborating. But I think a lot of us are trying. That's what's exciting" (Guy interview 2002).

Whether or not CALFED survives in its current form, the lessons its participants and others learn from the CALFED process may contribute to ecosystem programs in other watersheds across the country. From an "issue perspective," the lessons are many and detailed—ranging from governance shared among federal, state, and local agencies to the incorporation of sound science.

From an overall perspective, the lesson CALFED offers is more fundamental: Successful watershed ecosystem restoration programs require leaders who recognize that they will never be able to defeat all the other stakeholders in the watershed. They will never get everything they want at the expense of others. Neither litigation nor unilaterally sponsored legislation offers the answer. Even when the conflicts arise and return to the courtroom, leaders need to continue to strive to resolve the conflicts, not to create winners and losers but to achieve a balanced outcome.

Notes

1. It is important to clarify at the outset of this chapter that the CALFED Bay-Delta Program, renamed and reconstituted as the California Bay-Delta Authority, was never a state program with federal cooperation. Rather, from the beginning, it has been a joint state–federal program, hence the name "CALFED."

2. Much of the background information contained in this case study can be obtained from the documents CALFED issued over the years of its development. For more information on CALFED, see www.calwater.ca.gov.

3. Former U.S. assistant secretary of the Interior for Water and Science Elizabeth Ann Rieke wrote a law review article after Delta agencies and stakeholders achieved the 1994 Accord on water quality. See E. A. Rieke, 1996, "The Bay-Delta Accord: A Stride Toward Sustainability," *University of Colorado Law Review* 67 (Spring). This article provides an excellent overview of the Delta interest groups and the political underpinning of the 1994 Accord.

Former director of the CALFED Bay-Delta Program Patrick Wright also wrote an informative article about the 1994 Accord. See P. Wright, 1995, "The Federal Perspective on the Bay/Delta Standards: A Triumph of Common Sense over Politics as Usual, " *California Water Law & Policy Reporter 5 (February): 27.*

4. See *Town of Antioch v. Williams Irrigation District*, 205 P. 688, 689 (Cal. 1922).

5. *United States v. State Water Resources Control Board*, 182 Cal. App. 3d 82, 227, *California Reporter* 161 (1986). For legal citations and a more detailed legal history of the Delta's conflict, see A. W. Brandt, 2002. "An Environmental Water Account: The California Experience," *Water Law Review* (University of Denver) 5:426–56.

6. Central Valley Project Improvement Act, Public Law 102-575, Title 35 (1992).

7. *Framework Agreement between Governor's Water Policy Council of the State of California and the Federal Ecosystem Directorate*, 1994. See www.calwater.ca.gov/Archives/GeneralArchive/Framework1994.shtml.

8. See *Principles for Agreement on Bay-Delta Standards between the State of California and the Federal Government*, 1994 (December 15): 1, cited in Rieke, supra note xx, at 347; see also Wright 1995, 28; J. Kay, 1994, "U.S., State Unite with Plan to Save Bay," *San Francisco Examiner*, Dec. 16, A1; E. Diringer, 1994, "'Peace Has Broken Out' in Water Wars, Wilson Says," *San Francisco Chronicle*, Dec. 16, A1; J. H. Cushman, Jr., 1994, "U.S. and California Reach Pact to Regulate Flow of Fresh Water," *New York Times*, Dec. 16, A1.

9. For a more complete analysis of the EWA, see Brandt 2002.

10. See California Bay-Delta Authority Act, 2002, California Water Code Section 79400. At www.calwater.ca.gov.

11. The California Court of Appeals decision, 2005, *In re Bay-Delta Programmatic Environmental Impact Report Coordinated Proceedings*, 133 Cal. App. 4th 154 (Cal. App. 3d Dist. 2005), held that, under the California Environmental Quality Act, CALFED

should have considered alternatives that reduced water exports to protect the Delta. That decision is, as of August 2007, on appeal to the California Supreme Court.

12. In recent years, the litigation has shifted from secondary legal issues to challenging the foundations of Delta water management, such as recent federal and state litigation on compliance with the federal ESA and the California Endangered Species Act (CESA). See *Natural Resources Defense Council v. Kempthorne*, Civ. No. 1:05-CV-01207 OWW (E.D. Cal. May 25, 2007); *Watershed Enforcers v. CA Dept. of Water Resources*, No. RG 06292124 (Alameda Co. March 22, 2007).

13. *Tulare Lake Basin Water Storage Dist. v. United States*, 49 Fed. Cl. 313 (2001); 59 Fed. Cl. 246 (2003) (reduced Delta pumping was a taking of farmer property rights). Subsequent decisions from the same court that have rejected or departed from the *Tulare* analysis include *Klamath Irr. Dist. v. U.S.*, 74 Fed. Cl. 677, (2007); *Casitas Mun. Water Dist. v. U.S.*, 76 Fed. Cl. 100, (2007).

14. In a state court ruling with potentially far-reaching implications, a trial court ruled that SWP did not have authorization required for "take of listed species" (those listed under the California ESA), and, therefore, its Delta pumping operations were unlawful. In March 2007, the court issued an order requiring SWP pumping to cease in sixty days unless "take authorization" under CESA were obtained. The ruling has been appealed and has been stayed, pending appeal. See *Watershed Enforcers v. CA Dept. of Water Resources* (2007).

References

Bobker, G. 2002. Interview by D. Nawi and A. W. Brandt (hereafter, authors).

Brandt, A. W. 2002. "An Environmental Water Account: The California Experience." *Water Law Review (University of Denver)* 5:426–56.

CALFED Bay-Delta Authorization Act. 2004. Title 1, Pub. L. 108-361. U.S. Congress.

CALFED Programmatic Record of Decision (CALFED ROD). 2000. At www.calwater.ca.gov/Archives/GeneralArchive/RecordOfDecision2000.shtml.

California Bay-Delta Authority Act. 2002. California Water Code, Sections 79400 *et seq*. At www.calwater.ca.gov.

California Bay-Delta Program and California Bay-Delta Authority (CALFED). 2000. At www.calwater.ca.gov.

California Court of Appeals. Decision *In re Bay-Delta Programmatic Environmental Impact Report Coordinated Proceedings*, 133 Cal. App. 4th 154 (Cal. App. 3d Dist. 2005).

California Water Code (CWC). 2002. Sections 79403, 79452, 79470, and 79471. At www.calwater.ca.gov.

California's Water Future: A Framework for Action. 2000. June 9. At www.calwater.ca.gov/Archives/GeneralArchive/adobe_pdf/new_final_framework.pdf.

California State Budget (CSB). 2006–2007. Senate Bill (SB) 1574 (Kuehl). At www.leginfo.ca.gov.

CSB. 2007–2008. At www.leginfo.ca.gov.

Central Valley Project Improvement Act. Public Law 102-575, Title 35 (1992).

Clark, T. 2002. Interview by authors.

Clean Water Act (CWA). 1977. See below Federal Water Pollution Control Act.

Cushman, J. H., Jr. 1994. "U.S. and California Reach Pact to Regulate Flow of Fresh Water." *New York Times*, Dec. 16, A1.

"Delta Vision." 2007. At www.deltavision.ca.gov.

Diringer, E. 1994. "'Peace Has Broken Out' in Water Wars, Wilson Says." *San Francisco Chronicle*, Dec. 16, A1.

Federal Advisory Committee Act (FACA). Public Law 92-463, 86 Stat. 770 (1972).

Federal Ecosystem Directorate. 1993. December 15. Federal Register notice. The 1993 creation of this Directorate is cited in the *Framework Agreement Between Governor's Water Policy Council of the State of California and the Federal Ecosystem Directorate*. 1994. At www.calwater.ca.gov/Archives/GeneralArchive/Framework1994.shtml.

Federal Water Pollution Control Act (Clean Water Act [CWA] as of 1977). [1948.] 1972. 1977. 33 U.S.C Sec. 1251–1387 (2007). At http://uscode.house.gov/search/criteria.shtml.

Framework Agreement Between Governor's Water Policy Council of the State of California and the Federal Ecosystem Directorate. 1994. At www.calwater.ca.gov/ Archives/GeneralArchive/Framework1994.shtml.

Guy, D. 2002. Interviews by authors. October–December.

Kay, J. 1994. "U.S., State Unite with Plan to Save Bay." *San Francisco Examiner*, Dec. 16, A1.

Luoma, S. 2002. Interviews by authors. October–December.

Metropolitan Water District of Southern California (MWDSC). 2007. "Delta Action Plan." June. At www.mwdh2o.com/mwdh2o/pages/board/highlights01.html.

Nat'l Audubon Society v. Superior Court 33 Cal. 3d 419 (1983).

Natural Community Conservation Planning (NCCP). 1991. California Public Fish & Game Code, Sec. 2800 et seq. California state law. At www.dfg.ca.gov/nccp/.

Natural Resources Defense Council v. Kempthorne Civ. Cal. No. 1:05-CV-01207 OWW (E.D. Cal. May 25, 2007).

Nichols, M. 2002. Interview by authors.

People v. Murrison 101 Cal. App. 4th 349, 359 (2002).

People v. Weaver 147 Cal. App. 3d. Supp 23, 30 (1980).

Principles for Agreement on Bay-Delta Standards between the State of California and the Federal Government. 1994. Dec. 15. At http://calwater.ca.gov/Archives/General Archive/SanFranciscoBayDeltaAgreement.shtml.

Quinn, T. 2002. Interviews by authors. October–December.

Rieke, E. A. 1996. "The Bay-Delta Accord: A Stride toward Sustainability." *University of Colorado Law Review* 67 (Spring):341–69.

Town of Antioch v. Williams Irrigation District 205 P. 688, 689 (Cal. 1922).

Tulare Lake Basin Water Storage Dist. v. United States 49 Fed. Cl. 313 (2001).

Tulare Lake Basin Water Storage Dist. v. United States 59 Fed. Cl. 246 (2003) (reduced Delta pumping was a taking of farmer property rights).

United States v. State Water Resources Control Board 182 Cal. App. 3d. 82, 100 (1986).

U.S. v. SWRCB 182 Cal. App. 3d at 129–30 (1986), citing Cal. Const. Art. X, Sec. 2.

U.S. v. SWCB 182 Cal. App. 3d 82, 227 (1986). Cited in *California Reporter* 161, 1986.

Watershed Enforcers v. CA Dept. of Water Resources, Alameda Co. Superior Court No. RG06292124; First District Court of Appeal No. A117750 (March 22, 2007).

Wolk, L. 2007. *Congressional Briefing Book*. February. Files of the California Assembly, Water, Parks & Wildlife Committee.

Wright, P. 1995. "The Federal Perspective on the Bay/Delta Standards: A Triumph of Common Sense over Politics as Usual." *California Water Law & Policy Reporter* 5 *(February)*.

Wright, P. 2002. Interview by authors.

Chapter 8

The Ecology of Bay-Delta Restoration

An Impossible Dream?

THOMAS L. CRISMAN

The two largest river basins of California, Sacramento (70,567 square kilometers—km^2) and San Joaquin (32,000 km^2), discharge approximately 40 percent of runoff for the state through their combined Delta (1,153 km^2), the largest wetland of the western United States and a critical ecosystem component of the largest estuary of the Pacific coast, San Francisco Bay (CDWR 2005).

Prior to European settlement (preimpact) of this part of California, the ecological structure and function of the Sacramento and San Joaquin rivers represented a longitudinal continuum from the mountains to the Delta (Vannote et al. 1980) with strong, two-way lateral interactions with their riparian zones (area adjacent to rivers) along their lengths and with wetlands of the middle and lower river segments. The headwaters in the mountains exerted a strong control over seasonal flow dynamics in the rivers via the timing and extent of rainfall and snowmelt. Floodplain wetlands and riparian zones of intermediate river reaches reduced flooding; transformed and stored inorganic chemicals, organic matter, and sediments; and regulated their release downstream. The Delta played a similar role because it regulated primary productivity of San Francisco Bay and was a critical link in the reproductive cycle of many fish and invertebrates.

Native Americans appear to have played a critical role in the management of these rivers, especially through their interactions with riparian zones and floodplain wetlands. They tended, harvested, and burned wetland plants and thereby created a shifting mosaic of vegetation types interspersed with open-water areas, which likely enhanced biodiversity of the area. This sizeable population's sustainable practices in caring for wetland plants (for example, white root) produced a lawnlike layer of vegetation within riparian forests that expanded channel capacity and reduced flood impacts for the riparian zone.

The Native Americans' sustainable management of the watershed was replaced by gradual environmental deterioration caused by European settlers' mining, land clearance, agriculture, and urbanization.

Recognizing the importance of this rich natural resource to nature conservation, expanding human demands, and the immediate need to protect and restore critical habitats, the California Bay-Delta Authority (CALFED) was created in 2002 as a cooperative effort of twenty-one state and federal agencies to ensure sustainable management of the Bay-Delta and its watershed for conservation and human needs. See chapter 7 of this volume for details on progress toward the program's integrated ecosystem approach for the watershed.

While recognizing the importance of longitudinal and lateral linkages as controlling factors over river structure and function and over the Delta and bay, this analysis follows current management scenarios and examines such relationships for major geographic units of the watershed. This chapter clarifies how these units could be managed to achieve sustainability—as envisioned by CALFED.

Upper Watershed

Headwater streams of the Sacramento and San Joaquin rivers set the stage for the hydrology and water quality of the watershed. Ultimate control factors over upper watershed discharge are the timing and extent of snowmelt, mediated annually by vegetation impact on evapotranspiration. Any modifications in forest management practices prescribed by state and federal agencies, including forest thinning and width of the riparian zone, can have significant impacts on evapotranspiration regimes and leaf input to streams. Ecologists have expressed growing concerns that projected climate changes in headwater regions of this watershed will alter snowmelt and associated discharge of freshwater to the Delta (Gleick and Chalecki 1999). Throughout the length of the rivers, species composition and life histories of invertebrate communities are closely timed with discharge periodicity and magnitude; significant changes in either of these parameters are likely to alter aquatic food webs profoundly.

Baseline water quality in headwater streams reflects patterns in regional geology and soils and the role of vegetation in storing or releasing sediment and chemicals to receiving waters. Although heavy metals are common components of the igneous and metamorphic geology of headwater areas and can enter local streams, acid mine drainage and leaching from waste disposal sites are the major contributors to stream waters (CDWR 2005). The problem is especially acute for creeks draining from areas of historic gold mining, with high concentrations of bioavailable heavy metals reported at least 120 km downstream from the source area (Cain et al. 2000).

Several small reservoirs have been constructed on headwater streams to store snowmelt for generating electricity (CDWR 2005). These structures interrupt the longitudinal linkages outlined by the river continuum concept (Vannote et al. 1980), especially for downstream movement of organic matter, sediments, and nutrients. Ward and Stanford (1983) developed the serial discontinuity concept to explain the role of dams on longitudinal structure and function of streams. Dams disrupt longitudinal patterns in the evolution of stream structure and function, but expected patterns can often return below the channel segment affected by human-altered structure (the discontinuity distance). While environmentalists have great concerns that headwater dams interfere with upstream movements of threatened and endangered fish species (CDWR 2005), the dams' potential impacts on other components of the food web—particularly those associated with isolation of cold- and warm-water fauna and with altered flow and sedimentation of terrestrially derived organic matter necessary to drive heterotrophic stream segments (those populated by organisms that do not manufacture their own food via photosynthesis but must obtain matter for food) downstream—are poorly known. To make effective, sustainable, adaptive management decisions, water managers must understand how constantly changing conditions of headwaters affect downstream ecosystem structure and function.

Middle Watershed

Prior to extensive modification, the floodplain of the midreaches of the Sacramento–San Joaquin river system was a mosaic of habitats consisting of wetlands, channels, and riparian forests (Mount 2003). Sedimentation on the floodplain increased by an order of magnitude during the period from 1849 to 1920 and subsequently declined as levee construction blocked connections with the river channel and as progressively increasing numbers of instream reservoirs trapped sediments behind dams. Coupled with increasing agricultural activities, the floodplain lost habitat heterogeneity, and sediment filled the floodplain channel network, rendering it largely nonfunctional. While the upper montane reaches of the watershed are mainly in national forests, most of the central watershed is privately owned and devoted to agriculture. Formerly extensive floodplain wetlands have been converted into intensive agricultural areas, causing problems associated with pesticides, herbicides, and fertilizers. Invasion by nonnative plants has reduced soil moisture in floodplain soils (Gerlach 2004), and agricultural floodplains are from 80 to 150 percent more erodible than comparable areas of riparian forest (Micheli et al. 2004). Floodplain ponds are often dominated by alien fish species and extremely eutrophic (containing excessive aquatic plant growth in response to particular nutrients) (Feyrer et al. 2004).

Pronounced changes have been made to the channels of the Sacramento and San Joaquin rivers. Of the forty-three reservoirs of the region, most are considered multipurpose but used primarily for water supply and flood control (CDWR 2005). Much of the stored water is conveyed to the Delta, then pumped south for municipal, industrial, and agricultural uses. It is also used to maintain delta outflow standards. State and federal agencies have made joint plans to expand many of the instream reservoirs for greater storage capacity to meet increasing human needs.

The extensive reservoir network has altered river structure and function above and below dams. Vannote et al. (1980) developed the river continuum concept to explain the progressive, predictable changes in structure and function that rivers experience from their headwaters to their termini. For rivers like the Sacramento and San Joaquin that originate in high mountains, when benthic invertebrates (cold-water fauna without backbones) that depend on a detritus-based food web in alpine areas travel to lower elevations, they must change gradually into warm-water fauna, whose food web is driven more by autotrophic (able to make their own food via photosynthesis) production and sediment processes. Ward and Stanford (1995) developed the serial discontinuity concept to explain how dams and other channel alterations can disrupt the progression of evolving patterns in downstream structure and function, creating disconnects within the system. Dams disrupt seasonal hydrological patterns and associated sediment grain size, transport, and deposition. Current velocity is one of the most important factors structuring benthic invertebrate communities in the Sacramento River, with alterations in flow regimes from dams of special concern (Nelson and Lieberman 2002).

Fish communities are also structured in large part by hydrological impacts on the size of sediment substrates and by the quality of instream and riparian habitat (Coutant 2004; May and Brown 2002). While biologists clearly recognize that dams block upstream movement of many fish, including salmon, and thus can interfere with reproductive success, plans to remove small dams to assist in restoration of salmon populations (CDWR 2005) are of questionable value in light of a major proposed expansion in both the number and size of existing mid-basin dams.

A major disconnect exists between the Sacramento and San Joaquin rivers and their floodplains as a result of levee construction, loss of riparian forests, and drainage of floodplain wetlands (CDWR 2005). More than 95 percent of riparian forests have been destroyed during the past 130 years, ruining critical heterogeneous habitat for salmon and migratory birds (Larsen and Greco 2002).

Invasive plant species have altered additional riparian habitat significantly (CDWR 2005). Aquatic habitats of the floodplain often develop elevated algal biomass—associated with shallow water and longer water residence times—that, when exported to the river channel, enhances productivity of higher trophic levels of the food web, including invertebrates and fish (Sommer et al. 2004). These positive interactions are threatened by pesticides and herbicides from agricultural sources and heavy metals from acid mine drainage and local geology discharged from the floodplain (CDWR 2005).

Restoration of riparian forest and its connectivity as well as the floodplain to the river channel is considered critical for management success of the midbasin region and ultimately the Delta (Holl and Crone 2004). Allowing reflooding of the floodplain will hold 80 percent of floodwater, increase species diversity, and facilitate export of phytoplankton (microscopic algae suspended in the water column) and detritus from the river channel for incorporation into the food web. The California legislature recognized the importance of riparian forests for river and delta management and called for creation of the Upper Sacramento River Fisheries and Riparian Habitat Management Plan in 1986 but did not enact it (CDWR 2005). The plan seeks to preserve the remaining patches of intact riparian forests and to reestablish a continuous riparian forest corridor along the Sacramento River. Several thousand hectares of riparian area are targeted for protection or restoration.

Restoration of Central Valley wetland habitats is also considered essential for management of fish and wildlife in the San Joaquin River Basin (CDWR 2005). While it is critical to understand the extent to which wetland structure and function are essential for specific management goals and to acknowledge that both are important, sometimes achieving function is sufficient. In particular, the role of wet agriculture, especially rice, in essential wetland functions must be examined, together with how such a role can be incorporated at the landscape scale to provide additional wetland services and to serve as a buffer for protection of natural wetlands.

Integration of surface and groundwater resources via pumping and storage is crucial to attempt to meet future water demands of the Sacramento–San Joaquin Basin while fulfilling obligations to export water elsewhere in California (CDWR 2005). Most stream theory has considered streams mainly as surface water systems operating within a longitudinal and horizontal context. Stanford and Ward (1993), however, provided a conceptual framework for the vertical dimension—interactions between the stream channel and the hyporheic zone (area beneath the stream channel that interacts vertically and laterally with groundwater). Their hyporheic corridor concept considered how vertical and lateral interactions between the channel and the hyporheic corridor change along the length of the stream or river, proposing that the influence of the hyporheic zone on stream structure and function was lowest at the headwaters and near the river terminus and greatest in intermediate reaches.

The midbasin portions of both the Sacramento and San Joaquin rivers are planned for maximization of water storage and potential reallocation. It is critical that scientists, water managers, and all stakeholders make a major effort to understand not only interactions of river channels and the hyporheic zone, but also those of the riparian and the extensive floodplain areas with the hyporheic zone. Failure to do so could result in irreparable damage to the structure and function of the rivers, riparian zone, and floodplain water bodies.

Lower Watershed and Sacramento–San Joaquin Delta

The lower Sacramento and San Joaquin rivers and associated Delta are the most ecologically complicated portions of the basin. They integrate inputs of water, sediments, nutrients, and contaminants from the entire basin and, in turn, regulate conditions within San Francisco Bay by their ability to store, transform, and release physical, chemical, and biological parameters received from upstream. Longitudinal, vertical, and lateral interactions within this area are complicated further by the dynamic nature of the freshwater–saltwater transition and its importance for native flora and fauna. The area is important for migratory birds and fish and as a nursery ground for many commercially important fish and invertebrates.

Freshwater inputs to the lower rivers and Delta have varied significantly over the past few millennia (Ingram et al. 1996a). The upper Delta gradually came under tidal influence between 3,000 and 800 years ago and between 1,900 and 1,670 years before current freshwater inflows to San Francisco Bay were substantially greater than those estimated for the prediversion period (before human alteration). Over the past 750 years, stream flow to the bay has varied with alternating wet and dry intervals lasting from forty to sixty years. Thus biota (animal and plant life) associated with this region adapted to a dynamic freshwater–saltwater regime prior to human intervention.

A number of biotic changes have been recorded over the last century, especially after freshwater diversion from the Delta began. Phytoplankton production in the lower San Joaquin River is limited by levels of sunlight that can penetrate the water, while benthic algae production is limited by an interaction between salinity and nitrogen (Leland et al. 2001). Increased clarity associated with elevated river flow during spring leads to increased phytoplankton activity (Leland 2003).

By the 1980s, only eleven of the nineteen fish species present in the San Joaquin River prior to European settlement of the area remained, five of which were not common (Brown and Moyle 1994). Striped bass populations of the Sacramento River in 1994 totaled only 20 percent of those reported in the mid-1970s, with the loss attributed to water diversion, reduced food organisms, and potential agricultural pollution. Several fish species have experienced elevated selenium tissue concentrations in the vicinity of agricultural discharges into the San Joaquin River (Saiki et al. 2001), and extreme oxygen stress is evident in the lower 14 km of the river as a result of nitrogen transformations of organic matter carried by the flow (Lehman et al. 2004). While the impact from nonnative species on fish species is unclear, the declining population of endemic delta smelt in the saltwater–freshwater mixing zone and changes in its maintenance since 1983 are clearly associated with altered size and timing of the mixing zone due to freshwater flow diversion (Moyle et al. 1992).

The Sacramento–San Joaquin Delta is a tidal freshwater ecosystem whose food web is based primarily on heterotrophic processing of detrital organic matter delivered by inflowing rivers (Jassby and Cloern 2000). Phytoplankton are a secondary source of organic matter that can easily be converted for use by metazoans (all animals other than protozoans and sponges). The Delta is a net accumulator of organic matter, with transport of material from the Delta to the bay always less than the input to the Delta from upstream, as a result of engineered water diversion. Although data are incomplete, it appears that invertebrates of portions of the lower Sacramento–San Joaquin system, the principal processors of detrital organic matter, may not have been significantly affected by human modifications of the channels (Leland and Fend 1998).

The Delta is a fragmented ecosystem of approximately 1,130 km of waterways within a complex of numerous islands, many of which have been effectively isolated from the river and bay by more than 1,770 km of levees (CDWR 2005). Such isolation has resulted in significant subsidence (the wetland surface is now more than 6 meters below sea level) and significant reductions in the delivery of organic matter to San Francisco Bay. In addition, many tidal channels within the Delta have experienced lateral erosion, which complicates flow and sediment deposition patterns (Gabet 1998).

Between 1972 and 2002, approximately 940 hectares (2,322 acres) of leveed, former salt marsh were reconnected to the bay; via tidal action they accreted sediment to depths permitting reestablishment of marsh vegetation (Williams and Orr 2002). The time required to reach the restoration goal throughout this vast area varied significantly, influenced by the supply of sediment from the estuary and its deposition patterns within the wetland. Dredged materials were used to restore a tidal marsh at Sonoma Baylands, effectively reducing the time to reach restoration goals by decades (Marcus 2000). Most projects appear to have met their restoration goals; the most successful are larger sites and those connected to existing wetlands (Breaux et al. 2005). Whether water managers allow the marsh system, following its reconnection with the bay, to self-organize or to engineer sedimentation actively, reconnection to the bay is critical for the supply of organic matter for food web processing and ecosystem productivity to increase (Jassby and Cloern 2000).

The biomass of phytoplankton reaching nearshore areas of San Francisco Bay for utilization by the food web is heavily influenced by diversion of estuarine water for human use and river inflow (Jassby and Powell 1994). The food web's need for phytoplankton must be balanced by the desire to increase water extraction from the Delta for human use and the salinity levels required by nontidal invertebrates of the river (Leland and Fend 1998).

Future restoration scenarios need to include two aspects to promote ecosystem sustainability. First, it is critical that current climate change models for the Sacramento–San Joaquin Basin (Vanrheenen et al. 2004) be revisited as new data emerge. Second, McCreary et al. (1992) suggested that farmed wetlands in the Delta and North Bay should be protected as part of any landscape-level management plan. Although the biotic structure of agricultural wetlands does not approximate that of natural wetlands, such systems can function effectively for wetland processes, including production, transformation, storage, and export of organic matter (Crisman et al. 1996), critical to heterotrophic food webs.

San Francisco Bay

From a pre-European volume of approximately 2,088 km^2 of water at high tide, San Francisco Bay has been reduced by 37 percent to 1,323 km^2 via drainage and infilling (CDWR 2005). Based on isotopic analyses of subfossil mollusks, average annual river discharge to San Francisco Bay is 1.8 times less than values calculated for the five millennia immediately prior to European colonization, with salinities in parts of the bay lower than today (Ingram et al. 1996b). Sedimentation in the bay increased significantly about one hundred years ago, in some areas by an order of magnitude, reflecting increased delivery of erosion-derived, inorganic sediments from the watershed associated with industrialization, urbanization, mining, agriculture, deforestation, and changes in freshwater discharge from diversion projects (van Geen et al. 1999).

The structure and function of San Francisco Bay's food web are constantly self-organizing in response to alterations in organic matter input (both detrital and algal) from the river and Delta, phytoplankton and benthic algal production in response to nutrient loading and internal recycling, sediment resuspension, variations in salinity and turbidity, and spread of invasive species. Climate change and human diversion have reduced stream inflow to the bay (Jassby et al. 2002), and elevated salinity has been noted in the bay during the wet season because water is trapped higher in the watershed for human use (Knowles 2002).

Productivity of San Francisco Bay is controlled by a delicate balance between autochthonous (formed in its present position) and allochthonous (derived from terrestrial sources entering the aquatic ecosystem) organic matter availability. Watershed delivery of organic matter to the bay has been reduced via sedimentation in upstream reservoirs; fragmentation and destruction of delta marshes; and consumption by invasive species, especially clams (Jassby et al. 2002). The main carbon sources within the bay are phytoplankton and sediment resuspension of benthic algae and detritus (Canuel et al. 1995). It is crucial that these carbon sources are kept in balance because carbon is the driver of the food web, the currency of life.

Management of San Francisco Bay is a temporally moving target, reflecting watershed export and alterations in the food web via introduction of invasive species of plants and animals, which is why adaptive management techniques have been part of every CALFED program element since 1998 (see chapter 7, this volume). San Francisco Bay purportedly has the greatest number of exotic species of any aquatic system in North America, 234 as of 1998 (Cox 1999). Alterations in food web structure and function for both saltwater and freshwater portions of the bay have been observed for all trophic levels from primary producers to predators. Native species have suffered population declines, and perhaps the most worrisome nonnative species are those that consume phytoplankton and detritus, the energy basis of the food web. Although many species have been introduced into the bay and Delta purposefully, those entering via ballast water from commercial shipping are a serious and possibly expanding threat to the ecosystem. Without elimination of further foreign introductions and a detailed understanding of the impacts to food web structure and function from the current pool of invasive species, management of San Francisco Bay is a moving target that may not be sustainable.

Perhaps the great missing link in understanding ecological interactions as part of the CALFED program is the relationship of San Francisco Bay with the Pacific Ocean. Jassby et al. (1993) noted that exchange between the bay and the ocean likely plays an important but unknown role in the organic carbon balance of the system. Such interactions are two way, with the bay exporting carbon (detritus, phytoplankton, living biota) and the ocean providing migratory and breeding species. A complete picture of CALFED must include these components.

Conclusions

In light of mandated allocations for human consumption, wildlife conservation, ecosystem restoration, and the functioning of San Francisco Bay, the situation reminds one of Don Quixote pursuing the impossible dream. A sustainable water management plan for the Sacramento–San Joaquin Basin and Northern California should reflect less emphasis on obtaining more water and more effort toward determining the minimum amount of water required to meet human and ecosystem needs. In Mediterranean climates with progressively increasing human demands for diminishing surface and groundwater resources, it is critical to determine not only the minimum quantity of water needed for the functioning of aquatic systems but also the timing and duration of stage required (Mitraki et al. 2004). As is becoming clearer regarding interactions between surface and groundwater within the Sacramento–San Joaquin Basin, it is critical that regional ecosystem management not only take a traditional vertical approach to water management but also incorporate a horizontal approach that evaluates lateral consequences of vertical changes in water level (Crisman et al. 2005). While it would be ideal to address all aspects of structure and function of ecosystems as part of adaptive management plans, the reality of water management in Mediterranean climates is that water is so limited we can only address overall ecosystem function, not the more complicated structural aspects, such as species present and their relative importance within the biotic community.

Two factors work against restoration of this ecosystem. The first is determination of the baseline conditions to which the system is to return. Is it pre-European colonization, pre-mining, pre-intensive irrigation, or pre-diversion of water? Selection of the proper time frame for the restoration goal will ultimately control the probability of success for any effort. For such large landscapes, it is likely that the baseline for restoration/rehabilitation goals will vary regionally, given the nature, timing, and intensity of past environmental disturbances. All restoration efforts, such as dam removal and riparian and delta reconnection, must keep in mind that ecosystems maintain a memory of past abuses in their sediments, that, if disturbed during a restoration effort, can have a profound ecological impact.

The second factor working against restoration is fundamental: the quantity of water available. Minimum water mandates have been established for river flow and water diversion for human use. Given the reality of climate change, associated alterations in the timing and quantity of water available, and a progressively increasing human population within the basin and beyond who will depend on this water, it will be extremely difficult to meet the expectations of humans and needs of ecosystems in the face of diminishing water resources.

Restoration implies a return to original ecosystem structure and function, while rehabilitation is more concerned with returning an ecosystem to a state-functional goal, with far more emphasis on allowing the ecosystem structures, both biotic and abiotic, to self-organize over time. For the Sacramento–San Joaquin Basin, perhaps it is more realistic to plan in terms of rehabilitation of the ecosystem.

References

Breaux, A., S. Cochrane, J. Evens, M. Martindale, B. Pavlik, L. Suer, and D. Benner. 2005. "Wetland Ecological and Compliance Assessments in the San Francisco Bay Region." *California Journal of Environmental Management* 74:217–37.

Brown, L. R., and P. B. Moyle. 1994. "Distribution, Ecology and Status of the Fishes of the San-Joaquin River Drainage, California." *California Fish and Game* 79:96–114.

Cain, D. J., J. L. Carter, S. V. Fend, S. N. Luoma, C. N. Alpers, and H. E. Taylor. 2000. "Metal Exposure in a Benthic Invertebrate, *Hydropsyche californica*, Related to Mine Drainage in the Sacramento River." *Canadian Journal of Fisheries and Aquatic Sciences* 57:380–90.

California Department of Water Resources (CDWR). 2005. *California Water Plan: Update 2005*. At www.waterplan.water.ca.gov/.

Canuel, E. A., J. E. Cloern, D. B. Ringelberg, J. B. Guckert, and G. H. Rau. 1995. "Molecular and Isotopic Tracers Used to Examine Sources of Organic Matter and Its Incorporation into the Food Webs of San Francisco Bay." *Limnology and Oceanography* 40:67–81.

Coutant, C. C. 2004. "A Riparian Habitat Hypothesis for Successful Reproduction of White Sturgeon." *Reviews in Fisheries Science* 12:23–73.

Cox, G. W. 1999. *Alien Species in North America and Hawaii*. Washington, DC: Island Press.

Crisman, T. L., L. J. Chapman, and C. A. Chapman. 1996. "Conserving Tropical Wetlands through Sustainable Use." *Geotimes* (July): 23–25.

Crisman, T. L., C. Mitraki, and G. Zalidis. 2005. "Integrating Vertical and Horizontal Approaches for Management of Shallow Lakes and Wetlands." *Ecological Engineering* 24:379–89.

Feyrer, F., T. R. Sommer, S. C. Zeug, G. O'Leary, and W. Harrell. 2004. "Fish Assemblages of Perennial Floodplain Ponds of the Sacramento River, California (USA), with Implications for the Conservation of Native Fishes." *Fisheries Management and Ecology* 11:335–44.

Gabet, E. J. 1998. "Lateral Migration and Bank Erosion in a Saltmarsh Tidal Channel in San Francisco Bay, California." *Estuaries* 21:745–53.

Gerlach, J. D. 2004. "The Impacts of Serial Land-Use Changes and Biological Invasions on Soil Water Resources in California, USA." *Journal of Arid Environments* 57:365–79.

Gleick, P. H., and E. L. Chalecki. 1999. "The Impacts of Climatic Changes for Water Resources of the Colorado and Sacramento–San Joaquin River Basins." *Journal of the American Water Resources Association* 35:1429–41.

Holl, K. D., and E. E. Crone. 2004. "Applicability of Landscape and Island Biogeography Theory to Restoration of Riparian Understory Plants." *Journal of Applied Ecology* 41:922–33.

Ingram, B. L., J. C. Ingle, and M. E. Conrad. 1996a. "A 2000 Year Record of Sacramento–San Joaquin River Inflow to San Francisco Bay Estuary, California." *Geology* 24:331–34.

Ingram, B. L., J. C. Ingle, and M. E. Conrad. 1996b. "Stable Isotope Record of Late Holocene Salinity and River Discharge in San Francisco Bay, California." *Earth and Planetary Science Letters* 141:237–47.

Jassby, A. D., and J. E. Cloern. 2000. "Organic Matter Sources and Rehabilitation of the Sacramento–San Joaquin Delta (California, USA)." *Aquatic Conservation—Marine and Freshwater Ecosystems* 10:323–52.

Jassby, A. D., and T. M. Powell. 1994. "Hydrodynamic Influences on Interannual Chlorophyll Variability in an Estuary—Upper San Francisco Bay-Delta (California, USA)." *Estuarine and Coastal Shelf Science* 39:595–618.

Jassby, A. D., J. E. Cloern, and B. E. Cole. 2002. "Annual Primary Production: Patterns and Mechanisms of Change in a Nutrient-Rich Tidal Ecosystem." *Limnology and Oceanography* 47:698–712.

Jassby, A. D., J. E. Cloern, and T. M. Powell. 1993. "Organic-Carbon Sources and Sinks in San Francisco Bay—Variability Induced by River Flow." *Marine Ecology Progress Series* 95:39–54.

Knowles, N. 2002. "Natural and Management Influences on Freshwater Inflows and Salinity in the San Francisco Estuary at Monthly to Interannual Scales." *Water Resources Research* 38:1289.

Larsen, E. W., and S. E. Greco. 2002. "Modeling Channel Management Impacts on River Migration: A Case Study of Woodson Bridge State Recreational Area, Sacramento River, California, USA." *Environmental Management* 30:209–24.

Lehman, P. W., J. Sevier, J. Giulianotti, and M. Johnson. 2004. "Sources of Oxygen Demand in the Lower San Joaquin River, California." *Estuaries* 27:405–18.

Leland, H. V. 2003. "The Influence of Water Depth and Flow Regime on Phytoplankton Biomass and Community Structure in a Shallow, Lowland River." *Hydrobiologia* 506:247–55.

Leland, H. V., and S. V. Fend. 1998. "Benthic Invertebrate Distributions in the San Joaquin River, California, in Relation to Physical and Chemical Factors." *Canadian Journal of Fisheries and Aquatic Sciences* 55:1051–67.

Leland, H. V., L. R. Brown, and D. K. Mueller. 2001. "Distribution of Algae in the San Joaquin River, California, in Relation to Nutrient Supply, Salinity and Other Environmental Factors." *Freshwater Biology* 46:1139–67.

Marcus, L. 2000. "Restoring Tidal Wetlands at Sonoma Baylands, San Francisco Bay, California." *Ecological Engineering* 15: 373–83.

May, J. T., and L. R. Brown. 2002. "Fish Communities of the Sacramento River Basin: Implications for Conservation of Native Fishes in the Central Valley, California." *Environmental Biology of Fishes* 63:373–88.

McCreary, S., R. Twiss, B. Warren, C. White, S. Huse, K. Gardels, and D. Roques. 1992. "Land-Use Change and Impacts on the San Francisco Estuary: A Regional Assessment with National Policy Implications." *Coastal Management* 20:219–53.

Micheli, E. R., J. W. Kirchner, and E. W. Larsen. 2004. "Quantifying the Effect of Riparian Forest versus Agricultural Vegetation on River Meander Migration Rates, Central Sacramento River, California, USA." *River Research and Application* 20:537–48.

Mitraki, C., T. L. Crisman, and G. Zalidis. 2004. "Lake Koronia, Greece: Shift from Autotrophy to Heterotrophy with Cultural Eutrophication and Progressive Water-Level Reduction." *Limnologia* 34:110–16.

Mount, J. F. 2003. "Changes in Lowland Floodplain Sedimentations Processes: Pre-disturbance to Post-rehabilitation, Cosumnes River, CA." *Geomorphology* 56:305–23.

Moyle, P. B., B. Herbold, D. E. Stevens, and L. W. Miller. 1992. "Life-History and Status of Delta Smelt in the Sacramento–San Joaquin Estuary, California." *Transactions of the American Fisheries Society* 121:67–77.

Nelson, S. M., and D. M. Lieberman. 2002. "The Influence of Flow and Other Environmental Factors on Benthic Invertebrates in the Sacramento River, USA." *Hydrobiologia* 489:117–29.

Saiki, M. K., B. A. Martin, S. E. Schwarzbach, and T. W. May. 2001. "Effects of an Agricultural Drainwater Bypass on Fishes Inhabiting the Grassland Water District and the Lower San Joaquin River, California." *North American Journal of Fisheries Management* 21:624–35.

Sommer, T. R., W. C. Harrell, A. M. Solger, B. Tom, and W. Kimmerer. 2004. "Effects of Flow Variation on Channel and Floodplain Biota and Habitats of the Sacramento River, California, USA." *Aquatic Conservation—Marine and Freshwater Ecosystems* 14:247–61.

Stanford, J. A., and J. V. Ward. 1993. "An Ecosystem Perspective of Alluvial Rivers—Connectivity and the Hyporheic Corridor." *Journal of the North American Benthological Society* 12: 48–60.

Stevens, M. L. 2004. "Ethnoecology of Selected California Wetland Plants." *Fremontia* 32 (4): 7–15.

van Geen, A., N. J. Valette-Silver, S. N. Luoma, C. C. Fuller, M. Baskaran, F. Tera, and J. Klein. 1999. "Constraints on the Sedimentation History of San Francisco Bay from C-14 and Be-10." *Marine Chemistry* 64:29–38.

Vannote, R. L, G. W. Minshall, K. W. Cummins, J. R. Sedell, and C. E. Cushing. 1980. "The River Continuum Concept." *Canadian Journal of Fisheries and Aquatic Sciences* 37:130–37.

Vanrheenen, N. T., A. W. Wood, R. N. Palmer, and D. P. Lettenmaier. 2004. "Potential Implications of PCM Climate Change Scenarios for Sacramento–San Joaquin River Basin Hydrology and Water Resources." *Climatic Change* 62:257–81.

Ward, J. V., and J. A. Stanford. 1983. "Serial Discontinuity Concept of Lotic Ecosystems." In *Dynamics of Lotic Systems*, eds. T. D. Fontaine and S. M. Bartell. Ann Arbor, MI: Ann Arbor Science, 29–42.

Ward, J. V., and J. A. Stanford. 1995. "The Serial Discontinuity Concept: Extending the Model to Floodplain Rivers." *Regulated Rivers: Research and Management* 10:159–68.

Williams, P. B., and M. K. Orr. 2002. "Physical Evolution of Restored Breached Levee Salt Marshes in the San Francisco Bay Estuary." *Restoration Ecology* 10:527–42.

Chapter 9

Water Fights

The Economics of Allocating Scarce Water and Bay-Delta Restoration

Stephen Polasky

In California, as in much of the western United States, water is destiny. Aside from the Pacific Northwest and high mountain regions, most of the western United States is arid. Los Angeles receives on average 13 inches of precipitation per year and Las Vegas only 4.5 inches, compared with 30 to 60 inches in the eastern United States. Without adequate local precipitation, most agriculture in the West depends on irrigation, and most cities depend on water supplies piped in from afar. With water, the desert blooms, cities expand, and fortunes are made. Without water, hardship and dissension overwhelm entire communities. Gaining and maintaining access to water spawn intense political fights among rival economic interests, as depicted in books like Marc Reisner's *Cadillac Desert* (1986) and in movies such as *Chinatown*. As Mark Twain reportedly said, "Whiskey is for drinking. Water is for fighting over" (n.d.).

At the heart of the water supply system in California are the federal Central Valley Project and California's State Water Project. These projects link relatively abundant water supplies from snowmelt in the Sierra Nevada and more plentiful rainfall in northern California with water demands from agriculture in the Central Valley and large cities in central and southern California. The two projects were designed to move water from where it flowed naturally to where people wanted to use it. By moving water to where people wanted it, the projects allowed a vast expansion of highly productive agriculture in the Central Valley. Agriculture in California generated $31.8 billion in revenue in 2004 (NASS 2005), making it the top agricultural state in the nation. Fresno County, in the middle of the Central Valley, led all counties in the nation in agricultural revenues, generating $4.7 billion in 2004 (NASS 2005). Moving water to where people wanted to use it also facilitated the growth of large metropolitan areas. Both Los Angeles and the Bay Area rank among the five most populous metropolitan areas in the country. The San Francisco Bay-Sacramento-San Joaquin Delta (Bay-Delta) watershed supplies water to 22 million people (CALFED 2004).

Although redirecting water to irrigate crops and supply cities has helped the economy of California to grow and develop, it has also disrupted natural water flow regimes and caused large changes in important ecosystems. Of particular concern is the effect of water diversions on the Bay-Delta ecosystem, a rich estuarine system at the base of the Sacramento and San Joaquin rivers as they flow into San Francisco Bay. Environmental concerns over water quality, commercial fish species (chinook salmon), and endangered species in the Bay-Delta area led to calls for changes in water management in California. In 1992, the Central Valley Project Improvement Act (CVPIA) formally added protection of fish and wildlife as a goal of the Central Valley Project.

Placing further new demands on the water management system, in this case, environmental protection, without adding any new water to the system increases the potential for conflict among competing interests. Leaving more water for instream flows to help fish means less water may be diverted for use in agriculture or urban water supply and vice versa. Though some actions can improve both human uses of water and environmental outcomes, such as steps to increase efficiency of water conveyance and changing the timing of water deliveries, high levels of competition among interest groups remain. Since the early 1990s, legal and political battles over water allocation among various groups representing agricultural, environmental, and urban interests have multiplied. The joint federal–state CALFED Bay-Delta Program and its successor, the California Bay-Delta Authority (CALFED), as described in chapter 7 (this volume), is an attempt to surmount conflicts over water allocation and devise a water management plan that is acceptable to all major interests. While the CALFED program has had some success in coordinating actions of federal and state governmental agencies, it faces major challenges in trying to please all of the competing interests for water in California.

Water Management

Water management can be accomplished via several routes. One route is to use science and economics to allocate water to uses that generate the greatest net benefit for society, sometimes called "efficient" or "optimal" management. Though real-world water management may contain elements of efficient management, there is no reason to expect such ideal management to be dominant. Water management is governed by a complex web of laws and regulations pertaining to water rights, property rights, land use, and environmental laws at both the federal and state levels. These laws and regulations are written by legislatures subject to lobbying pressure from interest groups and administered by agencies facing similar political pressures. Laws and regulations related to water use—typically complex, often unclear, and not mutually consistent—are regularly challenged in court by various parties seeking more advantageous interpretations. Amid this political cauldron of competing interests, as well as complex rules and numerous institutions, CALFED is trying to develop cooperative, or at least acceptable, water management plans.

Efficient Water Management

Before plunging into water politics, let us step back and first consider how one might allocate water in an efficient manner using economic principles. At its core, economics is about how to allocate scarce resources. One of the most basic principles in economics is that scarce resources should be allocated to their "highest and best use," defined as the use that generates the greatest net benefit. The net benefit of allocating an additional unit of water is equal to the "addition benefits" minus the "addition costs" when water is allocated versus when it is not. For example, suppose that additional water is allocated to a farmer to irrigate crops. The additional water increases crop yields, leading to greater revenue for the farmer. This increase in revenue represents the marginal benefit of additional water measured in dollar terms. If there were increased costs associated with allocating additional water to the farmer, these costs would be deducted from the benefits to generate net benefits.

Estimating the net benefits of water use is not without difficulties; however, methods exist for undertaking such benefits estimates (Young 1996). In the case of the farmer given earlier, it is fairly easy to see what the net benefits are, at least in principle. In practice, getting an estimate of the change in benefits with and without water requires estimating benefits for a counterfactual situation. The farmer either received water or did not, meaning that yield for one of the cases is not observed. Estimating what would happen in the counterfactual requires using evidence from different time periods and/or different places where other differences (weather, soil conditions, and so on) may obscure the relationship between additional water and crop yields. While perfect estimates are unlikely, the use of multiple regression analysis with observations across many farms over many time periods can generate fairly robust estimates of the net benefit of water in agriculture. A number of studies have also estimated the net benefits of water in other uses, such as allocation for municipal and industrial use (Renzetti 2002).

Water availability in California is dependent on weather patterns that can result in plentiful winter rains with large snowpack in the mountains or in drought. When there is a drought, not all demands for water can be met, and some mechanism must be in place for deciding which demands to satisfy. One way to handle variable supply is to establish a queue among water users. Water is allocated in descending order in the queue until the water is fully allocated. Having a position near the front of the queue is of great value because the likelihood of receiving water is increased. Other mechanisms for allocating water are also possible, such as having water users bid for water or using some sort of lottery, though the latter, purely random mechanism may not deliver water to those who value or need it the most.

For any allocation system, there is value in knowing ahead of time whether you will receive water. For example, if farmers in a particular area know at the beginning of the season that they will be unlikely to get irrigation water that year, they can leave the land fallow or plant crops that need little water, thereby avoiding the wasted effort and expense of planting crops that will fail because of lack of water.

Evidence on net benefits from agricultural and urban water uses can be combined to analyze the highest and best uses for water. An analysis of the optimal water allocation system for California was recently undertaken by a team of researchers based at the University of California at Davis (Jenkins et al. 2001). Using an economic-engineering model of the water allocation system, these researchers analyzed such questions as the value of making various investments in efficiency of transporting water, increasing available supply or storage capacity, as well as the optimal allocation of water among various uses. This type of analysis can inform water managers of the likely consequences of decisions and can potentially lead to better water allocation and investment. In an earlier study, the Congressional Budget Office (CBO 1997) estimated the benefits and costs of implementing the CVPIA of 1992. This study estimated that changes brought about under the act would generate additional benefits of $11 million annually for municipal users but that agricultural users would lose an estimated $43 million annually compared with the status quo. Environmental benefits, which were the primary reason for changes in water management, were not estimated.

There are two major challenges to thinking about an efficient water allocation system designed to send water to the highest and best use. The first challenge has to do with incorporating environmental benefits of increased instream flows and more natural water flow regimes. The CBO study already mentioned did not quantify environmental benefits—not because they were not deemed important but because doing so is difficult. The second challenge has to do with institutional arrangements, water rights, and vested economic and political interests.

Balancing the Value of Use and Instream Flow

The Central Valley Project and the State Water Project were designed with the primary purpose of supplying water for agricultural and urban uses. Over time, however, it became apparent that what was good for agriculture and urban water supplies was not good for the Bay-Delta ecosystem. The adverse effects of water projects on populations of salmon, steelhead, and delta smelt, species listed under the Endangered Species Act (ESA), as well as the state's failure to meet clean water standards under the Clean Water Act (CWA), compelled federal and state agencies to address environmental consequences of California's water supply system. The current objective for federal and state agencies is to satisfy both water supply goals and environmental goals. The mission of CALFED is "to improve water supplies in California and the health of the San Francisco Bay–Sacramento–San Joaquin River Delta Watershed" (CALFED 2004).

In some important ways, water supply goals and environmental goals are at odds with each other. Supporting fish populations and healthy aquatic ecosystems requires keeping water in rivers and flowing through the Bay and Delta. Supporting agriculture and urban water needs requires diverting water out of rivers, in many cases out of the Bay-Delta Basin, so that this water does not flow through the Bay and Delta. How much water should be kept instream for environmental goals and how much should be diverted to use in agricultural and municipal water supplies are basic questions facing CALFED.

One way to attempt to answer the question of the proper balance between environmental goals and water supply goals is to extend the calculations of net benefits to include the environmental benefits of instream flows and natural flow regimes. Environmental benefits include the value of protecting fish populations and populations of other aquatic species, of providing a healthy aquatic ecosystem, and of improving water quality. What is difficult, however, is to measure these values in dollar terms that are comparable to measures of net benefits of water use in agriculture or municipal systems (NRC 2004). Economists have developed an extensive literature on "nonmarket valuation," the purpose of which is to measure environmental benefits in dollar terms (Freeman 1993). For some environmental benefits, particularly those that affect the value of residential property (for example, living near a lake with clean water or having a beautiful view) or recreational opportunities (such as fishing, hiking, or skiing), economists have generated credible estimates of the dollar value of environmental benefits. But there is little agreement on the dollar value for many other environmental benefits. For example, how can the value of saving an endangered species or the value of preventing eutrophication of coastal estuaries be determined? These environmental benefits clearly have positive and lasting values, but assessing those values in dollar terms is problematic.

Rather than attempting to assess the dollar value of environmental benefits, policy makers have found another way to incorporate environmental goals to specify directly the environmental standard to be met. Typically, this is the approach for creating environmental policy. For example, the Clean Water Act and Clean Air Act set standards for water quality and ambient air quality.

When there are clear thresholds beyond which environmental systems or populations of species would be in trouble, setting environmental standards is entirely appropriate. However, many water allocation issues are not so black and white. A little more water kept instream would be beneficial for fish populations and the health of aquatic ecosystems. A little more water diverted to agriculture would be beneficial for farmers. When the world is gray rather than black and white, how should CALFED, or society at large, decide how much water to devote to each use? Making these decisions requires knowledge of the relative impact of a little more water to achieving a goal and a value judgment about the relative importance of the goal. Whether we decide on the stringency of environmental standards or attempt to compare the dollar value of environmental benefits with benefits of water use in agriculture or municipal supply, making trade-offs between different and sometimes incompatible goals requires difficult choices.

Of course, not every decision involves trade-offs between environmental and water use objectives. Some ways of operating the system can increase efficiency and allow progress in meeting both environmental and water use goals. A good example of this is CALFED's Environmental Water Account (EWA). The EWA buys water from willing sellers, stores this water, and releases the water when it is needed to protect fish populations. Stored water can also be used to trade with water users when it is important to shut off diversions of water at particular points in the system at particular times. An example of this is action to protect delta smelt: "In late spring 2004, when concern for delta smelt was high and real-time monitoring indicated that young delta smelt were in the vicinity of Delta export pumps, water managers reduced pumping to improve survival and allow fish to move out of the interior Delta. EWA water was used to offset export reductions that would otherwise have reduced project supplies" (CALFED 2004).

When negotiators have the ability to find creative solutions to help meet one goal without making it more difficult to meet other goals, then agreements can be made, and conflicts can be avoided. In some cases, the timing of water diversions can be changed to improve the efficiency of water conveyance so that less water is lost between the point of diversion and the end user. In such cases, progress on environmental goals can be made without harming water users. In other cases, when water diversion means less water for instream flows, conflict among different interests is almost inevitable.

Institutional Arrangements: Water Rights, Water Markets, and the Courts

The tasks of CALFED are to satisfy agricultural and urban water users and to protect endangered species and the Bay-Delta ecosystem. These tasks would be complex even when trying, as objectively as possible, to allocate water to generate the greatest net benefits. However, because the economic and environmental stakes in water management are so large, interest group politics are never far below the surface. All interests—agricultural water users, urban water users, fishermen, and environmentalists—pay close attention to water management decisions to make sure that their interests are well represented. Close scrutiny and political pressures make it difficult for government agencies participating in CALFED to sustain cooperation among all parties.

Unlike water itself, there has been no shortage of court cases over water. Litigation has been spurred largely by the clash between traditional water users and environmental interests. Large-scale water projects in California and throughout the West largely predate modern environmental laws. Rightly or wrongly, many water users view water delivery from federal and state projects as an entitlement. According to this viewpoint, if water is not delivered, perhaps because a federal or state agency is now mandated by an environmental law to keep more water instream, then the government has not fulfilled its obligations and should pay water users for failure to deliver. The alternative view is that water allocation from government projects is not an entitlement to private water users. Government agencies responsible for water management should allocate water to best serve the public, which may mean keeping the water instream to satisfy environmental goals. According to this view, when water is needed instream for environmental protection, as dictated by environmental laws like the ESA, then government agencies must comply, and compensation does not need to be paid to private water users.

In a controversial court decision in 2001, *Tulare Lake Basin Water Storage District v. U.S.*, Judge John P. Wiese sided with farmers in the Central Valley who claimed that they were owed compensation when water deliveries were cut off because of requirements to protect species listed under the ESA. The Bush administration, generally supportive of private property rights arguments, decided to settle rather than appeal and paid $16.7 million to the farmers in 2004. Others, though, think the case was wrongly decided and that no compensation should be required (Benson 2002; Echeverria 2005). The debate on water law and compensation is far from settled.

Given all the litigation and debate, it is easy to lose hope that cooperative solutions to water management can be worked out. There are, however, some grounds for optimism. The current water fights focus on the issue of whether compensation is necessary when water is shifted from traditional water users to meet environmental mandates. If the legal issues surrounding rights and compensation can be settled clearly one way or the other, then negotiations among parties can proceed, and reasonable decisions about water use may yet occur. According to Nobel laureate Ronald Coase, efficient solutions to economic and environmental disputes can be achieved among parties through bargaining as long as property rights are clearly established up front (Coase 1960). Coase explained his insight using an example of a rancher whose cattle may roam and destroy a neighboring farmer's crops. If the laws of the land are that cattle are free to roam, then it is up to the farmer to pay the rancher to keep his cattle contained or suffer crop damage. The farmer will choose to pay the rancher to contain the cattle if that is less expensive than suffering crop damage. Conversely, if the laws of the land are that damages to crops must be compensated, then the rancher will have to contain the cattle or pay the farmer. The rancher will choose to contain the cattle if that is less expensive than paying for the crop damage. Either way, an incentive to find an efficient solution that minimizes the costs of conflict between the parties exists. What differs between the two sets of rules is the distribution of benefits and costs among the parties.

In the case of water allocation, a similar logic could apply. The same water allocation outcome could be achieved regardless of whether water is a public good or a private entitlement. If water is an entitlement, federal or state agencies could decide to keep water instream and pay compensation for doing so if this were viewed to be in the public interest. In contrast, if water is a public good, federal and state agencies might sell water for agricultural or urban users if that were consistent with managing water resources in the public interest, or they could keep water for instream use if necessary to meet environmental objectives. What differs between these two alternative sets of rules is the distribution of benefits and costs. Protecting the environment can occur under either approach but will be more expensive if water allocations are viewed as a private entitlement rather than a public good.

If legal issues related to water use and environmental laws can be sorted out, one possible institutional arrangement for allocating water among competing uses is to use water markets (Carey and Sunding 2001). Water markets that allow trading water among users already exist in some parts of the West. Extending water markets to allow government agencies to hold or purchase water for "nonuse" (for example, protecting instream flows) does not require much of an extension to this already existing institution. In fact, the EWA discussed earlier is similar in nature to a water market and provides many of the benefits of a functioning water market. Many economists view expansion of water markets as holding promise for promoting more efficient use of water, so that it truly does flow toward its highest and best uses (Howitt and Hansen 2005; Saleth and Dinar 2004).

Clearly, establishing policies for water allocation among water users and environmental protection will require the courts and society at large to decide what the parameters for fair and equitable rules are, as well as which needs constitute high priorities. However, at present, intense battles over these rules continue. For example, a spokesperson for the U.S. Department of the Interior said, in regard to the Tulare case, "If we have a public good which is reflected in the Endangered Species Act, it's fair that the act should be supported by all taxpayers as opposed to only particular individuals having to pay for that public good" (quoted in Murphy 2005). Alternatively, Joseph Sax, a prominent environmental lawyer who wrote an amicus brief for the government in the Tulare case, summarized the debate over water use and environmental protection as "a change in terms of society's priority and a need to figure out how to adapt people with traditional uses and expectations to that change" (quoted in Coyle 2004).

The impacts of past water management decisions together with the California economy's growth have placed greater burdens on current water managers "to improve water supplies in California and the health of the San Francisco Bay–Sacramento–San Joaquin River Delta Watershed" (CALFED n.d.). Increasing urbanization, continued demands from agriculture, potential impacts of climate change, declining populations of endangered species, declines in catch of commercial fish species, and low water quality all will make it difficult to meet environmental objectives while improving water supplies for various uses. The best chance for being able to meet water user demands and environmental objectives simultaneously will occur if the legal issues over water rights are settled relatively soon and CALFED is able to work cooperatively among government agencies, water user groups, and environmental groups. If that happens, there is some hope that the CALFED Bay-Delta Authority will achieve its stated aims.

References

Benson, M. B. 2002. "The Tulare Case: Water Rights, the Endangered Species Act, and the Fifth Amendment." *Environmental Law* 32:551–88.

CALFED Bay-Delta Program. 2004. *Annual Report 2004*.

CALFED website. n.d. At http://calwater.ca.gov/AboutCalfed/AboutCALFED.shtml.

Carey, J., and D. Sunding. 2001. "Emerging Markets in Water: A Comparative Institutional Analysis of the Central Valley and Colorado-Big Thompson Projects." *Natural Resources Journal* 41 (2): 283–328.

Central Valley Project Improvement Act (CVPIA). Public Law 102-575, Title 35 (1992).

Coase, R. 1960. "The Problem of Social Cost." *Journal of Law and Economics* 3: 1–44.

Coyle, M. 2004. "Water Suits Flow Like H_20." *National Law Journal* (December 13, 2004).

Congressional Budget Office (CBO). 1997. *Water Use Conflicts in the West: Implications of Reforming the Bureau of Reclamation's Water Supply Policies*. Washington, DC: CBO.

Echeverria, J. D. 2005. "Why *Tulare Lake* Was Incorrectly Decided." Fall Meeting of the Section on Environment, Energy, and Resources, American Bar Association, September 2005, Nashville, Tennessee.

Freeman, A. M., III. 1993. *The Measurement of Environmental and Resource Values: Theory and Methods*. Washington, DC: Resources for the Future.

Howitt, R., and K. Hansen. 2005. "The Evolving Western Water Markets." *Choices: The Magazine of Food, Farm and Resource Issues* 20 (1): 59–63.

Jenkins, M. W., A. J. Draper, J. R. Lund, R. E. Howitt, S. Tanaka, R. Ritzema, G. Marques, S. M. Msangi, B. D. Newlin, B. J. Van Lienden, M. D. Davis, and K. B. Ward. 2001. *Improving California Water Management: Optimizing Value and Flexibility*. Report for the CALFED Bay-Delta Program. At http://cee.engr.ucdavis.edu/faculty/lund/CALVIN/Report2/CALVINExecSumm2001.pdf.

Murphy, D. E. 2005. "In Fish vs. Farmer Cases, the Fish Loses Its Edge." *New York Times*, February 22. At www.nytimes.com/2005/02/22/national/22west.html.

National Agricultural Statistical Service (NASS), U.S. Department of Agriculture. 2005. (Revised January 2006). *California Agricultural Statistics 2004*. At www.nass.usda.gov/ca.

National Research Council (NRC). 2004. *Valuing Ecosystem Service: Towards Better Environmental Decision-Making*. Washington, DC: National Academies Press.

Reisner, M. 1986. *Cadillac Desert: The American West and Its Disappearing Water*. New York: Viking.

Renzetti, S. 2002. *The Economics of Water Demand*. Norwell, MA: Kluwer Academic Publishers.

Saleth, R., and A. Dinar. 2004. *The Institutional Economics of Water: A Cross-Country Analysis of Institutions and Performance*. Cheltenham, UK: Edward Elgar Publishing Ltd.

Twain, Mark. n.d. At www.twainquotes.com. The quotation in this chapter has been attributed to Mark Twain, not verified.

Young, R. A. 1996. *Measuring Economic Benefits for Water Investments and Policies*. World Bank Technical Paper 338. Washington, DC: The World Bank.

PART IV

The Chesapeake Bay

The magnificent Chesapeake Bay is the nation's largest and most biologically diverse estuary, home to more than 3,000 species of animals, fish, and plants. While the bay is great in area (2,300 square miles), it is extremely shallow, with an average depth of only 21 feet. The vastness of its watershed, which covers 64,000 square miles across six states and the District of Columbia, as well as its shallow depth, make the bay particularly vulnerable to the effects of human activity, such as agriculture, sewage treatment, and urban development. The primary environmental problems of the bay are the results of nutrient pollution, coming from sewage treatment plant discharge and farm runoff. The bay has also suffered damage from sedimentation due to urban development in the watershed and the subsequent loss of forests and wetlands.

Overharvesting of native oysters has virtually eradicated once abundant populations, and agencies and stakeholders have been at work on developing policies to conserve the dwindling blue crab fishery. Thomas L. Crisman's chapter 11 considers how, from an ecological viewpoint, sustainable management of the Chesapeake Bay must take into account both the structure and the function of the ecosystem. He analyzes bay management issues in the context of vertical interactions within the ecosystem (interrelations between the atmosphere and the water, between sediment and water, and within the water column) and horizontal interactions (those between the water and its surrounding areas, including the watershed, the littoral zone, coastal wetlands, and the pelagic zone). In chapter 12, Stephen Polasky analyzes the economic costs of water pollution and overharvesting in the bay and identifies and quantifies some of the benefits of remedying these problems. He notes the need to put in place regulations and incentives for farmers, commercial fishermen, and others whose work impacts the bay, as well as the necessity of educating policy makers and the public about the benefits of saving the bay if they are expected to summon the will to bear the costs.

Established for a generation, the Chesapeake Bay Program is one of the oldest large ecosystem restoration projects in the United States, with much to teach regarding reaching decisions by consensus. Since the Chesapeake Bay Agreement of 1983, the states in the bay watershed and the U.S. Environmental Protection Agency (EPA), with stakeholder approval, have been successful in articulating goals and timetables for saving the bay. The apotheosis of their effort is the *Chesapeake 2000* agreement (C2K), with water quality improvement its main objective. The product of a two-year negotiation, C2K contains dozens of specifically defined goals, covering not only water quality but also protection and restoration of living resources, sound land use, and robust community engagement—all to be achieved by 2010.

By contrast, negotiations on the California Bay-Delta, Platte River, and Everglades ecosystems have struggled for agreement on more abstract, conceptual approaches. While the collaborative effort that led to adoption of C2K is commendable, the agreement is without regulatory force, and there is no plan for funding the estimated $19 billion cost of its implementation. At this point, federal funding on such a large scale does not appear likely. In contrast with the high priority accorded to restoring the Everglades, demonstrated by the $7.8 billion Comprehensive Everglades Restoration Plan authorization, the federal government has not chosen to view cleanup of the Chesapeake Bay as a matter of national priority. The Chesapeake's major federal partner, EPA, is a research and regulatory agency with relatively small annual budgets compared with the large sums commanded by projects of the U.S. Army Corps of Engineers. In further contrast with the Chesapeake Bay, the U.S. Department of the Interior has provided a measure of federal funding to the California Delta and Platte River programs because of the presence of U.S. Bureau of Reclamation water facilities and compelling Environmental Species Act issues there. Whether the states with major responsibility for the Chesapeake Bay—Virginia, Maryland, and Pennsylvania—will summon the political will to fund C2K goals and whether they will have the capabilities to do so are the critical political issues facing the Chesapeake Bay program.

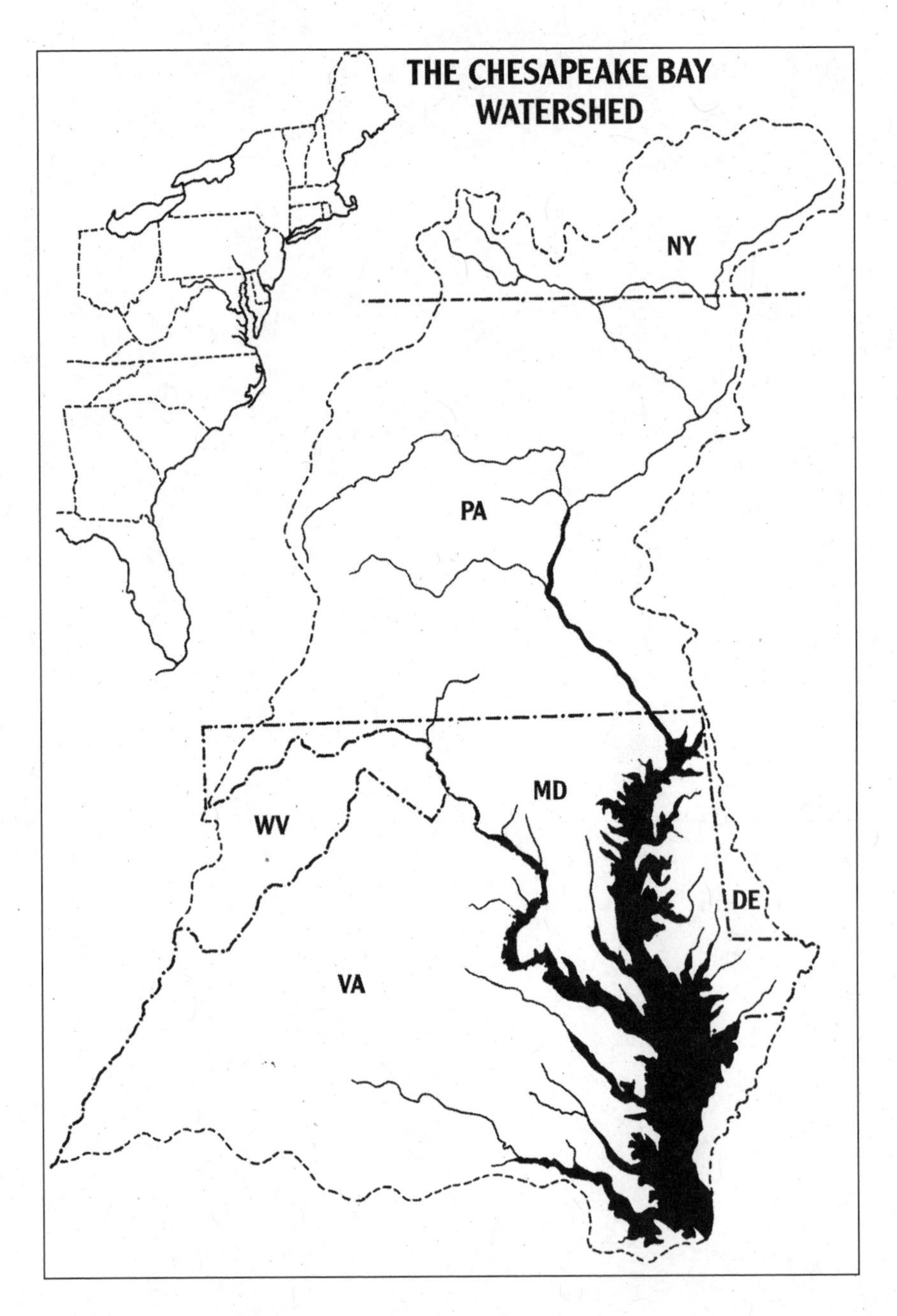

FIGURE 10.1. The Chesapeake Bay Watershed (Chesapeake Bay Program 2000b).

Chapter 10

The Culture of Collaboration in the Chesapeake Bay Program

Mary Doyle and Fernando Miralles-Wilhelm

The Chesapeake Bay is the nation's largest and most biologically diverse estuary, home to more than 3,000 species of animals, fish, and plants. The bay and the rivers and wetlands of its watershed host a network of interconnected habitats and food webs that form a complex ecosystem. Certain species may belong to a variety of communities and use a number of habitats during their life cycles. Because these habitats in the bay and its watershed are connected and communities overlap, biological, chemical, or physical changes in one species' habitat will affect not only the community it supports but also habitats and communities of other species.

The Chesapeake Bay is approximately 200 miles long, stretching from Havre de Grace, Maryland, to Norfolk, Virginia. It varies in width from about 3.4 miles to 35 miles at its widest. Including its tidal tributaries, the bay has approximately 11,684 miles of shoreline. Fifty major tributaries pour into the Chesapeake Bay every day. Eighty to 90 percent of the freshwater entering the bay comes from waters flowing from the Appalachian Mountains to the north and west. The Susquehanna River runs through heavily agricultural counties in Pennsylvania and supplies 45 percent of the bay's freshwater. The flow of the Susquehanna into the bay, at a rate of 40,000 cubic feet per second, has a flushing effect that allows the bay to renew itself continually. Four other major rivers, the Rappahannock, York, James, and Potomac, supply another 45 percent of freshwater in the bay. Ten to 20 percent of its freshwater supply enters the bay from its eastern shore, which consists of low, flat, marshland areas cut by rivers running to the bay. Nearly half the water in the bay is saltwater that enters from the Atlantic Ocean.

Remarkably, while the bay is very great in area, 2,300 square miles or approximately 5,180 square kilometers (km^2), it is extremely shallow, with a depth averaging only about 21 feet. Because it is so shallow, the Chesapeake is extremely sensitive to climate factors, such as temperature fluctuations and wind. The bay has a huge watershed for the amount of water it contains, almost 3,000 km^2 of land for every cubic kilometer of water, due to the shallow water depth of the bay (Horton and Eichbaum 1991, 5). The vast watershed of Chesapeake Bay covers 64,000 square miles (or 40.96 million acres). It extends from the headwaters of the Susquehanna River near Cooperstown, New York, to the Atlantic Ocean, encompassing significant portions of Pennsylvania, Maryland, and Virginia; smaller portions of West Virginia and Delaware; and the District of Columbia. The Chesapeake Bay watershed constitutes one-sixth of the eastern seaboard of the United States.

Environmental Problems and Challenges

Chesapeake Bay and its tributaries endure an array of assaults from air, water, and land. The bay's primary environmental problem is nutrient pollution of its waters. Other adverse effects on water quality are attributable to toxic chemicals, sediment loads, air pollution, and land use changes. These interrelated environmental water quality problems can create harmful, costly synergistic effects. Therefore, analyzing the issues and developing solutions require an ecosystem-wide approach that recognizes the complex interactions of the bay's systems and elements. Thomas L. Crisman provides a more detailed ecological analysis of the bay in chapter 11. Stephen Polasky analyzes how economic activities, such as harvesting, farming, and transportation, have changed the ecological composition of the bay in chapter 12.

Nutrient Overload

Most experts agree that the major water quality problem facing the Chesapeake is nitrogen and phosphorous overload. Before nutrient reduction goals were set in 1987, 330 million pounds of nitrogen and 20 million pounds of phosphorous entered the bay every year. The primary sources of nutrient pollution in the bay watershed are the nitrogen and phosphorous present in manure and chemical fertilizers. These nutrients leach from farmlands into the ground and run off into nearby waterways. Because of the large agricultural operations in southern Pennsylvania, the Susquehanna River is the largest contributor of nutrient pollution in the bay (Walsh and Kohn 1998).

Another major cause of nutrient overload is discharge from sewage treatment plants in the region. Sixty million pounds of nitrogen flow into the bay's mainstem and tributaries every year from these facilities (CBF 2002a). Upgrading sewage treatment plants to reduce the discharge of nutrients is an important component of restoration. In February 2007, the Virginia General Assembly authorized $250 million in bonds for such upgrades to assist the state in meeting the goals of the *Chesapeake 2000* agreement (or C2K). This bond funding will allow for upgrades of wastewater plant improvements that will lead to a reduction of approximately 4 million pounds of nitrogen pollution entering Virginia waterways (CBF 2007b). About 6 percent of nitrogen entering the bay is contributed by septic systems built for residential development, a 20-percent increase since 1990 (CBF 2000, 6). A significant percentage of nitrogen runoff is also contributed by air pollution from automobiles, trucks, power plants, and industry (CBF 2002a).

These nutrients spur the growth of algae and the occurrence of large algal blooms. The algal blooms blanket the surface of the water, preventing sunlight from reaching submerged aquatic vegetation (SAV) and causing the vital underwater plants to die. When algae die, they sink to the bottom of the bay, where their decomposition uses up dissolved oxygen, causing death or severe stress for fish and shellfish populations (DCR, SWCP 2006).

Urban Growth

The land surface of the Chesapeake Bay watershed is fifteen times greater than the area of the bay itself. How land is used within the watershed has a direct and profound effect on the bay. In the bay's natural state, forest, wetlands, and SAV served to regulate the flow of water moving across the land to the bay and its tributaries and filtered contaminants from the runoff. The filtering function, particularly as performed by forest and wetlands adjacent to waterways, is critical to the bay's ability to sustain its living resources (CBF 2000, 2).

Following European settlement, the bay area's population increased, and great parts of the land were converted to agricultural and urban uses. Until the mid-twentieth century, development was located primarily around central urban areas. However, in the last several decades, a pattern of low-density land use has emerged, moving away from existing communities and increasing the rate of conversion of natural lands. Housing is located farther away from existing infrastructure, such as schools, commercial areas, and wastewater treatment facilities. This pattern of urban sprawl fosters increasing rates of land consumption: current development uses four to five times more land per person than just forty years ago. Population growth also increases dependence upon automobile travel, resulting in additional road building requirements and air pollution effects (CBP 2004). Less than 60 percent of the historic forest in the Chesapeake Bay watershed remains today. Riparian forests protect only 53 percent of rivers and streams in the watershed. Urban expansion is consuming forestlands at the rate of one hundred acres per day. Sixty percent of wetlands in Maryland, Virginia, and Pennsylvania have been destroyed; now only 12 percent of this historic acreage is covered by SAV. The combined effects of development deprive the bay of the vital filtration function performed by these lost resources, increasing significantly the daily pollution loads reaching the bay (CBP 2004).

The population of the bay area increased from 8.1 million in 1950 to 16 million in 2000. By 2020, 17.6 million people are estimated to be living in the watershed and 19.4 million by 2030. Each year, approximately 20,000 people move from large cities, such as Baltimore and Washington, D.C., into the suburbs. More land is being used than is needed because, although average household population has decreased, lot sizes have increased by 60 percent, and average house sizes increased from 1,500 to more than 2,200 square feet between 1970 and 2000. Thus the number of households in the watershed has grown faster than the population (CBP 2004). One effect of increased development is the change of natural lands into impervious surfaces, such as pavement, that do not allow water to flow into the ground. As water moves across the land, runoff carries nonpoint source pollution into tributaries of the bay (CBP 2004). Between 1990 and 2000, impervious surfaces increased from 611,017 to 860,004 acres in the bay area; if that rate continues, 250,000 more acres will become impervious by 2010. One of the major goals of the C2K agreement is to reduce the rate of development in the watershed by 30 percent before 2012, saving 175,000 acres from impervious construction (CBP 2006a).

Another goal of the C2K agreement is to preserve from development 20 percent (approximately 8.2 million acres) of the watershed's land area by 2010 (CBP 2000a). Land can be protected permanently with a perpetual conservation or open space easement or fee ownership held by a federal, state, or local government or a nonprofit organization. An estimated 6.7 million acres had been preserved by July 2000 (CBF 2002c, 48). In the watershed areas of Maryland, Pennsylvania, Virginia, and Washington, D.C., 527,329 acres of land were preserved between July 2000 and July 2004. To achieve the agreement's goal of preserving 20 percent of the land area, an additional 292,963 acres must be preserved (CBP 2005).

The Bay Partner Communities Program, started in 1997, works with municipalities to employ practices that will make accomplishing the goals of C2K possible. Local communities are recognized for their accomplishments in four areas: improving water quality, promoting sound land use, protecting and restoring living resources and habitat, and engaging the community. Through coordination of programs such as this and the cooperation of all levels of government, the goals set out by C2K to ensure the health of the bay will come closer to reality as 2010 fast approaches.

Sediment

Loss of natural lands and destruction of wetlands increase the rate and volume of runoff, washing more sediment and pollution directly into the bay and its tributaries. Increases in impervious construction on the land, such as rooftops and paving, cause water to rush into and scour streams, pulling tons of sediment into great clouds. These sediment clouds block sunlight, inhibiting SAV growth. When the sediment loads settle, they smother bottom-dwelling species such as oysters and clams (CBF 2000, 3, 7).

The Chesapeake Bay Program

By the early 1970s, it was obvious that the Chesapeake Bay had suffered serious environmental harm: "Fear of fecal contamination shut down oyster beds. Rivers were closed to swimming. Shad, rockfish, and other freshwater spawners appeared in greatly reduced numbers or disappeared altogether. . . . Farm fertilizers and other sources of nitrogen were causing massive blooms of algae that blocked sunlight and smothered underwater grasses. Herbicides and chemicals entered the bay in record amounts, contaminating the water and accumulating in river and bay sediment" (Wennersten 2001, 211).

U.S. Senator Charles Mathias of Maryland, a devoted sailor and fisher on the bay, decided in the mid-1970s to make "saving the bay" a national priority. In 1975, at Senator Mathias's instigation, Congress directed the U.S. Environmental Protection Agency (EPA) to conduct a study of the Chesapeake Bay, for which Congress appropriated $25 million. In 1983, as a result of the study, the EPA formed and launched the Chesapeake Bay Program (CBP), authorized by Section 117 of the Clean Water Act (CWA). The first mandate of the Program was to carry out a massive, in-depth study of the health of the bay and to report its findings and recommend improved ways of managing environmental quality in the Chesapeake (Ernst 2003, 148–49).

Creation of the Chesapeake Bay Commission

While the EPA study was under way, legislative leaders in Maryland and Virginia, also alarmed at the condition of the bay, convened a joint Maryland–Virginia Chesapeake Bay Legislative Advisory Commission to evaluate the existing management of the bay and to make recommendations for its improvement. After considering a number of possible devices for cooperative management by the state governments and the alternative of seeking direct federal intervention, the Advisory Commission in 1978 recommended creation of a bi-state legislative commission to coordinate bay protection and restoration efforts. The Advisory Commission's leadership believed that the two state legislatures had to become more actively engaged in efforts to clean the bay because, without their ongoing commitment, restoration efforts would fail. In 1980, responding to the Advisory Commission's recommendations, the legislatures of Maryland and Virginia created the Chesapeake Bay Commission (CBC) to assist the states in managing the Chesapeake Bay cooperatively. In recognition of the importance of the Pennsylvania portion of the bay's watershed, it was added as a member state of the CBC in 1985 (CBC n.d.).

The CBC now has twenty-one members, seven from each state, of whom fifteen are legislators. The governors of Maryland, Virginia, and Pennsylvania, represented by their cabinet officials responsible for natural resources, are also Commission members, along with a citizen representative from each state. The Commission's charter directs it to "assist the legislatures . . . in evaluating and responding to problems of mutual concern relating to the Chesapeake Bay; to promote intergovernmental resource planning and action by the [states]; to provide . . . through recommendation to the respective legislatures, uniformity of legislative application; . . . and to recommend improvements in the existing management system for the benefit of the present and future inhabitants of the Chesapeake Bay region" (CBC 1981, A-2).

The CBC is rare among governmental coordinating bodies addressing ecosystem-wide environmental problems, in that most of its members are legislators. In the typical model, such a coordinating entity would be made up of representatives of executive branch agencies of government. The Commission's unusual membership has served a vital role because it has kept the three states' legislatures deeply educated and involved in the work of Chesapeake Bay restoration and protection for more than two decades. Several of the state legislative members have continued to serve during much of the Commission's life, providing continuity, institutional memory, and depth of understanding. These benefits are difficult to achieve in entities whose members are only executive branch representatives because they change with each administration. The Commission has a small staff, under the leadership of a highly respected Chesapeake Bay veteran, Anne Pesiri Swanson, who has served as executive director since 1988. The CBC has tackled its coordination functions with ingenuity and skill and has served as a catalyst for a number of important advances in fisheries management, reduction of toxics and nutrient load, and land use controls, among others (CBC n.d.).

The Chesapeake Bay Agreement of 1983

In December 1983, the CBC sponsored and organized a baywide conference to consider the results and recommendations of the seven-year EPA study. The governors of Virginia, Maryland, and Pennsylvania; the mayor of the District of Columbia; and the EPA administrator participated actively. Representatives of government entities, user and environmental groups, scientists, and concerned citizens evaluated "public policy related to the management of uses of the Chesapeake Bay and its watershed for the protection and enhancement of living resources in the tidal system" (CBC 1983, 9).

The lasting legacy of the 1983 Chesapeake Bay conference is a commitment to the bay's restoration and protection, expressed in the *1983 Chesapeake Bay Agreement.* The governors of Virginia, Maryland, and Pennsylvania; the mayor of the District of Columbia; the administrator of EPA; and the chairman of the CBC signed the agreement on December 9, 1983 (CBP 1983, 1). The 1983 agreement is brief; its preamble reads as follows: "We recognize that the findings of the Chesapeake Bay Program have shown an historical decline in the living resources of the . . . Bay and that a cooperative approach is needed among the participating jurisdictions and entities to fully address the extent, complexity, and sources of pollutants entering the Bay. . . . We further recognize that EPA and the States share the responsibility for management decisions and resources regarding the high priority issues of the Chesapeake Bay" (CBP 1983, 1). This clearly establishes the partnership between the EPA and the three states.

The remainder of the 1983 agreement consists of three short paragraphs laying out a rudimentary structure for the state–federal partnership. It establishes the Chesapeake Executive Council, made up of the signatory governors, mayor, and the EPA regional administrator, who "meet at least twice yearly to assess and oversee the implementation of coordinated plans to improve and protect the water quality and living resources of the Chesapeake Bay estuarine system" (CBP 1983, 1). The Council was to be chaired initially by the EPA administrator and was directed to report annually to the signatories. The Executive Council was directed to "establish an implementation committee of agency representatives who will meet as needed to coordinate technical matters and to coordinate the development and evaluation of management plans" (CBP 1983, 1). The agreement also calls for the establishment of "a liaison office for Chesapeake Bay activities" at an EPA facility in Annapolis, Maryland (CBP 1983, 1). The 1983 agreement, although accurately characterized as "little more than a straightforward paragraph professing a resolve to work together to restore the Chesapeake Bay," initiated the CBP as it exists and operates today (CBC 2000, 35).

The essence of the CBP is the partnership among the three states and between them and the federal government, represented by the EPA. While other agencies have roles to play relating to the bay, notably the Army Corps of Engineers (USACE) and the National Oceanic and Atmospheric Administration (NOAA), the EPA is the lead federal agency. To clarify whether the CBP is, in fact, a federal program, the EPA has stated: "Although EPA provides seed money for the Bay restoration effort and has a lot of responsibilities for coordinating and staffing the Chesapeake Bay Program, it is a unique regional partnership dominated by its state partners" (EPA n.d.).

The Chesapeake Bay Agreement of 1987

After the 1983 agreement was established, according to the CBC, "meetings and joint ventures among the jurisdictional participants were relatively rare" (CBC 1990, 1). Members of the commission, who continued to work actively to plan legislative initiatives, came to the conclusion that for the basinwide effort to be truly effective, it needed a new level of commitment from the partners and a more specific set of goals and timetables: "We have now moved from the general spirit of cooperation and coordination which was engendered by that earlier Agreement into a more specific goal-oriented complex of programs with specific, meaningful and measurable targets and timeframes" (CBC 1990, 19).

In writing the *1987 Chesapeake Bay Agreement*, CBC members focused on six priority areas and specified objectives and commitments in each category. The agreement placed first priority on the bay's living resources and then set out a series of objectives relating to restoration and conservation of water quality and habitat, as well as commitments with dates attached, to develop guidelines and management plans to achieve the goals. The other priority areas identified in the agreement include population growth and development; public information, education and participation; public access; and governance. The most significant and ambitious of the undertakings in the 1987 agreement related to water quality. They reflected the parties' serious concern that "excessive nutrient enrichment (i.e. eutrophication) is . . . the single most important factor in the declining water quality and productivity of the Chesapeake Bay" (CBC 1990, 21).

The 1987 agreement showed for the first time the parties' willingness to set measurable water quality goals. Most notably, the parties committed to achieve by the year 2000 "at least a 40 percent reduction of nitrogen and phosphorous entering the main stem of the Chesapeake Bay" (CBP 1987, 3). The nutrient reduction targets were adopted as the means to achieve the goal of improving and maintaining dissolved oxygen concentrations essential to maintain living resources. The 40-percent goal, however, did not stay in place for long: "Shortly after the 1987 Agreement was signed . . . the 40 percent goal was significantly changed as officials added the word 'controllable' to the goal. Subsequently, many of the nutrients reaching the Bay were written off as 'uncontrollable,' including those originating from headwater states such as New York, West Virginia and Delaware, as well as from air pollution and a variety of other sources. Ultimately, the '40 percent' goal became a 20 percent reduction for nitrogen and a 31 percent reduction for phosphorus for the watershed" (Blankenship 2002d).

As for population growth and development, the 1987 agreement committed the signatories to adopt by 1989 "development policies and guidelines designed to reduce adverse impacts on the water quality and living resources of the Bay, including minimum best management practices for development and to cooperatively assist local governments in evaluating land-use and development decisions within their purview, consistent with the policies and guidelines" (CBP 1987, 23). The parties further agreed to evaluate state and federal development projects for their potential impacts on water quality and to encourage local governments to "incorporate protection of tidal and non-tidal wetlands and fragile natural areas in their land-use and water and sewer planning, construction and other growth-related management processes" (CBP 1987, 23).

Chesapeake 2000 (C2K)

Many of the goals set in the 1987 agreement were set for achievement by the year 2000, including the 40-percent nutrient reduction target. As the millennium approached, the Chesapeake Bay partners agreed it would be a propitious time to measure progress and redefine goals in light of increased scientific understanding of the dynamics of the bay system. Where the earlier goals had not been met—nutrient reduction in particular—the parties reconsidered the targets and reassessed strategies for their achievement. The next agreement, entitled simply *Chesapeake 2000* (often called C2K), expanded upon the 1983 and 1987 agreements and was endorsed by the same signatories. The product of a two-year drafting effort led by the CBC, C2K involved extensive input from an array of experts and from the public. The 2000 agreement reiterates the top priority accorded in the earlier agreements to restoration of the bay's living resources. The key to restoration of the bay's living resources is improved water quality, which C2K terms "the most critical element in the overall protection and restoration of the Bay and its tributaries" (CBP 2000a, 6). Dozens of other goals in C2K are organized within five categories: (1) water quality protection and restoration, (2) living resource protection and restoration, (3) vital habitat protection and restoration, (4) sound land use, and (5) stewardship and community engagement.

Water Quality Restoration and Protection

The first of the goals in this category is to continue efforts to achieve and maintain the modified 40-percent nutrient reduction goal agreed to in 1987, as well as to adhere to living resource requirements affecting southerly bay tributaries (CBP 2000a, 6). The second goal is to correct the nutrient- and sediment-related problems in the bay and tidal tributaries in order to remove them from the list of impaired waters under the CWA by 2010 and to avoid the imposition of burdensome total maximum daily load (TMDL) requirements by EPA. To achieve this second goal required defining "the water quality conditions necessary to protect aquatic living resources and then assign[ing] load reductions for nitrogen and phosphorus to each major tributary" (CBP 2000a, 6).

Unlike the two earlier agreements, C2K "targets sediment removal as a central objective of its water quality improvement efforts" (CBC 2000, 40). Employing a process like the one established for nutrient reduction, C2K commits to determining the load reductions necessary to protect living resources and assigning them to each major bay tributary by 2001 (CBP 2000a, 7). The 2000 agreement addresses the issue of chemical contaminants in the bay and adopts a goal of "a Chesapeake Bay free of toxics by reducing or eliminating the input of chemical contaminants from all controllable sources to levels that result in no toxic or bio-accumulative impact on the living resources that inhabit the Bay or on human health" (CBP 2000a, 7). Strategies for attaining this goal include revising the "Chesapeake Bay Basinwide Toxics Reduction and Prevention Strategy"; going beyond traditional point source controls to include nonpoint sources, such as groundwater discharge and atmospheric deposition; continually improving pollution prevention measures to strive for zero release of chemical contaminants from point sources; and using education, outreach, and implementation of integrated pest management and best management practices on lands with a high potential for contributing pesticide loads to the bay (CBP 2000a, 7).

Living Resource Protection and Restoration

The 2000 agreement adopts a goal of achieving, by 2010, a minimum tenfold increase in native oysters, using a 1994 baseline. To achieve this, the parties will commit to "develop and implement a strategy . . . using sanctuaries sufficient in size and distribution, aquaculture, and continued disease research and disease-resistant management" (CBP 2000a, 2).

Two points also called for in C2K are establishing harvest targets for the blue crab fishery and managing the fishery to restore a healthy spawning biomass, size, and age structure (CBP 2000a, 4). In recent years, the three states and the EPA have acknowledged the negative impact of exotic species in the bay ecosystem, often introduced by accident or released in the ballast water of ships. The 2000 agreement commits partners to identify exotic species with potential to cause significant harm to the bay's aquatic ecosystem and to prepare management plans for these species. A task force, according to C2K, is to be created to work with the U.S. Coast Guard, the ports, the shipping industry, and environmental interests to reduce the introduction of exotics from ballast water (CBP 2000a, 2). The 2000 agreement also recognizes the need to develop integrated goals for the protection of species that share habitat.

Vital Habitat Protection and Restoration

The vital habitats of the Chesapeake Bay and its watershed, including open water, underwater grasses, marshes, wetlands, streams, and forests, "support living resource abundance by providing key food and habitat for a variety of species. Submerged aquatic vegetation [SAV] reduces shoreline erosion while forests and wetlands protect water quality by naturally processing the pollutants before they enter the water. Long-term protection of this natural infrastructure is essential" (CBP 2000a, 4). In 1993, the Executive Council of the CBP adopted a goal of restoring and protecting 114,000 acres of SAV by 2005 (Landy et al. 1999).

Recommitting the parties to this goal, C2K calls for stricter revisions: "The revised goals will include specific levels of water clarity which are to be met in 2010. Strategies to achieve these goals will address water clarity, water quality and bottom disturbance" (CBP 2000a, 4). The agreement also calls for accelerated efforts to restore SAV beds in areas of critical importance to the bay's living resources. Unrelenting population and development pressures are threatening wetland habitat in the entire Chesapeake region. In C2K, the parties commit to "achieve a no-net loss of existing wetlands acreage and function in the signatories' regulatory programs" (CBP 2000a, 5).

By 2010, the parties seek to achieve a net gain of 25,000 acres (more than 39 square miles) of wetlands. The agreement relies on local efforts to achieve this goal, promising "information and assistance to local governments and community groups for the development and implementation of wetlands preservation plans as a component of a locally based integrated watershed management plan" (CBP 2000a, 5). The goal is to implement "the wetlands plan component in 25 percent of the land area of each state's Bay watershed by 2010. The plans would preserve key wetlands while addressing surrounding land use so as to preserve wetland functions" (CBP 2000a, 5). The goals adopted by the CBP's 1996 Riparian Forest Buffer Initiative, to "restore 2,010 miles of riparian forest by 2010" and "to conserve existing forests along all streams and shorelines," are endorsed by C2K (CBP 2000a, 5). Using such devices as conservation easements, greenways, and land acquisition, the parties will link forested areas, counteracting the current fragmentation of habitat (CBP 2000a, 5).

Sound Land Use

The Chesapeake Bay partners acknowledged, "[F]uture development will be sustainable only if we protect our natural and rural resource land, limit impervious surfaces and concentrate new growth in existing population centers or suitable areas served by appropriate infrastructure" (CBP 2000a, 8–9). They renewed their commitment to "appropriate development standards" yet acknowledged that much authority rests not with them but with local governments: "Local jurisdictions have been delegated authority over many decisions regarding growth and development which have both direct and indirect effects on the Chesapeake Bay system and its living resources. . . . States will engage in active partnerships with local governments in managing growth and development in ways that support the goal of sustainable development. We will also strive to coordinate land-use, transportation, water and sewer and other infrastructure planning so that funding and policies at all levels of government do not contribute to poorly planned growth and development or degrade local water quality and habitat" (CBP 2000a, 9).

The 2000 agreement establishes the goal of permanently preserving from development 20 percent of the land area in the watershed (estimated at 1.1 million acres) by 2010 (CBP 2000a, 9). This goal is to be achieved by land acquisition efforts, including acquisition of easements and transfer of development rights, with willing sellers. No specific plans for funding land acquisition are included, though the agreement says generally that the parties will "provide financial assistance or new revenue sources" to achieve the goal (CBP 2000a, 9). A number of strategies and commitments geared toward controlling development and sprawl are included in C2K. Most ambitious is the commitment, by 2012, "to reduce the rate of development of forest and agricultural land in the Chesapeake Bay watershed by 30 percent measured as an average over five years from the baseline of 1992–1997, with . . . progress reported regularly to the Chesapeake Executive Council" (CBP 2000a, 10).

Stewardship and Community Engagement

The partners state that they will "promote individual stewardship and assist individuals, community-based organizations, businesses, local governments and schools to undertake initiatives to achieve the goals and commitments of this agreement" (CBP 2000a, 11). The agreement outlines a number of education and outreach strategies. The parties also commit to "strengthen partnerships with Delaware, New York and West Virginia by promoting communication and by seeking agreements on issues of mutual concern" (CBP 2000a, 12–14).

Organization and Operation of the Chesapeake Bay Program

The Chesapeake Executive Council, created in the 1983 agreement, establishes policy direction for the CBP. Its Principals Staff Committee, made up of state and federal officials with responsibility for environment and natural resources, the CBC executive director, and an EPA representative, provides policy advice to the Executive Council; sets agendas for the Council's annual meetings; and gives direction to and coordinates the work of the Implementation Committee, the Executive Council, and its advisory committees. The Implementation Committee is composed of representatives from the signatories, ten federal agencies, and other program participants, such as the University of Maryland and Virginia Tech. This committee is responsible for implementing the decisions and technical studies of the Executive Council and coordinating activities carried out under C2K. Chaired by the director of the EPA CBP Office, the Implementation Committee provides staff support to the Council and is responsible for an annual work plan and budget. It receives advice from the Scientific and Technical Advisory Committee, the Local Government Advisory Committee, and the Citizens Advisory Committee, whose chairs are also members of the Implementation Committee (CBP 1983).

The Scientific and Technical Advisory Committee (STAC) is made up of scientists and technical experts from state and federal agencies, universities, research institutions, and private industry. Created in 1984, it provides scientific and technical advice through reports and position papers, reviews of Bay Program initiatives, and annual conferences and workshops. Members of the STAC serve on CBP subcommittees and workgroups and provide liaison between the scientific/engineering committee and the Bay Program. The Chesapeake Research Consortium (CRC), an association of six institutions, has assembled a team of scientists and engineers to work with policy makers to design and carry out multidisciplinary studies and find solutions to problems affecting the bay.

The CRC provides staff support to STAC and other Bay Program subcommittees (CRC n.d.).

Section 117 of the CWA authorizes EPA to participate in the CBP "as a member of" and "in cooperation with the Chesapeake Executive Council" (1977). Funded at about $20 million annually, the EPA office of the CBP has no regulatory authority. Instead, it provides technical information and support, watershed models, water quality assessments, and policy analysis to the Program's partners. The EPA is authorized to provide technical assistance grants to state and local governments and nonprofit or educational organizations to fund activities that support the CBP (CWA 1977).

Implementation and monitoring grants are authorized for signatory jurisdictions to carry out resource management mechanisms called for in the agreement or for monitoring the bay ecosystem (CWA 1977). The EPA is directed "in coordination with other members of the Chesapeake Executive Council" to "ensure that management plans are developed and implementation is begun . . . to achieve and maintain . . . " nutrient goals as to nitrogen and phosphorous levels; toxin reduction; and habitat restoration. As of 2003, the EPA must report to Congress every five years on the state of the bay ecosystem, assessing the effectiveness of management strategies implemented pursuant to C2K and making recommendations for improved management (CWA 1977).

The Role of Science in the Chesapeake Bay Program

Scientific research on the Chesapeake Bay seems to lack organization and focus. The CBP's Executive Council and STAC have not provided a comprehensive research agenda for the bay, and they have not established processes for setting research priorities. The CBP does not have a budget to fund research, though it has attempted to coordinate scientific efforts in various ways, including through its scientific and technical subcommittees. From time to time, STAC appoints standing and ad hoc working groups from the region's scientific and technical community to address Bay Program issues. Observers of the Chesapeake Bay science effort have expressed concern that high-level policy makers are not receiving the best scientific input from these working groups because the communication channels between them are weak. These observers are troubled by what they see as a lack of established forums through which scientists can communicate with nonscientists who set policy for the CBP. Concerns have been expressed in that STAC is not organized well enough to provide the most timely and relevant scientific and technical advice to the CBP. Members of STAC were selected on the basis of their institutional affiliations rather than their experience and expertise.

In 2001, the STAC executive board reorganized its membership structure. Gaps in expertise were identified, and new members were appointed (CBP STAC 2001). Since the adoption of C2K, STAC has developed a scientific and technical needs assessment, demonstrating the importance of "immediate and continuous support for research and monitoring as well as incorporation of progressive view technologies" if the goals are to be met (CBP STAC 2004, 3). The committee notes the danger of failing to integrate sound science into C2K implementation strategies: "STAC strategy believes that undertaking the restoration activities in the watershed without consideration of the best scientific judgment possible could result in unsuccessful restoration from potentially expensive outlays of money and labor" (CBP STAC 2004, 3). The reconstituted STAC is committed to provide the means for scientists to proceed in a coordinated fashion and to speak coherently to policy makers.

Chesapeake science efforts have been characterized more by a lack of communication and clear priorities than by chronic dispute and disagreement among experts. Naturally, bay scientists and others occasionally disagree on scientific and technical issues. The most notorious of the few publicized cases of serious scientific dispute in the bay region stems from the *Pfiesteria piscicida* outbreak of 1997. In April 1997, *Pfiesteria* was blamed for killing more than 50,000 fish, mostly menhaden, and sickening dozens of people on the Pocomoke River in Maryland (CBP 2006b).

Emergence of the allegedly toxic strains of *Pfiesteria* was attributable to high nutrient loads in the water (nitrogen and phosphorous). Because agricultural runoff was the leading contributor of nutrients added to the water, "[m]any environmentalists and scientists [came to] believe that agricultural runoff from poultry farms in Maryland and hog farms in North Carolina [was] responsible for the *Pfiesteria* outbreaks" (Bueschen 1998). Based on this view, the Maryland legislature imposed upon farmers mandatory restrictions on the use of fertilizers and the requirement that they adopt best practice fertilizer management plans (Bueschen 1998).

Since the dramatic events of 1997, a number of scientists have done more extensive work on *Pfiesteria*, leading them to raise questions about its toxicity. Some of the later studies have failed to find toxin-producing genes that could harm fish. Joann Burkholder, of North Carolina State University, whose research was the basis for the prior prevailing view on the toxic properties of *Pfiesteria*, dismissed "the newspapers as part of a coordinated 'press blitzkrieg attack' on the nearly 50 papers by herself and others describing *Pfiesteria* and its impacts" (Blankenship 2002b). What emerged was what the *Bay Journal* called "one of the ugliest disputes in environmental science" (Blankenship 2002b).

The *Pfiesteria* conflict remains unresolved. One widely respected Chesapeake Bay scientist, Donald Boesch, of the University of Maryland, said, "'It has been polarized for some time, and it's not getting better, and in some ways it's getting worse. . . . It's not the kind of professional conduct I condone, on either side.' . . . [H]e and others worry that the dispute has caused people to question the need for the nutrient control legislation that emerged" (Blankenship 2002b).

Assessing the Chesapeake Bay Program

The CBP is a regional partnership of peers; there is no hierarchy in its structure. Thus consensus is the guiding principle for policy making, and for action to take place, the partners must all agree on how and when to proceed. Swanson, executive director of the CBC, expressed in quantitative terms the political complexity of the Program when she pointed out that Chesapeake Bay lies within the jurisdiction and influence of "three governors, 40 members of Congress, hundreds of state legislators and local elected officials, 13 federal agencies, four interstate agencies, more than 700 citizen groups, and hundreds of businesses. . . . The Bay Program has established more than 50 subcommittees and workgroups to ensure that all of these interests are represented and that the goals of the Program are ultimately achieved" (Swanson 2003). These entities and individuals have interests, mandates, and constituencies that are often divergent, and securing their consensus on all major policy decisions can be an overwhelming challenge.

In this light, the Chesapeake Bay agreements of 1983, 1987, and 2000 are products of consensus and must be acknowledged as notable achievements. Of course, the agreements will ultimately be judged by their effectiveness in restoring and protecting the bay. The 2000 agreement is an elaborate document, containing close to 120 commitments by the signatories and accurately reflecting the parties' increased understanding of the complexity of the task they face. Over halfway to its target date of 2010, C2K has proved to be a useful catalyst, guide to action, and benchmark for bay restoration projects. However, adequate funding has not been forthcoming to see its goals met (MGA 2006).

A Culture of Courtesy

Perhaps because consensus is essential for action, a tradition of courtesy and comity, with very little acrimony, exists among the partners of the CBP. Many of the participants have known each other and worked together for a long time on the restoration effort; some have moved from government employment to work with a nongovernmental partner or vice versa. Except in the area of water quality, the program has generated almost no litigation. This lack of litigation may be attributable to the area's culture of civility as well as to the fact that the consensus-based program has no regulatory bite. In the bay region, one environmental group—the venerable Chesapeake Bay Foundation (CBF)—dominates the roles of outside watchdog, public interest advocate, and grassroots organizer.

Elsewhere, these roles are typically occupied by any number of environmental groups who focus on different issues and are sometimes at odds with one another. The CBF, with 177,000 members, 170 full-time employees, a $20-million annual budget, and a multiplicity of educational and advocacy initiatives (CBF 2006b) has overshadowed the efforts of other bay area environmental organizations and now occupies the field virtually alone. The focus of its approach to policy makers on important issues is typically educational—seeking to persuade them with good scientific information. Its annual State of the Bay report, with its health index scoring improvements or lack of progress, is widely looked to by decision makers and the public to assess the effort's trends on a regular basis. Litigation is not a widely practiced device for dispute settlement in Chesapeake Bay. Perhaps the Bay Program partners and other stakeholders would achieve more if they were to engage more frequently in open criticism and dissent, similar to participants in other regional restoration efforts. However, in light of the area's traditions, this sort of change is unlikely.

Maryland and Virginia: The Partnership

The principal partners in the CBP are the states fronting the bay, Maryland and Virginia. The two states historically have represented very different political cultures, one progressive and one conservative. Maryland has a regulatory government, a heavy tax burden, and a constitutionally powerful governor. Virginia has a tradition of limited government, a reluctance to tax, and a governor whose powers are constitutionally quite confined. At times, these historical differences have made for an uneasy partnership. Since 2000, however, these differences have blurred, with the election of a Republican governor in Maryland in 2002, succeeded by a Democrat in 2006, and the elections of Democratic governors in Virginia in 2001 and 2005.

The Partnership in Action: Land Conservation Efforts

The historical contrast between the two states can be seen in their approaches to land conservation, long a program priority now permanently enshrined in the C2K commitment to preserve from development 20 percent of the land area in the bay watershed by 2010. Maryland was one of the first states to fund a land conservation program through a dedicated real estate transfer tax. An average of

19,000 Maryland acres were protected each year from fiscal year 1992 to 1999, using these revenues, federal grants, county matching funds, and private donations of land. During this period, the state spent $325 million on land conservation in the Chesapeake Bay watershed. In contrast, Virginia relied heavily on private donations of land as the basis of its conservation program. From 1992 to 1999, an average per year of 11,500 acres were protected in Virginia, 60 percent of them donated. During these years, the state of Virginia spent $23.6 million on land protection in the bay watershed (TPL and CBC 2001).

Cooperation and Consensus on the Blue Crab Crisis

The blue crab, of the genus *Callinectes*, Latin for "beautiful swimmer," is the icon of the Chesapeake Bay. In 1996, serious concern about the dwindling blue crab fishery in the bay led the CBC and its constituents, the governors and legislatures of Maryland and Virginia, to form the Bi-State Blue Crab Advisory Committee (BBCAC). The mission of the BBCAC was to provide specific focus on issues related to the overharvesting of the blue crab and to give independent advice to the three regional management jurisdictions with responsibility for the blue crab—Maryland, Virginia, and the Potomac River Fisheries Commission. A technical work group was created to deliver scientific advice on the biological and economic consequences of alternative management scenarios.

From the outset, BBCAC worked simultaneously on a number of issues, including (1) an analysis of commercial crabbers, including a survey of the gear and equipment they use, the profits needed to sustain their business, and their views on managing the fishery; and (2) development of recommendations for reasonable thresholds and targets for the blue crab fishery that would define how many crabs could be taken without threatening the stock (CBC BBCAC 2001, 4). In 1999, Maryland and Virginia funded BBCAC to conduct a two-year blue crab management study and develop recommendations for action to protect the fishery. The report, entitled *Taking Action for the Blue Crab*, was published in January 2001. The BBCAC stated its key findings: "Blue crab stocks are near the lowest point measured since fisheries-independent surveys began. There is a limit to fishing pressure that blue crab stocks can stand at any given level of abundance. As the stocks rise, fishing level can also rise. As stocks fall, fishing pressure must also decrease" (CBC BBCAC 2001, 3). The BBCAC announced bay-wide goals to reduce crab harvest pressure and to double the size of the spawning stock from the 1997–1999 average level (CBC BBCAC 2001, 11). This strengthened a goal stated in the 2000 agreement.

Reducing Blue Crab Harvest Pressure

The 2001 report issued by the BBCAC contains specific recommendations, including a fishing threshold that would preserve 10 percent of the blue crab's spawning potential and a minimum stock size threshold. At 2001 fishing levels, the report, in effect, called for a 15-percent decrease in the crab harvest over the ensuing three-year period (CBC BBCAC 2001, 3). Maryland, Virginia, and the Potomac River Fisheries Commission accepted the BBCAC's recommendations, and each jurisdiction vowed to impose the reductions before the onset of the 2001 crabbing season in April. Watermen and seafood packers in Maryland and Virginia resisted and put pressure on regulators and legislators in each state to go slowly and to take no action until the other jurisdiction had acted. This resulted in a stalemate.

In Maryland, the legislature adopted a licensing requirement for recreational crabbers and limited their harvests but rejected regulations on commercial crabbers. A coalition of Lower Eastern Shore watermen and seafood packers argued to the legislature that the rules were unnecessary and would drive them out of business. While agreeing that the crab fishery was in crisis, the watermen maintained that the causes of the crisis were uncertain and disagreed with the position that overharvesting was to blame. They pointed to poor water quality and its harmful effects on SAV, rendering young crabs more susceptible to predators (Stoppkotte 2001).

In late April 2001, Maryland Governor Parris Glendening took action to promulgate rules limiting the crab harvest, both recreational and commercial, despite the legislature's resistance (Wilkinson n.d., 3). Following the Maryland governor's action, in 2001 the Virginia Marine Resources Commission enacted a series of regulations aimed to reduce the crab catch in Virginia by 5 percent (CBF 2002b). Late in 2001, Governor Glendening announced that Maryland would move to achieve the 15-percent reduction by the end of 2002, and Virginia followed with new actions intended to reduce the total harvest by almost 15 percent for the 2002 crabbing season (Maryland DNR 2001).

Lessons Learned about Consensus in the Blue Crab Crisis

Following the BBAC's recommendations, Maryland and Virginia put the crab harvest reductions into place, demonstrating that the partnership can take coordinated action to achieve consensus. At least four factors account for their success in handling the blue crab crisis. First, under the leadership of the CBC, the BBCAC gave the problems of the blue crab fishery high priority on the bay's agenda.

Second, science was well integrated into the process, and virtually all the scientists agreed that the blue crab fishery was on the brink of collapse, as expressed in the BBCAC's report early in 2001 (CBC BBCAC 2001). Virginia's and Maryland's representatives and their constituents, including watermen and seafood processors, shared this view, recalling accounts of the once abundant oyster fishery's precipitous collapse in the 1930s. Accepting the scientists' dire assessment, parties responsible for the Chesapeake acted with a sense of urgency: the possibility of losing the iconic blue crab symbolized losing the bay itself.

Third, while disagreement as to the causes of the catastrophic threat to the crab fishery existed, none of the parties developed a serious alternative plan of action to counter the move to reduce the crab harvest. Likewise, the BBCAC's recommendations for a 15-percent reduction over three years, supported by the technical committee of scientists, were not seriously challenged.

Fourth, because the BBCAC successfully involved political leadership at high levels in Maryland and Virginia to develop recommendations, the two states ultimately accepted responsibility. Maryland's governor took the necessary action, clearing the way for Virginia to adopt parallel restrictions. Had the issue not been focused and publicized compellingly by the BBCAC and its technical committee, these officials may not have acted.

Though the blue crab crisis led the responsible jurisdictions to act, this story also demonstrates the challenges of a consensus-based program. Partner jurisdictions often face controversy among their own constituents, clouding their ability to make decisions. Actions like these, the adoption and enforcement of regulations restricting the crab harvest, require cumbersome, time-consuming dual processes of enactment, with each state pursuing its own constitutionally required process. Additionally, no centralized enforcement mechanism ensures that the regulated conduct conforms to the adopted mandates, so accountability remains in question.

Developments since the Blue Crab Crisis

Due to insufficient state funding, the BBCAC was disbanded in July 2003, slowing interstate collaboration spurred by the committee. However, BBCAC's Bi-State Blue Crab Technical Advisory Committee (BBTAC), made up of leading bay scientists and technical experts, continues to monitor the fishery and to advise the Commission, the states, and regional fishery managers. Following the 2004 crabbing season, BBTAC's first annual report characterized the year as "mixed" in terms of fishery goals and called the future "uncertain" (CBC BBTAC 2004, 2).

Subsequent surveys have shown a rise in crab harvests throughout the bay since 2002, although the market-sized crab population remains below the long-term average. Survey results as to the quantity of female crabs have been inconsistent, making an accurate determination of the spawning stock difficult. The BBTAC's second annual report stated, "2005 can be reported as a slightly above average year in nearly a decade of low abundance. The lower stock levels of the winter dredge survey in 2005–2006 offer a preliminary indication that modest improvements seen in 2005 may not mean the beginning of a long-term trend. Cautious management should continue" (CBC BBTAC 2005, 3). According to the CBF's *2006 State of the Bay* report, the blue crab harvest remained below targets; if this trend continues, the reproducing female population will double in 2007 and 2008. The report recommends that the BBCAC reconvene to ensure that the goal of an increased spawning stock is achieved (CBF 2006a, 8).

Funding C2K

The Chesapeake 2000 agreement was adopted without estimating the costs of implementing its commitments and without a strategy for funding those costs. Late in 2001 and early in 2002, Maryland and Virginia proffered their estimates of the costs of implementing nutrient and sediment reduction and other elements of C2K. Their cost estimates differed markedly because the two states assumed different water quality goals. Maryland officials estimated that the cost of achieving desired nutrient and sediment reductions would be approximately $4 billion. They based this estimate on assumptions that statewide maximum nutrient reductions would be required and that all septic systems would be replaced, at a cost to the public of $1 billion. Virginia officials assumed a level of effort "halfway between today's level of implementation and the maximum possible control effort" that would cost between $2 and $3 billion (Blankenship 2002c).

In January 2002, the secretary of the Maryland Department of Natural Resources announced a more comprehensive assessment of the costs of reaching the C2K goals. He put the cost of nutrient reduction, habitat restoration, improved storm water management, and other commitments at about $7 billion. Virginia officials announced a much lower comprehensive cost figure. They estimated that meeting the major C2K goals would total a maximum cost of $3.9 billion. Again, this estimate did not include, as Maryland's did, the cost of advanced storm water treatment and replacement of septic systems in the state. Maryland officials estimated that the cost to Virginia, based on Maryland's budgetary and ecological assumptions, would be close to $9 billion (Blankenship 2002a). In the years since C2K was promulgated, Maryland's and Virginia's state governments have faced serious budget constraints, limiting their ability to fund bay improvement efforts.

In 2002, the CBC was the first organization to put a price tag on full implementation of C2K goals by 2010. In its 2003 report, *The Cost of a Clean Bay: Assessing Funding Needs throughout the Watershed*, the CBC assessed the cost at $18.7 billion over eight years. Acknowledging the difficulty of estimating future costs on such a large scale, the report notes that its findings "are vulnerable to changing conditions and technologies," and that while "these are the best numbers we have . . . they should not be the last" (CBC 2003, 3). Achieving the C2K nutrient and sediment reduction goals was the most costly component of the estimate, at $11 billion.

Besides estimating the cost of implementing C2K, *The Cost of a Clean Bay* addressed what it termed a "funding gap," meaning the difference between the cost estimate and funds projected to be forthcoming from existing or anticipated state and federal programs. The overall funding gap was estimated to be nearly $13 billion. The report broke down the cost and the funding gap among the three states—Maryland, Virginia, and Pennsylvania—where most of the C2K obligations fell. Implementation costs were attributed relatively evenly per state, at more than $6 billion each. The largest estimated funding gap was Virginia's, at $5.1 billion; Maryland's was $2.9 billion and Pennsylvania's $4.8 billion (CBC 2003, 6). Observing that its cost and funding gap estimates were "staggering," the report stated, "understanding the price tag should not diminish our drive or our expectations" (CBC 2003, 2, 10). The report urged government officials to "target all available funds to maximize environmental benefits" and to establish "a logical spending sequence over time" (CBC 2003, 11).

In December 2004, the CBC published a follow-up report, *Cost-Effective Strategies for the Bay: Smart Investments for Nutrient and Sediment Reduction*, which addressed issues raised by *The Cost of a Clean Bay*, in particular, how to prioritize spending to maximize progress toward the C2K goals. The report was intended to answer the question, Which practices will deliver the largest nutrient and sediment load reductions for the least cost? (CBC 2004, 2). It recommended six measures "representing a wide range of specific actions associated with wastewater treatment plants, agriculture, urban stormwater, land preservation and air pollution" selected for their "cost effectiveness and nutrient reduction potential" (CBC 2004, 3).

The report predicted that these measures would achieve approximately three-quarters of the 2010 C2K goal for nitrogen and phosphorous reduction at a cost of about $1 billion per year. In promulgating this "menu of most cost-effective measures," the CBC was taking the practical approach of identifying and advocating specific steps to achieve a measure of progress despite seriously limited resources (CBC 2004, 6). The report concluded on an encouraging note: "[T]his report should hearten us all—it demonstrates that significant water quality benefits can be had at reasonable cost" (CBC 2004, 19).

Implementing Nutrient and Sediment Reductions

In the late 1990s, EPA listed the Virginia waters of the bay, including tidal portions of tributary rivers, as "impaired waters," pursuant to Section 303(d) of the CWA (1977), meaning that they failed to meet state water quality standards. (Maryland had placed its bay waters and tidal tributaries on the "impaired" list without waiting for EPA to act.) The main reasons for the listings were the presence of excess levels of the nutrients phosphorous and nitrogen, which meant the states were not meeting standards for dissolved oxygen, and high levels of sediment, which resulted in failure to satisfy water clarity criteria. Under the CWA, states are required to adopt strategies for reaching TMDLs for each listed body of water. If the states fail to act, EPA must step in to set enforceable TMDLs.

In 1998, two environmental groups, the American Canoe Association and the American Littoral Society, sued EPA to compel the setting of TMDLs for listed waters (*American Canoe* 1999). In 1999, the parties entered a consent decree in federal court, which provided that unless Virginia adopted enforceable TMDLs for nutrients and sediments by 2010, EPA would do so by the end of 2011. As it turned out, the culture of collaboration, though invoked this time in response to possible regulatory intervention, continued to guide EPA, the states, and the stakeholders as they approached the task of establishing enforceable water quality standards for nutrients and sediment. At the time the consent decree was entered in 1999, they were negotiating and drafting the C2K agreement. In the section on "Water Quality Restoration and Protection," arguably the most important aspect of the agreement, the parties worked with a purpose of "TMDL avoidance," by which they meant avoiding the onerous intervention of EPA. For the water quality standards to be met in Virginia and Maryland waters, all the states in the watershed would have to control nutrient discharges and sediment-producing activities on their bay tributaries. In C2K, the signatory states committed themselves to meet water quality standards so that the Maryland and Virginia waters could be removed from the "impaired list" by 2010, and EPA pledged its full support for the effort.

After a difficult process of revising the general state water quality standards to levels that EPA and the states agreed to be scientifically sound and achievable, they established their basic approach to setting specific water quality criteria. The parties developed a zone approach to fixing dissolved oxygen level requirements, so that standards would be less stringent for deep waters, which are naturally low in oxygen, and stricter for shallow waters, which require higher oxygen levels to serve as spawning habitats. They also produced water clarity criteria to restore and protect SAV areas and developed chlorophyll criteria to address excess nutrients that degraded the fish food and exacerbated harmful algal blooms. This marked the first time that EPA had incorporated stakeholder consultation in its process of setting water quality criteria (EPA et al. 2007). States with tidal waters—Virginia, Maryland, Delaware, and the District of Columbia—revised their water quality standards to incorporate the agreed-upon criteria.

The next step of the collaboration was equally difficult. The states, stakeholder groups, and EPA developed baywide load limits, similar to TMDLs, for nitrogen and phosphorous, and then allocated them to sub-basins divided by state lines. Through this process, the parties created thirty-six "watershed areas," for which the states agreed to develop tributary strategies to allocate loads and ultimately to assign discharge limitations and other duties to individuals within the watershed areas whose conduct affects the tributaries. Whether the result of this huge effort will be enough to avoid EPA-imposed TMDLs is not yet known. However, if the tributary strategies are proved effective in 2011, even if they are not yet sufficient, the impact of the TMDLs will be merely incremental.

With the imposition and enforcement of these water quality standards by 2010 or 2011, the Bay Program partners will move from making commitments to act for the bay's protection to imposing costly regulations to save the bay's living resources. The process of consultation and negotiation they have employed should serve to minimize the controversy that would otherwise be expected from imposing new, watershed-wide water quality regulations that may seem costly and onerous to those affected. In this next phase of the long-lived Chesapeake Bay restoration, it will be interesting to see whether consensus and courtesy will overcome controversy and litigation.

Federal Funding Sought to Meet C2K Goals

Responding to the CBC challenge to meet the heavy funding requirements associated with C2K, the Chesapeake Executive Committee appointed the Chesapeake Bay Watershed Blue Ribbon Finance Panel (CBWBRFP) in December 2003. The Panel's charge was "to identify funding sources sufficient to implement basinwide cleanup plans so that the bay and its tidal tributaries would be restored sufficiently by 2010 to remove them from the list of impaired waters under the Clean Water Act" (CBWBRFP 2004, ii). After seven months of briefings, study, and deliberation, the Panel reached several conclusions: "[W]hile it is difficult to determine the full cost of restoring Bay water quality, it is clear that current funding does not begin to meet financing needs for restoring Bay water quality. . . . The restoration of the Bay will only become more expensive with time; and . . . it will be difficult to achieve a fully integrated approach for funding and implementation, given the number of the jurisdictions . . . along with the large presence of federal facilities and operations in the watershed" (CBW-BRFP 2004, 2).

The Panel's central recommendation was creation of a Chesapeake Bay Financing Authority, with an initial capitalization of $15 billion, to be funded by the federal and state governments to make loans and grants to state and local cleanup projects. The Panel's report estimated that the actions necessary to meet water quality standards throughout the watershed would require $28 billion in upfront capital costs and $2.7 billion in annual funding (CBWBRFP 2004, 8). The Panel took the view that funding the new Financing Authority in the amount of $15 billion would provide "funding for the most cost effective areas of wastewater treatment and agriculture. . . . Addressing these two sectors alone will provide major benefits to water quality" (CBWBRFP 2004, 10). To say that the Panel's report was disappointing is an understatement, in that it did not answer the tough question of where the billions of dollars required would come from. Although the Panel did recommend that the states impose development fees of various kinds to raise revenues for water quality projects, it calls generally for the fund "to be capitalized by the federal and state governments," without specifics on what sources might be found together with strategies for success in securing funds (CBWBRFP 2004, 11).

Is it realistic for the CBP to look to the federal government to provide substantial funding to help meet the C2K goals? The Program has mounted a thoughtful and energetic effort to secure funding increases for water quality improvement projects in the bay's watershed through the federal Farm Bill of 2007 (CBF 2007a). Even if the Panel's legislative hopes for the Farm Bill were realized, the Program would secure only a small percentage of the estimated $19 billion cost to clean up the bay. Historically, neither the U.S. president nor Congress has made Chesapeake Bay restoration a national priority that would justify significant funding by the federal government. This fact, along with the high costs of funding the Iraq war, the large federal deficit, and national attention to matters other than environmental restoration, militate against the prospect of federal funding in the amounts needed.

Conclusion

The CBP has achieved broad and deep understanding of the problems facing the bay, clearly reflected in the Chesapeake 2000 agreement's comprehensive, detailed goals and plans addressing those problems baywide on a scientific basis. The Program's participants have the experience, knowledge, and competency to implement C2K. What is lacking is the sustained allocation of financial resources on a scale that makes implementation feasible to approach the ecosystem as a whole. Until these resources are forthcoming, the Program must make do with localized, incremental actions that fall far short of the vision and ambition represented in thirty years of effort to save the bay.

References

American Canoe Ass'n, Inc. v. U.S. Envtl. Prot. Agency, 54 F. Supp. 2d 621 (E.D. Va. 1999).

Blankenship, K. 2002a. "Maryland Tab for Bay Goals Put at $7 Billion." *Bay Journal* 11 (January/February): 10. Annapolis, MD: Alliance for the Chesapeake Bay for the CBP. At www.bayjournal.com/article.cfm?article=157.

Blankenship, K. 2002b. "*Pfiesteria* May or May Not Be Toxic, but the Dispute over the Issue Is." *Bay Journal* 12 (October): 7. Annapolis, MD: Alliance for the Chesapeake Bay for the CBP. At www.bayjournal.com/article.cfm?article=785.

Blankenship, K. 2002c. "VA Comes Up with Cost for Cleanup; Now It Must Come Up with Money." *Bay Journal* 12 (March): 1. Annapolis, MD: Alliance for the Chesapeake Bay for the CBP. At www.bayjournal.com/article.cfm?article=210.

Blankenship, K. 2002d. "Was 40 Percent Right?" *Bay Journal* 12 (May): 3. Annapolis, MD: Alliance for the Chesapeake Bay for the CBP. At www.bayjournal.com/article.cfm?article=519.

Bueschen, E. 1998. "*Pfiesteria piscicida*: A Regional Symptom of a National Problem." *Environmental Law Reporter* 28 (June): 10317–30.

Chesapeake Bay Commission (CBC). n.d. "History." Annapolis, MD: CBC. At www.chesbay.state.va.us/history.html.

CBC. 1981. *The Chesapeake Bay Commission's 1980–1981 Annual Report*. Annapolis, MD: CBC.

CBC. 1983. *The Chesapeake Bay Commission's 1982 Annual Report*. Annapolis, MD: CBC.

CBC. 1990. *The Chesapeake Bay Commission's 1988 and 1989 Annual Report*. Annapolis, MD: CBC.

CBC. 2000. *Policy for the Bay: Annual Report 1999*. Annapolis, MD: CBC.

CBC. 2001. *The Chesapeake Bay Commission's 2000 Annual Report: Keeping the Agreement*. Annapolis, MD: CBC. At www.chesbay.state.va.us/Publications/cbc2000annreport.pdf.

CBC. 2003. *The Cost of a Clean Bay: Assessing Funding Needs throughout the Watershed*. January. Annapolis, MD: CBC. At www.chesbay.state.va.us/Publications/C2Kfunding.pdf.

CBC. 2004. *Cost-Effective Strategies for the Bay: Smart Investments for Nutrient and Sediment Reduction*. December. Annapolis, MD: CBC. At www.chesbay.state.va.us/Publications/cost%20effective.pdf.

CBC. 2006. *The Chesapeake Bay Commission's 2005 Annual Report: Milestones: 25 Years of Policy for the Bay*. Annapolis, MD: CBC. At www.chesbay.state.va.us/Publications/CBC-AR-2005.pdf.

CBC Bi-State Blue Crab Advisory Committee (BBCAC). 2001. *Taking Action for the Blue Crab: Managing and Protecting the Stock and Its Fisheries*. January. Annapolis, MD: CBC. At www.chesbay.state.va.us/Publications/Archive%20Publications/BBCACReportfinal.pdf.

CBC Bi-State Blue Crab Technical Advisory Committee (BBTAC). 2004. *Blue Crab 2004: Status of the Chesapeake Population and Its Fisheries*. Annapolis, MD: CBC. At www.chesbay.state.va.us/Publications/Blue%20Crab%202004.pdf. Or at www.chesbay.state.va.us/. Under Publications, see Blue Crab Reports.

CBC BBCTAC. 2005. *Blue Crab 2005: Status of the Chesapeake Population and Its Fisheries*. Annapolis, MD: CBC. At www.chesbay.state.va.us/Publications/CBC-CRAB-05.pdf.

Chesapeake Bay Foundation (CBF). 2000. "Land and the Chesapeake Bay." Annapolis, MD: CBF. July 13. At www.cbf.org/site/DocServer/Lands_Book.pdf?docID=118.

CBF. 2002a. "CBF Refines Its Bay-Saving Plan." *Bay Weekly* 28 (Spring): 2. Annapolis, MD: CBF.

CBF. 2002b. "Crab Catch." *Bay Weekly* (September 17). At www.bayweekly.com/year02/issue9-17/chesout9-17.html.

CBF. 2002c. *The State of the Chesapeake Bay: A Report to the Citizens of the Bay Region*. Annual Report for 2002. June. Annapolis, MD: CBF. At www.chesapeakebay.net/pubs/sob/sob02/sotb_2002_final.pdf.

CBF. 2004. *State of the Bay Report 2004*. Annual Report. November. Annapolis, MD: CBF. At www.cbf.org/site/DocServer/2004_sotb2.pdf?docID=6763.

CBF. 2006a. *2006 State of the Bay*. Annual Report. November. Annapolis, MD: CBF. At www.cbf.org/site/DocServer/SOTB_2006.pdf?docID=6743.

CBF. 2006b. "President's Profile." Annapolis, MD: CBF. At www.cbf.org/site/PageServer?pagename=about_sub_leadership_profile.

CBF. 2007a. *CHESSEA: Chesapeake's Healthy and Environmentally Sound Stewardship of Energy and Agriculture Act of 2007* (H.R. 1766/S. 1346). April 15. Annapolis, MD: CBF. At www.cbf.org/site/DocServer/CHESSEA_fact_sheet.pdf? (docID=8843).

CBF. 2007b. "Funding for Clean Water." Annapolis, MD: CBF. At www.cbf.org/site/PageServer?pagename=state_sub_va_legislation_water.

Chesapeake Bay Program (CBP). 1983. *1983 Chesapeake Bay Agreement*. December 9. Annapolis, MD: CBP. At www.chesapeakebay.net/pubs/1983ChesapeakeBayAgreement.pdf.

CBP. 1987. *1987 Chesapeake Bay Agreement*. December 15. Annapolis, MD: CBP. At http://www.chesapeakebay.net/pubs/1987ChesapeakeBayAgreement.pdf.

CBP. 2000a. *Chesapeake 2000* (C2K). Annapolis, MD: CBP. At http://www.chesapeakebay.net/agreement.htm.

CBP. 2000b. "The Watershed." *Chesapeake Bay: An Introduction to an Ecosystem*. Annapolis, MD: CBP. At www.chesapeakebay.net/.

CBP. 2004. "Development and the Bay Watershed." Annapolis, MD: CBP. At www.chesapeakebay.net/info/development.cfm.

CBP. 2005. Land Growth and Stewardship Subcommittee (LGSS). "LGSS Digest Regarding Land Conservation Keystone Commitment for IC April 2005." Annapolis, MD: CBP.

CBP. 2006a. "Bay Trends and Indicators." June 26. At www.chesapeakebay.net/status.cfm.

CBP. 2006b. "*Pfiesteria*." At www.chesapeakebay.net/pfiesteria.htm.

CBP Scientific and Technical Advisory Committee (STAC). 2001. "Meeting Summary." March 29. Annapolis, MD: CBP. At www.chesapeake.org/stac/MeetInfo/march2001mins.PDF.

CBP STAC. 2004. *Scientific and Technical Needs for Fulfilling Chesapeake 2000 Goals. 2004 Update*. At www.chesapeake.org/stac/STNeeds/04C2KSTNeeds.pdf.

Chesapeake Bay Watershed Blue Ribbon Finance Panel (CBWBRFP). 2004. *Saving a National Treasure: Financing the Cleanup of the Chesapeake Bay*. A Report to the Chesapeake Executive Council. At www.chesapeakebay.net/pubs/blueribbon/Blue_Ribbon_fullreport.pdf.

Chesapeake Research Consortium (CRC). n.d. At www.chesapeake.org/crc/crc.html.

Clean Water Act (CWA). 1977. Section 117 (33 U.S.C. 1267); (33 U.S.C. 1276) (b)(1); (d); (e); (h). Federal Water Pollution Control Act [1948] 1972. 33 U.S. C. Sec. 1251-1387 (2007). At http://uscode.house.gov/search/criteria.shtml.

Department of Conservation and Recreation, Soil and Water Conservation Programs (DCR, SWCP). 2006. "Soil and Water Conservation." Richmond, VA. At www.dcr.state.va.us/sw/nutenbay.htm.

Environmental Protection Agency (EPA). n.d. *Waterdrops*. At www.epa.gov/region03/waterdrops/CBP.pdf.

EPA. 2007. E-mail correspondence between program director, Chesapeake Bay Program Office, Washington, DC, and C. Rotchford, research assistant to M. Doyle. January.

EPA and R. Beusse, C. Blair, H. Canes, B. Chuong, J. Hatfield, and G. Pierce, contributors. 2007. *Evaluation Report: EPA Relying on Existing Clean Air Act Regulations to Reduce Atmospheric Deposition of the Chesapeake Bay and Its Watershed*. Office of the Inspector General. Report No. 2007-P-00009. February 28. Washington, DC: EPA. At www.epa.gov/oig/reports/2007/20070228-2007-P-00009.pdf.

Ernst, H. R. 2003. *Chesapeake Bay Blues: Science, Politics, and the Struggle to Save the Bay*. Lanham, MD: Rowman & Littlefield Publishers, Inc.

Horton, T., and W. M. Eichbaum. 1991. *Turning the Tide: Saving the Chesapeake Bay*. Washington, DC: Island Press.

Landy, M. K., M. M. Susman, and D. S. Knopman. 1999. *Civic Environmentalism in Action: A Field Guide to Regional and Local Initiatives*. January. Washington, DC: Progressive Policy Institute. At www.ndol.org/documents/Civic_Enviro_Full_Report.pdf.

Maryland Department of Natural Resources (DNR). 2001. "Governor Glendening Announces New Action to Protect Chesapeake Bay Blue Crabs—New Regulations Achieve 15 Percent Reduction in Harvest Effort Ahead of Schedule." December 7. Annapolis, MD: Maryland DNR. At http://dnrweb.dnr.state.md.us/dnrnews/pressrelease2001/120701.html.

Maryland General Assembly (MGA). 2006. "Fiscal and Policy Note, Chesapeake Bay—Cleanup and Restoration—Federal Funding." Department of Legislative Services. HJ7. At http://mlis.state.md.us/2006rs/fnotes/bil_0007/hj0007.pdf.

Stoppkotte, K. 2001. "Chesapeake Bay Watermen Question Limits on Crab Harvests." *National Geographic Today*. May 10. At http://news.nationalgeographic.com/news/2001/05/0510_crabbing.html.

Swanson, A. 2003. *The Chesapeake Bay: Lessons Learned from Managing a Watershed*. Annapolis, MD: CBC. An abridged copy of this report is available at http://usinfo.state.gov/journals/itgic/0404/ijge/gj07.htm.

Trust for Public Land and Chesapeake Bay Commission (TPL and CBC). 2001. *Keeping Our Commitment: Preserving Land in the Chesapeake Watershed*. Washington, DC, and Annapolis, MD: TPL and CBC. At www.tpl.org/content_documents/chesapeake_report.pdf.

Walsh, K. L., and R. A. Kohn. 1998. "Pollution in the Chesapeake Bay." Site maintained by R. A. Kohn, Department of Animal and Avian Sciences, University of Maryland, College Park. At www.agnr.umd.edu/Nutrients/Bay/#Ref.

Wennersten, J. 2001. *The Chesapeake: An Environmental Biography*. Baltimore: Maryland Historical Society.

Wilkinson, J. n.d. *Managing the Blue Crab: Seeking a Solution to an Uncertain Crisis*. Charlottesville, VA: University of Virginia Institute for Environmental Negotiation. At www.virginia.edu/ien/vnrli_update/docs/briefs/blue%20crab.pdf.

Chapter 11

An Ecological Perspective on Management of the Chesapeake Bay

THOMAS L. CRISMAN

The Chesapeake Bay is the largest and best studied estuary in North America. Since 1970, approximately 3,408 refereed papers have been published about the Chesapeake, while the record for other major North American bays/estuaries, including San Francisco (1,005), Delaware (614), Tampa (502), Florida (410), Galveston (368), Mobile (228), and Biscayne (182), pales by comparison. Still, a comprehensive, sustainable management plan for the ecosystem eludes managers because of historic biases in the focus of scientific research and differences in management programs among the three states comprising the Chesapeake watershed that are potentially at odds with one another.

This is not surprising, given the following conditions: (1) horizontal gradients from upper to lower parts of the watershed in geology, chemistry, biology, and degree of urbanization; (2) salinity gradients from freshwater to fully marine along the length of tributary streams and rivers and a north–south axis within the bay proper; and (3) vertical gradients reflecting spatial patterns in atmospheric loading (quantity) of nutrients, acidity, and contaminants, as well as vertical differences within the water column of the bay for light, dissolved oxygen, salinity, and nutrient availability. Horizontal and vertical heterogeneity in the physical and chemical environment have laid the foundation for the high biotic (related to living flora and fauna) productivity and diversity of the bay; however, sustainable management of any aquatic ecosystem must consider both ecosystem structure and function within an integrated horizontal and vertical management approach (Crisman et al. 2005).

Ecosystem Structure versus Function

To understand the structure of an ecosystem, its biota must be identified and described, and then the system's physical and chemical properties, which control the composition and distribution of plant and animal species, are assessed.

Biological Structure

Given the economic importance of fish and seafood harvests from Chesapeake Bay, studies of biotic oscillations within the food web, especially for commercially and recreationally important species, are useful. Within the fish fauna, American eels, river herring, and red drum continue to be threatened, but croaker, striped bass, white perch, and summer flounder are among the species that have rebounded or are slowly recovering, following catastrophic declines during the past century (Horton 2003). For commercially important invertebrates, however, the situation is bleaker. Both blue crabs (Hines et al. 2003) and native oysters, *Crassostrea virginica* (Jordan et al. 2002), have declined precipitously within the past two decades. Only hard clams (*Mercenaria mercenaria*) in the lower bay increased dramatically during the same period (Arnold et al. 2004).

Such changes reflect significant alterations in both the taxonomic composition (number of species of plants and animals in an area) and productivity of primary producers and energy transfer within the food web of Chesapeake Bay. Since at least the early 1980s, the bay's pelagic zone (open water offshore) has experienced a dramatic increase in biomass of phytoplankton (microscopic algae suspended in the water column) (Marshall et al. 2002). Harmful algal blooms, especially by the dinoflagellates (single-celled organisms with two flagella, whiplike appendages enabling them to swim) *Pfiesteria piscicida* and *Prorocentrum minimum*, have been noted throughout the bay since 1997 (Gilbert et al. 2001). See chapter 10 for a discussion of the deleterious effects of *Pfiesteria piscicida*.

In addition to increased phytoplankton biomass in tidal areas of several rivers (Burchardt and Marshall 2004), nearshore littoral (between high and low watermarks) areas have experienced a significant reduction in the aerial (overall) extent of sea grasses and submerged aquatic vegetation since the 1970s (Moore et al. 2003). Closer to shore, native emergent macrophyte taxa (*Spartina*, *Typha*, *Panicum*), visible aquatic plants that produce oxygen and provide food and cover for fish and other wildlife, are being replaced by the invasive reed *Phragmites australis* (Rooth et al. 2003). When native macrophytes are driven out by hardier plants, fish and waterfowl that depend on the native plants are often forced to find new habitats. Fortunately, in the case of this reed, no negative results have been observed.

The *Phragmites australis* reed is one of only 6 percent of the bay's 196 invasive plant species for which ecological studies have been conducted (Ruiz et al. 1999). Invasive species have entered the food web at all levels from microbes to fish, but the greatest attention has been paid to the possible purposeful introduction of a Pacific oyster species to enhance depleted stocks of the native *Crassostrea virginica* (Lutz 2003) and to the accidental introductions of two mollusks, the Asiatic clam (*Corbicula fluminea*) and zebra mussel (*Dreissena polymorpha*), along with their observed and potential environmental impacts (Phelps 1994).

Physical and Chemical Structure

Ecosystem structure also includes physical and chemical attributes, those factors controlling the composition and distribution of plant and animal species. The physical structure of the pelagic zone of the bay is controlled by patterns of thermal stratification within the water column and associated vertical profiles for dissolved oxygen. The aerial extent of hypoxia (oxygen concentration stressful to an organism) or anoxia (absence of oxygen) in bottom waters of the central bay has increased markedly, to a point that the Chesapeake Bay is now the most hypoxic estuary of the Mid-Atlantic region (Uphoff 2003). Even nearshore areas and the mouths of rivers (Breitburg et al. 2003) are experiencing extensive hypoxia with associated impacts on food webs. Deep-water refugia (places where organisms can exist or be protected) from predation have been eliminated for vertically migrating zooplankton (microscopic, free-swimming animals). Detritivores (creatures that feed on dead organic material in bottom sediments) have been displaced to oxygenated, generally shallower waters.

Throughout the bay, the bottom configuration and sediment composition have been altered by increased delivery of watershed-derived erosion products and progressive cultural (modified by humans) eutrophication (excessive growth response of algae and other aquatic plants to elevated loadings of particular nutrients). Highly turbid river water has displaced macrophytes to shallower locations near deltas, where they can receive the sunlight they need for survival (Moore et al. 2003). After the Chesapeake region was colonized, progressive deforestation of its watershed increased delivery of inorganic sediments to nearshore areas of the bay. Since 1920, sediments within those areas have been redistributed and delivered to deep water (Cronin and Vann 2003). Sediments in the more protected nearshore areas and in deep-water sections of the bay have become increasingly more organic (consisting of living organisms and compounds they produce) due to eutrophication. Bioaccumulation of contaminants, such as heavy metals and organic chemicals exported from agricultural and urban areas of the watershed, directly affect the food web (Marshall et al. 2002), as well as salinity changes in the bay, reflecting alteration of the hydrological cycle through land use changes and sea level rises associated with climate change.

Ecosystem Function

One of the most important functions of ecosystems is the transformation and storage of chemical and, to a lesser extent, physical loadings from terrestrial sources. Nutrients entering the ecosystem are rapidly incorporated into the autotrophic (able to make their own food via photosynthesis) biomass and made available for consumption by heterotrophs (organisms that consume organic matter and compounds for energy); they, in turn, constitute numerous pathways in the food web. Available for export to the open ocean are all the ecosystem's incorporated nutrients that are not parts of the living biomass of plants and animals or the accumulating sediment pool.

Unfortunately, the quantity and rate of delivery of nutrients to the Chesapeake from urban and agricultural runoff have resulted in cultural eutrophication (Boesch et al. 2000) and a short-circuiting of the food web. Energy transfer within the open water has been hindered by the presence of algal taxa not favored by zooplankton grazers (Gilbert et al. 2001) and a marked reduction in one of the main fish grazers on algae, menhaden (Uphoff 2003). Progressively severe hypoxia in the area where the sediment and water meet has reduced the recycling of nutrients and direct consumption of organic detritus from deep areas of the bay (Dauer et al. 2000). Thus, while the bay's primary plant and animal life have increased steadily, much of this energy is not available to the food web; instead, it decomposes into organic sediments that increase oxygen demand and increase the aerial extent of hypoxia.

Nearshore areas of the Chesapeake were traditionally extremely important for the transformation of watershed-derived organic matter and thereby regulated the community metabolism of the pelagic zone of the bay. Depletion of the native oyster, the principal grazer of the near shore, has profoundly reduced the ability of the grazer community to transform suspended matter into invertebrate biomass via filter feeding (Jordan et al. 2002). While oysters in the bay proper have crashed, their role as processors of watershed-derived organic matter in rivers and deltas has been filled in part by filter-feeding activities of the invasive clam, *Corbicula fluminea* (Phelps 1994). The effects of the filter processing community's movement to freshwater tributaries on the collective metabolism of open water are unclear. As with open water, nearshore crab populations have also been affected by hypoxic conditions, limiting their ability to process organic detritus accumulating on the bottom.

Vertical versus Horizontal Management

The structure and function of Chesapeake Bay are controlled by vertical and horizontal interactions within the ecosystem and with adjacent ecosystems. Traditionally, management and restoration of aquatic ecosystems has been biased toward a vertical perspective of ecosystems, but effective management must also consider horizontal interactions and their linkage with vertically oriented processes (Crisman et al. 2005).

Vertical Interactions

Vertical interactions include those between the atmosphere and water, between sediment and water, and within the water column.

Atmosphere–Water Interactions

The atmosphere can be a significant loader of nutrients and toxics to aquatic ecosystems, leading to accelerating eutrophication and accumulation of contaminants within the food web and in the sediments, respectively. Entering as particulates or aerosols, such chemical substances can be derived from local, regional, or intercontinental sources. Acid rain from local and regional sources has reached Chesapeake Bay for decades (Horton 2003). Heavy metal, mercury (Mason et al. 1999), polychlorinated biphenyls, and particulate organic/elemental carbon (Brunciak et al. 2001) loadings can be high, especially in nearshore areas. The vast majority of atmospheric loadings, 95 percent of mercury, for example (Mason et al. 1999), remain in the bay and are ultimately deposited in sediments following cycling and bioaccumulation within the food web.

Sediment–Water Interactions

Management practices focused on sediment–water interactions seek to reduce nutrient release and increase permanent storage through either permanent burial or decreased nutrient availability for biotic production within the water column. Sealing of bottom sediments with physical barriers (silt and clay) and chemical barriers (alum) deposited on top of or injected into surface sediments have proved effective at reducing nutrient and contaminant recycling in small lakes but are not likely to be feasible for vast areas of the Chesapeake Bay (Cooke et al. 1993). The same applies to inactive chemical nutrients within the water column that form precipitates deposited on the lake bottom, promoting long-term nutrient storage in sediments. Paleoecological analyses of sediment cores have demonstrated that central bay sediments have become more organic and enriched with nutrients, heavy metals, and contaminants since the late nineteenth century (Zimmerman and Canuel 2002). Organic enrichment, reflecting high levels of cultural eutrophication, have led to hypoxia of bottom waters from increased decomposition and caused major changes in the benthic faunal community (organisms living on the bottom or in bottom sediments) of the bay (Zimmerman and Canuel 2000).

Water Column Interactions

The Chesapeake, the most hypoxic estuary in the Mid-Atlantic region (Uphoff 2003), has experienced a lessening of the pelagic food web to the upper, well-oxygenated portions of the water column in many areas. A method that has been effective at reoxygenating small lakes experiencing cultural eutrophication calls for selectively aerating the lower portion of the water column or the total water column during the annual thermal stratification period of the year (Cooke et al. 1993). Unfortunately, no similar strategy for large systems such as the Chesapeake exists. Deep-water oxygen levels in the bay can be appreciably improved only through reversal of the eutrophication process.

Use of top-down biomanipulation to alter the structure and biomass of pelagic food webs is a relatively recent alternative to traditional bottom-up management approaches, which focus on reducing nutrient availability to the primary production base of the ecosystem (Carpenter and Kitchell 1988). Most biomanipulation schemes seek to restructure fish communities either (1) to promote a cascading of interactions among trophic levels (organisms functioning comparably in the predator–prey relationship) of the food web, leading to altered taxonomic composition and/or biomass reduction of phytoplankton through enhanced predator fish populations and direct grazing on phytoplankton by plant-eating fish, or (2) to decrease nutrient cycling from sediments by elimination of the bioturbation actions (disturbance of the sediment) from benthivorous fish (bottom feeders). All biomanipulation schemes attempt to alter the biotic structure of fish communities directly, so that the function of pelagic food webs is changed independently from controls over nutrient availability. When scientists add to or subtract from numbers of certain species of plant-eating fish in a community, the fishes' consumption of particular plants changes how all elements of the food web operate—despite nutrient levels in the water.

The biomanipulation approach holds great promise for management of the Chesapeake Bay, where nutrient loading from its enormous watershed is continuing to be difficult to reduce and key components of energy transfer, such as the menhaden species, are threatened. If biomanipulation can produce a harvestable product, such as safe, edible fish, this approach will be considered successful.

Horizontal Interactions

Horizontal interactions are those between a body of water and its surrounding areas, including the watershed, the littoral zone, coastal wetlands, and the pelagic zone.

Watershed Interactions

The Chesapeake has the largest ratio of watershed area to water volume of any major estuary in the world (Horton 2003). Paleoecological analyses of sediment cores from the bay have shown that cultural eutrophication and hypoxia in deep waters are directly related to forest clearance and the subsequent partitioning of land into agricultural and urban use (Zimmerman and Canuel 2002). Local geology sets the background conditions for chemical release from the watershed (Liu et al. 2000), but anthropogenic point and nonpoint loadings far exceed the role of geology in most cases. Point loadings are specific, visible places where chemical wastes are released; nonpoint loadings cannot be traced to particular sources because they are generated by numerous activities and seep into the ground or into runoff. Although urban areas can be major sources of toxic contaminants (Hall et al. 2002), pesticide, heavy metal, and nitrogen loadings have been associated with agricultural activities (Dauer et al. 2000). Significant reductions in point source phosphorus (58 percent) and nitrogen (28 percent) loadings have been achieved (Boesch et al. 2000) but fewer for nonpoint sources (19 percent phosphorus, 15 percent nitrogen). Nonpoint nutrient loadings to the bay are likely to continue to increase as human populations grow throughout the watershed (D'Elia et al. 2003) and as intense developmental pressures increase on and near the shore (Dauer et al. 2000). Additionally, land use has profoundly altered the discharge of freshwater and sediments into the bay, leading to changes in autotrophic production through altered light regimes and the distribution and composition of food webs through salinity and sediment compositional changes (Horton 2003).

Littoral Zone and Coastal Wetland Interactions

The littoral zone, the region along the shore, is the key link determining the success of any management plan for the Chesapeake Bay. It acts to trap sediments, reduce erosion of the shore, and, most importantly, to transform nutrient loadings from the watershed into biomass and detritus that, in turn, are exported to the pelagic zone. The littoral zone and coastal wetland exercise ultimate control over the metabolism of the open water areas of the bay. Structural changes in the nearshore ecosystem, including loss of submersed aquatic vegetation and drastic declines in the principal filter feeder (oysters) and detritivore (blue crabs), have impaired its function as a transformer (much like a kidney), resulting in a direct pass-through of watershed nutrient, contaminant, and sediment loadings to the pelagic zone of the bay, causing cultural eutrophication, toxic algal blooms, alteration of the food web, and development of extensive hypoxia in deep waters.

While the littoral zone has witnessed the greatest alteration of biotic structure of any region of the Chesapeake Bay, it is not clear how its function will be affected in the long term. Aside from controversy about introduction of exotic oysters into the bay, the invasive clam (*Corbicula fluminea*) is a highly efficient filter feeder in the freshwater portions of the bay (Phelps 1994) and may have replaced, in part, the contribution of oysters at trapping suspended organic matter. A second invasive bivalve and highly efficient filter feeder, the zebra mussel (*Dreisenna polymorpha*), is moving down the Susquehanna River to the Chesapeake (Horton 2003). Where *Corbicula* clam populations have been able to reduce phytoplankton biomass significantly, macrophytes have become reestablished, and duck populations have increased partly because they eat the clams (Phelps 1994).

The invasive reed, *Phragmites australis*, has been spreading along the shore of the Chesapeake, colonizing areas where native aquatic vegetation has disappeared and actively replacing macrophytes *Panicum*, *Spartina*, and *Typha* (Rooth et al. 2003). While some have called for this reed's eradication, benthic microalgal biomass and benthic fauna were not significantly different between *Phragmites* and *Spartina* stands (Posey et al. 2003), and *Phragmites* appears to be more efficient at trapping both nutrients and sediments than native species (Rooth et al. 2003). With enhanced sedimentation in *Phragmites* stands, this plant may play an additional beneficial role by keeping pace with sea level rise and protecting the shoreline from erosion.

With 196 invasive plant species identified in tidal waters of the bay (Ruiz et al. 1999) and the likelihood of additional purposeful introduction of exotics for economic purposes, it is important that future management scenarios for the bay consider that such species may play major positive, as well as negative, roles in biomanipulation of food webs. As the original biotic structure of the bay has been changed irrevocably, future management must be based on ecosystem function and the realization that multiple species can perform similar functional attributes if managed properly.

Pelagic Zone Interactions

The pelagic zone of Chesapeake Bay displays a great deal of aerial heterogeneity in productivity and salinity, reflecting regional differences in watershed loadings of nutrients, contaminants, sediments, and freshwater. Interstate differences in management approaches for the bay largely reflect ecosystem structural and functional differences between the upper and lower bay. While sediments of the pelagic zone are the ultimate repositories of watershed inputs, the Chesapeake Bay itself interacts with the continent, the coastal ocean, and beyond. Nutrients and contaminants are exported from the bay to the continent through the feeding and bioaccumulation of substances in migrating and overwintering waterfowl and seafood harvest. Similar movements of such substances, including mercury (Mason et al. 1999), to the coastal and open ocean occur through tidal flushing and movements of biota for completion of their life cycles. Exotic species' introduction to the bay in ship ballast water has the potential for spreading invasive plant and animal taxa along the coastal Atlantic, just as ballast water from the Chesapeake might have been the inoculation source of the ctenophore *Mnemiopsis* (a transparent, free-swimming invertebrate), into the Black Sea (Purcell et al. 2001).

Rehabilitation versus Restoration

While seeking restoration of the Chesapeake Bay is a noble goal, rehabilitation is within the realm of reality. The biotic structure of the ecosystem has been altered irrevocably through changes in the physical and chemical attributes of the habitat and the introduction of invasive species. Above all, the memory of the system, the sediment record, has been altered to reflect (1) sediment, contaminant, and nutrient loadings from historical changes in land use; (2) the ability of coastal wetlands and the littoral zone to transform and store them; and (3) the ultimate response of the pelagic zone through eutrophication and development of hypoxic bottom waters.

As demonstrated through numerous paleoecological reconstructions from sediment cores (Cronin and Vann 2003), environmental change in Chesapeake Bay is not unidirectional or irreversible. It is a moving target, reflecting anthropogenic alterations surrounding the bay and within the bay proper, superimposed on cyclic climate change in hydrology, temperature, and sea level rise. Rather than attempting the impossible—restoration to conditions from some unspecified point in the past—it is better to set realistic goals for the bay that recognize the changed biological, physical, and chemical attributes of the system and then to set a realistic plan to reach them. Any long-term adaptive management plan for the Chesapeake will fail unless it recognizes that management and associated goals for the bay are constantly changing and that short-term reassessment of the ecosystem is needed regularly.

References

Arnold, G. L., M. W. Luckenbach, H. V. Wang, and J. Shen. 2004. "Predicting the Effects of a Changing Coastal Landscape on Large-scale Clam Aquaculture." *Journal of Shellfish Research* 23:281.

Boesch, D. F., R. B. Brinsfield, and R. E. Magnien. 2000. "Chesapeake Bay Eutrophication: Scientific Understanding, Ecosystem Restoration and Challenges for Agriculture." *Journal of Environmental Quality* 30:303–20.

Breitburg, D. L., A. Adamack, K. A. Rose, S. E. Kolesar, M. B. Decker, J. E. Purcell, J. E. Keister, and J. H. Cowan, Jr. 2003. "The Pattern and Influence of Low Dissolved Oxygen in the Patuxent River, a Seasonally Hypoxic Estuary." *Estuaries* 26:280–97.

Brunciak, P. A., J. Dachs, T. P. Franz, C. L. Gigliotti, E. D. Nelson, B. J. Turpin, and S. J. Eisenreich. 2001. "Polychlorinated Biphenyls and Particulate Organic/Elemental Carbon in the Atmosphere of Chesapeake Bay, USA." *Atmospheric Environment* 35:5663–77.

Burchardt, L., and H. G. Marshall. 2004. "Monitoring Phytoplankton Populations and Water Quality Parameters in Estuarine Rivers of Chesapeake Bay (USA)." *Oceanological and Hydrobiological Studies* 33:55–64.

Carpenter, S. R., and J. F. Kitchell. 1988. "Consumer Control of Lake Productivity." *BioScience* 38: 764–69.

Cooke, G. D., E. B Welch, S. A. Peterson, and P. R. Newroth. 1993. *Restoration and Management of Lakes and Reservoirs*. Boca Raton, FL: Lewis Publishers.

Crisman, T. L., C. Mitraki, and G. Zalidis. 2005. "Integrating Vertical and Horizontal Approaches for Management of Shallow Lakes and Wetlands." *Ecological Engineering* 24:379–89.

Cronin, T. M., and C. D. Vann. 2003. "The Sedimentary Record of Climatic and Anthropogenic Influence on the Patuxent Estuary and Chesapeake Bay Ecosystems." *Estuaries* 26:196–209.

Dauer, D. M., S. B. Weisberg, and J. A. Ranasinghe. 2000. "Relationships between Benthic Community Condition, Water Quality, Sediment Quality, Nutrient Loads, and Land Use Patterns in Chesapeake Bay." *Estuaries* 23:80–96.

D'Elia, C. F., W. R. Boynton, and J. G. Sanders. 2003. "A Watershed Perspective on Nutrient Enrichment, Science and Policy in the Patuxent River, Maryland: 1960–2000." *Estuaries* 26:171–85.

Gilbert, P. M., R. Magnien, M. W. Lomas, J. Alexander, C. Fan, E. Haramoto, M. Trice, and T. M. Kana. 2001. "Harmful Algal Blooms in the Chesapeake and Coastal Bays of Maryland, USA: Comparison of 1997, 1998, and 1999 Events." *Estuaries* 24:875–83.

Hall, L. W., Jr., R. D. Anderson, and R. W. Alden, III. 2002. "A Ten-Year Summary of Concurrent Ambient Water Column and Sediment Toxicity Tests in the Chesapeake Bay Watershed: 1990–1999." *Environmental Monitoring and Assessment* 76:311–52.

Hines, A. H., J. L. D. Davis, A. Young-Williams, Y. Zohar, and O. Zmora. 2003. "Assessing Feasibility of Stock Enhancement for Chesapeake Blue Crabs (*Callinectes sapidus*)." *Journal of Shellfish Research* 22:335.

Horton, T. 2003. *Turning the Tide*. Covelo, CA: Island Press.

Jordan, S. J., K. N. Greenhawk, C. B. McCollough, J. Vanisko, and M. L. Homer. 2002. "Oyster Biomass, Abundance and Harvest in Northern Chesapeake Bay: Trends and Forecasts." *Journal of Shellfish Research* 21:733–41.

Liu, Z. J., D. E. Weller, D. L. Correll, and T. E. Jordan. 2000. "Effects of Land Cover and Geology on Stream Chemistry in Watersheds of Chesapeake Bay." *Journal of the American Water Resources Association* 36:1349–66.

Lutz, C. G. 2003. "Asian Oysters in the Chesapeake Bay: A New Beginning?" *Aquaculture Magazine* 29:28–32.

Marshall, H. G., M. F. Lane, and K. K. Nesius. 2002. "Long-term Phytoplankton Trends and Related Water Quality Trends in the Lower Chesapeake Bay, Virginia, U.S.A." *Environmental Monitoring and Assessment* 81:349–60.

Mason, R. P., N. M. Lawson, A. L. Lawrence, J. J. Lerner, J. G. Lee, and G. R. Sheu. 1999. "Mercury in the Chesapeake Bay." *Marine Chemistry* 65:77–96.

Moore, K. A., B. A. Anderson, D. J. Wilcox, R. J. Orth, and M. Naylor. 2003. "Changes in Seagrass Distribution as Evidence of Historical Water Quality Conditions." *Gulf of Mexico Science* 21:142–43.

Phelps, H. L. 1994. "The Asiatic Clam (*Corbicula fluminea*) Invasion and System-level Ecological Change in the Potomac River Estuary near Washington, DC." *Estuaries* 17:614–21.

Posey, M. H., T. D. Alphin, D. L. Meyer, and J. M. Johnson. 2003. "Benthic Communities of Common Reed *Phragmites australis* and Marsh Cordgrass *Spartina alterniflora* Marshes in Chesapeake Bay." *Marine Ecology Progress Series* 261:51–61.

Purcell, J. E., T. A. Shiganova, M. B. Decker, and E. D. Houde. 2001. "The Ctenophore *Mnemiopsis* in Native and Exotic Habitats: U.S. Estuaries versus the Black Sea Basin." *Hydrobiologia* 451:145–76.

Rooth, J. E., J. C. Stevenson, and J. C. Cornwell. 2003. "Increased Sediment Accretion Rates following Invasion by *Phragmites australis*: The Role of Litter." *Estuaries* 26:475–83.

Ruiz, G. M., P. Fofonoff, A. H. Hines, and E. D. Grosholz. 1999. "Non-indigenous Species as Stressors in Estuarine and Marine Communities: Assessing Invasion Impacts and Interactions." *Limnology and Oceanography* 44:950–72.

Uphoff, J. H. 2003. "Predator–Prey Analysis of Striped Bass and Atlantic Menhaden in Upper Chesapeake Bay." *Fisheries Management and Ecology* 10:313–22.

Zimmerman, A. R., and E. A. Canuel. 2000. "A Geochemical Record of Eutrophication and Anoxia in Chesapeake Bay Sediments: Anthropologic Influence on Organic Matter Composition." *Marine Chemistry* 69:117–37.

Zimmerman, A. R., and E. A. Canuel. 2002. "Sediment Geochemical Records of Eutrophication in the Mesohaline Chesapeake Bay." *Limnology and Oceanography* 47:1084–93.

Chapter 12

Murky Waters and Murky Policies

Costs and Benefits of Restoring Chesapeake Bay

Stephen Polasky

Located in the heart of the Mid-Atlantic region, Chesapeake Bay, the largest estuary in the United States, has been a central feature in defining the region's sense of place, culture, and economic base. Chesapeake Bay has provided the surrounding area with rich sources of natural wealth, in terms of seafood harvests, transportation, recreation, and scenic beauty.

While Chesapeake Bay has greatly influenced the lives of people in the region, people have also greatly influenced the bay. Harvesting of oysters, crabs, and many fish species, often at unsustainable levels, has changed the biological composition of bay species. The chemical and biological compositions of the bay have been changed by nutrient and pesticide runoff from agriculture, deposition of nitrogen from air pollution generated from burning fossil fuels for electricity production and transportation, and household and industrial wastewater. The qualities that made Chesapeake Bay such a valuable natural resource for harvesting fish and shellfish, recreation, and scenic beauty are now threatened by human-induced biological and chemical changes to the bay.

Recognition that human actions pose a threat to Chesapeake Bay has led to efforts to reduce damage and restore the bay, as detailed in chapter 10. For the past twenty-five years, the Chesapeake Bay Commission (CBC) has spearheaded efforts to restore the bay to a more desirable condition by reducing pollution entering the bay and reducing pressures from overharvesting. Agreements negotiated by members of the CBC—the *1987 Chesapeake Bay Agreement* and *Chesapeake 2000* (C2K)—set forth a number of specific targets for water quality, land use, habitat, and species populations to enable recovery of the bay. These targets were often quite ambitious, such as the 1987 target that set a 40-percent reduction in the inflow of nitrogen into the bay by 2000. As it turned out, this ambitious target and others have not been met.

A large part of the reason for this failure is that the 1987 and 2000 agreements did not provide specific regulatory mechanisms to achieve these targets. Restoring Chesapeake Bay to conditions experienced before overharvesting and pollution caused major changes in the bay would yield large benefits. However, achieving such a restoration will require major changes in human activities that contribute nutrient or toxic inputs into the bay or to overharvesting, including fisheries management, agriculture, industry, and households. These changes will impose large costs on certain segments of society. Without policies in place that provide positive incentives for change or penalties for nonremediation of status quo pollution levels, the Chesapeake Bay is unlikely to be restored. In this case, agreements to restore Chesapeake Bay will remain idealistic statements whose lofty ambitions will not be matched by actual recovery of the bay.

Chesapeake Bay recovery, though difficult, is possible if citizens decide it is worth doing and devote sufficient resources to do so. Putting in place such a recovery effort requires two things. First, advocates must convince political leaders and the general public that recovery of the bay is worth the sacrifices necessary to make it happen, that the benefits of recovery are greater than the costs necessary to achieve it. Second, people whose decisions affect Chesapeake Bay—such as industrial firms and landowners whose land management practices contribute to pollution and boat owners who contribute to overharvesting—must have incentives to change their behavior in ways that are consistent with recovery of the bay. This second step requires an understanding of which human actions negatively impact the bay and which regulations or incentives will induce people to redirect their management practices from actions that harm the bay to actions that are consistent with recovery of the bay. The following sections address these two challenges.

Benefits Provided by Chesapeake Bay

Chesapeake Bay provides a broad range of benefits to humans, from various types of seafood to scenic beauty. The degree to which these benefits are provided depends in large part on the state of the ecosystem. Recent human-induced changes in the bay have reduced the quantity and quality of many of the benefits provided by the bay. For example, overharvesting and pollution have reduced the availability of many species, while declining water quality has reduced the bay's recreational benefits as well as aesthetic qualities. Historical records indicate what the bay once provided, and, if allowed to recover, what it possibly could provide again.

Some of the Chesapeake Bay's benefits, such as commercial fish harvests, are relatively easy to quantify and value in monetary terms. Other benefits, notably aesthetic values, are extremely difficult to pin down using any type of objective measure and very difficult to quantify in monetary terms. In addition, some people feel that humans should not alter ecosystems or harm species within the ecosystem. These people may place a higher value on having more natural or unaltered ecosystems. Such expressions of intrinsic value are also difficult to measure or value in monetary terms. Any attempt to provide a systematic accounting of the benefits of ecosystem recovery efforts will face such difficulties, making a complete accounting in a single metric of value problematic. Still, well-documented evidence about the bay's benefits, whether it is best expressed in qualitative terms, such as number of species or water quality measures, or in monetary values will be needed to convince decision makers and the general public that the costs to recover the bay are worth it.

Harvesting oysters, crabs, fish, and other seafood is probably the most readily apparent benefit provided by Chesapeake Bay. The first goal stated in *Chesapeake 2000*, is to "Restore, enhance and protect the finfish, shellfish and other living resources, their habitats and ecological relationships to sustain all fisheries and provide for a balanced ecosystem" (CBC 2000). Of course, "living resources" may be valued intrinsically, but in the case of Chesapeake Bay, harvesting of these living resources supports a large industry. From 1950 to 2004, the value of commercial harvests from Chesapeake Bay was $5.3 billion (National Marine Fisheries Service 2005). The top species in terms of value of harvest in 2004 were sea scallops ($92.6 million); blue crab ($60.9 million); menhaden ($24.4 million); clams ($6.4 million); flounder ($5.9 million); and striped bass ($5.2 million) (figures reported in constant 2004 dollars, National Marine Fisheries Service 2005). Total harvests of all species from the bay more than doubled from 1950 to 1990. However, since 1990, production has fallen dramatically. (See fig. 12.1.)

For certain species, such as oysters, the decline in production began much earlier. In fact, oyster production has been in decline since the late 1800s. It has been in continuous decline since 1950, when the National Marine Fisheries Service began record keeping (fig. 12.2). In 1950, oyster harvests made up 44 percent of the total harvest by value. By 2004, oyster harvesting had all but disappeared from Chesapeake Bay.

Other valuable species, such as blue crab, are also in decline. In 1990, blue crab harvests from the Chesapeake Bay were 46,493 metric tons. By 2000, blue crab harvests had declined to 23,448 metric tons. Blue crab harvests had recovered somewhat by 2004 to 27,882 metric tons.

Though it is impossible to predict exactly what additional value would be created in Chesapeake Bay commercial fisheries if populations of oyster, blue

Year	Metric Tons of Harvest	Value of Harvest (in millions of constant 2004 US$)
1950	172,771	196.0
1960	197,676	222.5
1970	285,970	197.4
1980	326,282	303.2
1990	398,286	240.0
2000	223,220	188.9
2004	240,888	209.8

FIGURE 12.1. Total Harvest of All Species from the Chesapeake Bay (NOAA 2005; USBL/BLS 2007). Note: Prices deflated using the consumer price index for all urban consumers (CPI-U).

crab, and other species were restored close to historical levels, it is safe to say that the harvest values have been severely depressed from reduced populations of these valuable species. If oysters and blue crab managed to recover to their population levels of fifty or one hundred years ago, the value of harvests would likely increase by tens or hundreds of millions of dollars from current levels. Given the large-scale changes in Chesapeake Bay in terms of species composition caused by overharvesting of some species, the introduction of exotic species, the presence of diseases (especially those affecting oysters), and changes in water quality, it may not be possible to recover population levels of some species. On the other hand, harvests of other species have increased in recent years. Scallop harvests have gone from insignificant levels in the 1950s and 1960s to becoming the most valuable species harvested in 2004. Of course, scallops may only be the latest species subject to severe harvest pressures and may follow the way of the oyster and blue crab if proper precautions in management are not taken.

Changes in species abundance also affect the values of recreation. Sport fisheries and shellfish harvesting in Chesapeake Bay also generate benefits, though statistics on sport fisheries are not as readily available as for commercial fisheries. Other important recreational activities on Chesapeake Bay

Year	Metric Tons of Harvest	Value of Harvest (in millions of constant 2004 US$)
1950	13,587	87.0
1960	12,297	123.2
1970	11,190	73.4
1980	10,338	67.2
1990	2,048	23.0
2000	1,148	8.4
2004	40	0.4

FIGURE 12.2. Harvest of Oysters from the Chesapeake Bay (NOAA 2005; USBL/BLS 2007). Note: Prices deflated using CPI-U.

include boating and swimming. Lipton (2003) estimated that the value of improved water quality to individual boaters in Maryland was $146. Declines in water quality diminish the value of these activities and are also linked to potential health problems. A 1997 outbreak of *Pfiesteria* that killed tens of thousands of fish triggered alarms about possible human health consequences, though human health consequences are a subject of some debate (see chapter 10, this volume).

Many people like to live near water, and property values on water typically reflect a large premium. This premium is increased with improved water quality. For example, Leggett and Bockstael (2000) estimated that benefits from reducing fecal coliform from current levels to state water quality standards for 494 residential waterfront properties in Anne Arundel County to be $12 million (with a 95 percent confidence interval ranging between $3 million and $20 million). This multi-million-dollar value was generated for one particular component of water quality for a small portion of properties in one county fronting on Chesapeake Bay. A more inclusive study, including additional aspects of water quality and more properties fronting Chesapeake Bay, would likely generate far higher estimates.

Chesapeake Bay is a natural asset that generates large benefits in terms of seafood production, recreation, and aesthetics. Overfishing, excessive nutrient inputs, and other forms of pollution diminish these values. While a precise estimate of the dollar value of damages caused by overfishing and pollution is not available, these losses clearly run into hundreds of millions of dollars. When aesthetic qualities, such as the role that the Chesapeake Bay plays in defining a sense of place and regional identity, are factored in, along with its intrinsic value, a clear case for Chesapeake Bay recovery can be made. However, this case must be very strong to outweigh the high costs of recovery and the political opposition that such costs engender.

Reducing Pollution and Overharvesting in Chesapeake Bay

The twin problems facing Chesapeake Bay, pollution and overfishing, can be corrected, but to do so, the people living near the bay will have to alter the ways they currently conduct their affairs. To make these changes, at least in the short run, some segments of society will have to pay. Addressing overfishing may require closing off parts of the bay to fishing for a period of time, which will cause commercial fishermen financial hardship. Farmers may be asked to use less fertilizer, and municipal sewage systems may need to upgrade their treatment plants to reduce nutrient inputs. Governor of Virginia Mark Warner recently announced plans to commit $150 million to reduce nitrogen and phosphorus emissions from sewage treatment plants. The Chesapeake Bay Foundation, however, estimates that Virginia will have to pay $2.3 billion to cover its share of pollution costs (Shear 2005). Overall, the CBC estimated that it would cost $18.7 billion to meet commitments established in the *Chesapeake 2000* agreement (CBC 2002). The prospect of such high costs makes some people and groups resist change, even when societal benefits far exceed societal costs. If groups who must bear the costs are politically powerful and if their concerns are not adequately addressed, then changes that will lead to restoration of the bay are unlikely to occur.

Of the two problems facing Chesapeake Bay, overfishing is, in principle, the easier one to address. The people who are the source of the problem are also the ones who would benefit most from correcting it. Income from fishing increases when populations of harvested species are allowed to recover. However, achieving long-term benefits typically requires short-term costs in the form of reduced harvests for a period of time.

Unregulated fisheries are classic examples of the "tragedy of the commons" (Gordon 1954; Hardin 1968). Open-access resources (those available to anyone who wishes to use them) are subject to overuse because individual resource users do not take account of the impacts that their uses have on the value of use of the resource to others. In fisheries, overharvesting can be particularly severe. Though all fishermen would be better off conserving some fish to allow for successful regeneration and healthy future fish stocks, no individual fisherman has an incentive to conserve. A fish not harvested by one fisherman will just be harvested by another.

Overcoming the tragedy of the commons typically requires some type of collective action to overcome individual incentives to overharvest. Some resources are successfully managed by resource users themselves (Ostrom 1990). Self-regulation can work when resource users are a relatively small, cohesive group who can restrict access of outsiders to the resource and can devise effective monitoring and enforcement mechanisms among members of the group. In other cases, overcoming the tragedy of the commons requires enforcement of harvest restrictions by government agencies. One means of doing this is to institute a system of individually transferable quotas (ITQs). The quota system limits total allowable harvest. Harvest rights can be traded among fishermen, which allows for greater flexibility and efficiency. Used correctly, ITQs or other regulatory or self-management schemes are good for ecosystems and for fishermen because they prevent overfishing, which allows fish stocks to increase and allows future harvests to increase. For successful management to operate in a depleted fishery, however, fishermen must get through lean times necessary to allow the resource stock to recover. Fujita et al. (2004) proposed a fishery investment fund that would make loans to fishermen to pay for lost income in the short term, which would be repaid from the additional income available from harvesting in a recovered fishery in the longer term.

Solutions for water quality problems in Chesapeake Bay will be tougher to implement than fishery management improvements because the groups responsible for creating the problem are often different from those who benefit from improved water quality. In addition, there are typically a large number of activities and sources that contribute to water quality problems, which complicates designing water quality policy as well as monitoring and enforcing the policy. Chesapeake Bay faces pollution from toxic chemicals and excessive inputs of nitrogen and phosphorus. In what follows, the focus will be on reducing inputs of nitrogen and phosphorus because this is the primary water quality concern and one that illustrates well the main policy issues.

Water Quality Concerns

Chesapeake Bay received an estimated 600 million pounds of nitrogen and 30 million pounds of phosphorus from 1990 to 1992 (USGS 1995). From

1985 to 2004, both nitrogen and phosphorus loadings have declined slightly, though levels of both nutrients remain above targeted reduction levels (CBP 2005). Both the *1987 Chesapeake Bay Agreement* and *Chesapeake 2000* targeted 40-percent reductions in nitrogen and phosphorus inputs into the bay. Meeting these targets will require much greater efforts than have been exerted to date and will impose greater costs. Despite some improvement in lowering nutrient inputs, water quality in Chesapeake Bay remains impaired. In August 2005, 41 percent of the bay had dissolved oxygen levels too low to support fish, which was the largest such dead zone ever recorded in the bay (CBF 2005).

Most estuaries, including Chesapeake Bay, are nitrogen limited; therefore, the excessive input of nitrogen is of primary concern. Reducing nitrogen inputs into the bay presents some formidable challenges. Sources of nitrogen inputs are numerous and diffuse. While some nitrogen inputs come from "point sources," such as municipal wastewater facilities, a majority of the nutrient inputs reaching the bay come from "nonpoint sources," such as leaching of nutrients from agricultural fields and atmospheric deposition. Much of the atmospheric deposition originates from burning of fossil fuels in power plants and vehicles outside the Chesapeake Bay watershed. Some reductions from air deposition will come from Clean Air Act regulations that tighten rules on nitrogen oxide (NO_x) emissions.

Reducing nitrogen leaching from agriculture and other nonpoint sources is also a major thrust of current efforts to improve water quality. Under the Clean Water Act, the Environmental Protection Agency has established water quality standards that require establishment of total maximum daily loads (TMDLs) for a variety of pollutants that cause water quality impairment. To meet TMDL requirements, nonpoint source reductions are necessary. However, relatively little success has been achieved in reducing loadings from nonpoint sources. It is difficult to measure nonpoint source loadings directly. Instead, indirect means of reducing loadings, such as adopting best management practices in agriculture (for example, buffer strips along streams), are typically used because such management practices can be monitored. To date, reducing nutrient loadings from point sources has met with far greater success. Point sources have reduced loadings by 26.5 million pounds per year between 1985 and 2003, despite increasing population in the region (CBP 2005).

If water quality goals for Chesapeake Bay are to be met, even larger reductions in nutrient loadings will need to be accomplished. Since the relatively easy steps to reducing loading have already been taken, remaining steps will be neither easy nor cheap.

Promoting Recovery of Chesapeake Bay

While progress has been made on limiting overfishing and loadings of nutrients and other pollutants, goals for Chesapeake Bay remain far from fulfilled. The *1987 Chesapeake Bay Agreement* and *Chesapeake 2000* established a framework for cooperation among various local, state, and national government authorities. These agreements have also established specific targets for water quality, submerged vegetation, species populations, and other important aspects that determine the condition of Chesapeake Bay. However, establishing a framework for cooperation along with goals and targets is not enough. Goals and targets have to be translated into actions that make a real difference on the ground and in the water. Action to improve the bay will only happen when there are regulations and incentive mechanisms in place that cause landowners, business owners, fishermen, and municipal officials whose decisions affect Chesapeake Bay aquatic life and water quality to face the impact that their decisions have on Chesapeake Bay. A number of innovative regulatory and incentive-based policy approaches exist, including ITQs approaches in fishery management and approaches to give incentives to reduce nonpoint source loadings of nutrients. The benefits of recovery of Chesapeake Bay are large, but so are the costs. If sufficient resources are to be devoted to recovery of Chesapeake Bay, political leaders and the general public will need to be aware of the benefits in order to have the will to face up to the costs. Only then will the tough steps required for a full recovery of Chesapeake Bay be put in place.

References

Chesapeake Bay Commission (CBC). 1987. *The Chesapeake Bay Agreement of 1987*. At www.chesapeakebay.net/pubs/1987ChesapeakeBayAgreement.pdf.

CBC. 2000. *Chesapeake 2000*. At www.chesapeakebay.net/agreement.htm.

CBC. 2002. *Cost of a Clean Bay*. At www.chesbay.state.va.us/Publications/C2Kfunding.pdf.

Chesapeake Bay Foundation (CBF). 2005. *Annual Report*. At www.cbf.org.

Chesapeake Bay Program (CBP). 2005. "Bay Trends and Indicators." At www.chesapeakebay.net/status.cfm?sid=115.

Fujita, R., K. Bonzon, J. Wilen, A. Solow, R. Arnason, J. Cannon, and S. Polasky. 2004. "Rationality or Chaos? Global Fisheries at the Crossroads." In *Defying Ocean's End: An Agenda for Action*, eds. L. K. Glover and S. A. Earle. Covelo, CA: Island Press.

Gordon, H. 1954. "The Economic Theory of a Common Property Resource: The Fishery." *Journal of Political Economy* 62:124ç42.

Hardin, G. 1968. "The Tragedy of the Commons." *Science* 162:1243–48.

Leggett, C. G., and N. E. Bockstael. 2000. "Evidence of the Effects of Water Quality on Residential Land Prices." *Journal of Environmental Economics and Management* 39 (2): 121–44.

Lipton, D. 2003. *The Value of Improved Water Quality to Chesapeake Bay Boaters*. Working Paper 03-16. College Park, MD: Department of Agricultural and Resource Economics, University of Maryland.

National Marine Fisheries Service, National Oceanic and Atmospheric Administration (NOAA). 2005. "Fisheries Statistics." At www.st.nmfs/st1/commercial /index.html.

Ostrom, E. 1990. *Governing the Commons: The Evolution of Institutions for Collective Action*. New York: Cambridge University Press.

Shear, Michael D. 2005. "Warner Plans Huge Bay Commitment: More than $150 Million to Accompany New Discharge Limits." *Washington Post*, November 22, B07.

U.S. Bureau of Labor, Bureau of Labor Statistics (USBL/BLS). 2007. "Consumer Price Indexes." At www.bls.gov/cpi/.

U.S. Geological Survey (USGS). 1995. "Chesapeake Bay: Measuring Pollution Reduction." U.S. Geological Survey Fact Sheet FS—55-95. At http://water.usgs.gov/wid/html/chesbay.html#HDR0.

PART V

The Upper Mississippi River

The Mississippi River is the third longest river in the world. Its Upper Mississippi River (UMR) Basin drains over 189,000 square miles (portions of seven states) in the northern part of the largest floodplain ecosystem in America. The Upper Mississippi River Basin (UMRB) contains over 4,000 square miles of diverse habitat—aquatic, wetland, forest, grassland, and agricultural—in its floodplain alone. Some three hundred species of birds, including 40 percent of North American migratory waterfowl and shorebirds, depend on the UMRB for food and shelter, as do 150 species of fish, 45 of amphibians and reptiles, 57 of mammals, and nearly 50 of mussels. The UMR's record of human history reaches back some 12,000 years; today, daily visits by recreationists total over 12 million per year. As Mark Twain said, "The Mississippi is well worth reading about. It is not a commonplace river, but on the contrary it is in all ways remarkable" (1863, 21). Not only for these reasons, but also for the following three, the analyses in the UMR portion of this volume encounter somewhat different challenges from those discussing the ongoing, large-scale restoration projects for the Everglades, Platte River, California Bay-Delta, and Chesapeake Bay.

First, the proposed large-scale ecosystem restoration project that chapter 13 analyzes—achieving ecological sustainability along one of the world's great river systems—must be done, if it is to be done at all, along the same storied river developed and managed for nearly two hundred years as a national transportation system (containing thirty-five locks at twenty-nine locations on the mainstem river and eight locks at eight locations on the Illinois Waterway connecting to it). No matter how many billions of dollars might be spent, the UMRB can never be restored to a state of nature—nor the river to the particularly remarkable one Twain knew.

Second, especially after the Great Mississippi Flood of 1927 that left some 700,000 homeless in Louisiana and Mississippi, human efforts to contain the mighty river's flow within its channel, by constructing ever-increasing levees for flood protection, isolated the main channel's ecosystem from that of its historic floodplain. Even before Hurricane Katrina in 2005, the rapid conversion of coastal Louisiana wetlands to open water in the late twentieth century had already proved to be a destructive externality placed upon the lower Mississippi River Basin, as a consequence of having met late nineteenth-century demands for flood control along the entire length of the river to protect cities such as St. Louis (located at the hinge point between the Upper and Lower Mississippi River Basins). However, post–Hurricane Katrina, enhanced flood control efforts are more likely to be pursued rather than existing projects removed. In this reality, how can the floodplain and river be reconnected to restore the health of the ecosystem?

Third, although the U.S. Army Corps of Engineers (USACE or Corps) has proposed the first fifteen-year, $1.463-billion UMR ecosystem restoration increment (of an eventual fifty-year, $5.3 billion program), funding of this plan awaits congressional action—without which the Corps' plan cannot be realized. For, in our constitutional democracy, USACE, like the Bureau of Reclamation working on the Platte River and the California Bay-Delta and the Environmental Protection Agency on the Chesapeake Bay, is simply "a federal agency—a creature of statute. It has no constitutional or common law existence or authority, but only those authorities conferred upon it by Congress" (*Michigan* 2000, 1081).

In studying how the Corps came to its present recommendation that Congress grant USACE new authority for and fund its proposed UMR ecosystem restoration program, Cynthia A. Drew analyzes how an agency under pressure found a different vision and a new path to it that encompassed greater transparency and stakeholder participation (chapter 13, this volume). Within this context, Thomas L. Crisman considers how the Corps' efforts to restore the UMR will need to develop scientific understanding and greater knowledge of the dynamics of complex river ecosystems (chapter 14, this volume). Stephen Polasky discusses how, given the large scale of economic and other analyses required by the Corps' attempts to conduct reasonable projections of possible future demand for lock expansions, USACE will need to reevaluate decisions continually, then be prepared to shift course in response to new data—a kind of "adaptive engineering" approach paralleling the adaptive management approach planned for the ecosystem restoration side of the sustainability program (chapter 15, this volume).

Of course, as with the Corps' first large-scale ecosystem restoration plan, that for the Florida Everglades (chapter 1, this volume), the ultimate success of the

UMR restoration plan depends upon receiving continued congressional funding. Moreover, the proposed UMR ecosystem sustainability program shares not only with the Everglades, but also the Platte River, California Bay-Delta, and Chesapeake Bay, the challenge to devise a plan to restore what has, over time, ineluctably become a managed system.

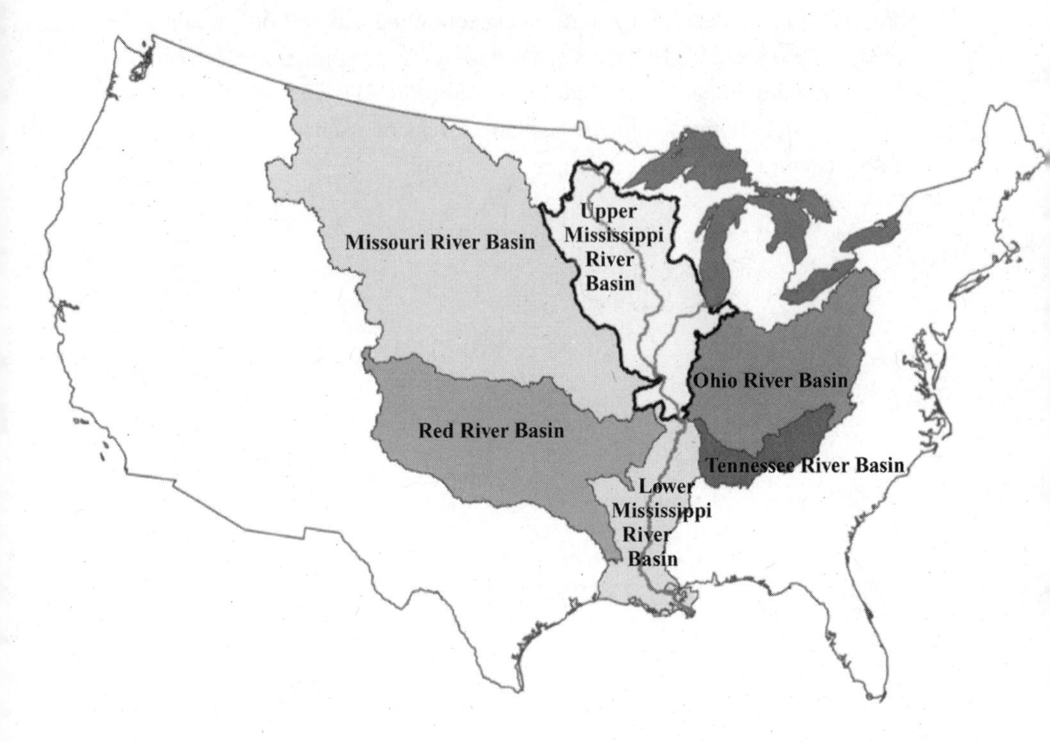

FIGURE 13.1. The Mississippi River Sub-basins (USACE 2004, 110).

Chapter 13

The Upper Mississippi River and the Army Corps of Engineers' New Role

Will Congress Fund Ecosystem Restoration?

CYNTHIA A. DREW

On February 13, 2000, the "old" way of approaching navigation feasibility studies by the U.S. Army Corps of Engineers (USACE or Corps) died with a bang, not a whimper, when the *Washington Post* broke Michael Grunwald's story, "How Corps Turned Doubt into a Lock; In Agency Where the Answer Is 'Grow,' A Questionable Project Finds Support."[1] The U.S. Army launched an investigation into the misconduct Grunwald alleged,[2] asking the National Research Council (NRC) to conduct an independent technical review of the *Upper Mississippi River–Illinois Waterway System Navigation Feasibility Study* (Study or Navigation Study). After reviewing NRC's findings on the Study's economics, environmental analysis, and engineering,[3] the Assistant Secretary of the Army "paused" the Study in February 2001. The Corps created an innovative Federal Principals Group to evaluate NRC's critical findings and to recommend the best way forward.[4]

In accord with new Environmental Operating Principles, the Corps chose a fresh path to resume the Study: ecosystem sustainability[5] (USACE 2002a). Between 2002 (Study's Interim Report) and 2004 (Final Report), USACE set for itself the extraordinary task of turning a "traditional" navigation study of the feasibility of enlarging or adding locks and dams into one striving to "integrate Federal river management activities to achieve sustainability of the [Upper Mississippi River] system" (USACE 2002c, 20). Remarkably, the Corps developed this groundbreaking effort in response to fallout from a whistleblower lawsuit, alleging that USACE was "cooking the books" to justify project benefits for expanding navigation facilities.

In adopting the Group's August 2001 recommendations, the Corps explicitly "restructured" the Study in 2002 to "give equal consideration of fish and wildlife resources along with navigation improvement planning"; to be "comprehensive and holistic" in considering the UMR system's multiple purposes; and to seek a "robust strategy that will work well under a variety of scenarios" (USACE 2002c, 1).[6] But could USACE, the agency long perceived by environmental nongovernmental organizations (NGOs) as aligned primarily with navigation interests,[7] ultimately bring forth recommendations that all stakeholders would support and Congress would fund?

By 2004, when USACE completed the Study's *Final Integrated Feasibility Report and Programmatic Environmental Impact Statement* (Final Report),[8] environmental NGOs had long contended that mitigation for past and continuing impacts of the existing navigation system should be "augmented by a comprehensive restoration effort."[9] These NGOs have not embraced all the ultimate Study recommendations to achieve an "integrated dual-purpose plan" (USACE 2004b, 2). Nonetheless, the Corps' recommendations in the Final Report, especially those providing for equal consideration of ecological needs, are widely deemed a large first step in the right direction.

The 2002 Restructured Study represented the first federal effort in decades to formulate a "comprehensive and holistic" plan (USACE 2002c, 1) for using and restoring the river Congress proclaimed both a "nationally significant ecosystem" and a "nationally significant commercial navigation system" (WRDA 1986). Moreover, if Congress funds future work as USACE's final 2004 Study recommended, with equal consideration of navigation and environmental needs designed to achieve ecosystem sustainability, this decision would be the harbinger for a whole new way of approaching ecosystem restoration, both for USACE and the nation.

Although the Corps lacks full congressional authority to move forward with its envisioned multipurpose project, Congress has appropriated initial funds for USACE to start planning how to achieve ecosystem sustainability. The Corps calls its ongoing post-Study planning activities, designed to achieve an "integrated" program of facilities improvements and environmental restoration, the Navigation and Ecosystem Sustainability Program (NESP). To appreciate this program's goals, one must understand the unique context facing those who make restoring the Upper Mississippi River Basin (UMRB) ecosystem a high priority.

Ecosystem Description, History

The Upper Mississippi River System (UMRS), a large, dynamic floodplain ecosystem, is part of the largest such river system in North America (McGuiness 2000, 2). The Mississippi River and its associated wetlands and forests are important habitat for fish and wildlife, including the most extensive continuous system of wetlands in North America (Wiener et al. 1998, 1). The Mississippi is the third longest river in the world, has the third largest watershed area (about one-eighth of North America and 40 percent of the United States), and has the seventh largest average discharge (Wiener et al. 1998, 1).

The UMRB drains over 189,000 square miles. The UMR floodplain ecosystem alone contains more than 4,000 square miles (over 2.6 million acres) of aquatic, forest, wetland, grassland, and agricultural habitats, including the Mississippi Flyway, used by 40 percent of migratory waterfowl that cross the United States (USACE 2002c, 2). Maps depicting the UMRB's national (fig. 13.1) and regional (fig. 13.2) extent are shown at the beginning of this chapter and following here.

The UMRB drains portions of seven midwestern states: Minnesota, Wisconsin, Iowa, Illinois, Missouri, Indiana, and South Dakota. It includes land drained by the Illinois Waterway as that water body flows from Chicago to Grafton, Illinois, at its confluence with the Mississippi River.[10] This waterway links the Great Lakes with the inland waterway system by allowing navigation to move from Lake Michigan at Chicago to the Mississippi River.[11] The study area encompasses the UMRS, defined in the Water Resources Development Act (WRDA) of 1986 as also the "navigable portions of the [1] Minnesota (15 river miles); [2] St. Croix (24 river miles); [3] Black (1 river mile); and [4] Kaskaskia rivers (36 river miles)" (USACE 2004a, 3). (A river mile is 1 mile measured from the centerline, the "sailing line," of the river.) These rivers flow, respectively, through Minnesota, along the border between Minnesota and Wisconsin, through Wisconsin, and through Illinois. The UMRB ecosystem plays the following significant roles for transportation, fish and wildlife, preservation, water supply, recreation, and cultural resources:

- A means for shippers to transport millions of tons of commodities within the study area, 122 million tons on the Mississippi River and 44 million tons on the Illinois Waterway in 2000
- Food and habitat for at least 485 species of birds, mammals, amphibians, reptiles, and fish (including 10 federally endangered or threatened species and 100 state listed species)
- Almost 285,000 acres of National Wildlife and Fish Refuge[12]
- Water supply for 22 communities and many farmers and industries[13]
- A multiuse recreational resource providing more than 11 million recreational visits each year[14]
- Cultural evidence of our Nation's past (USACE 2002c, 2)

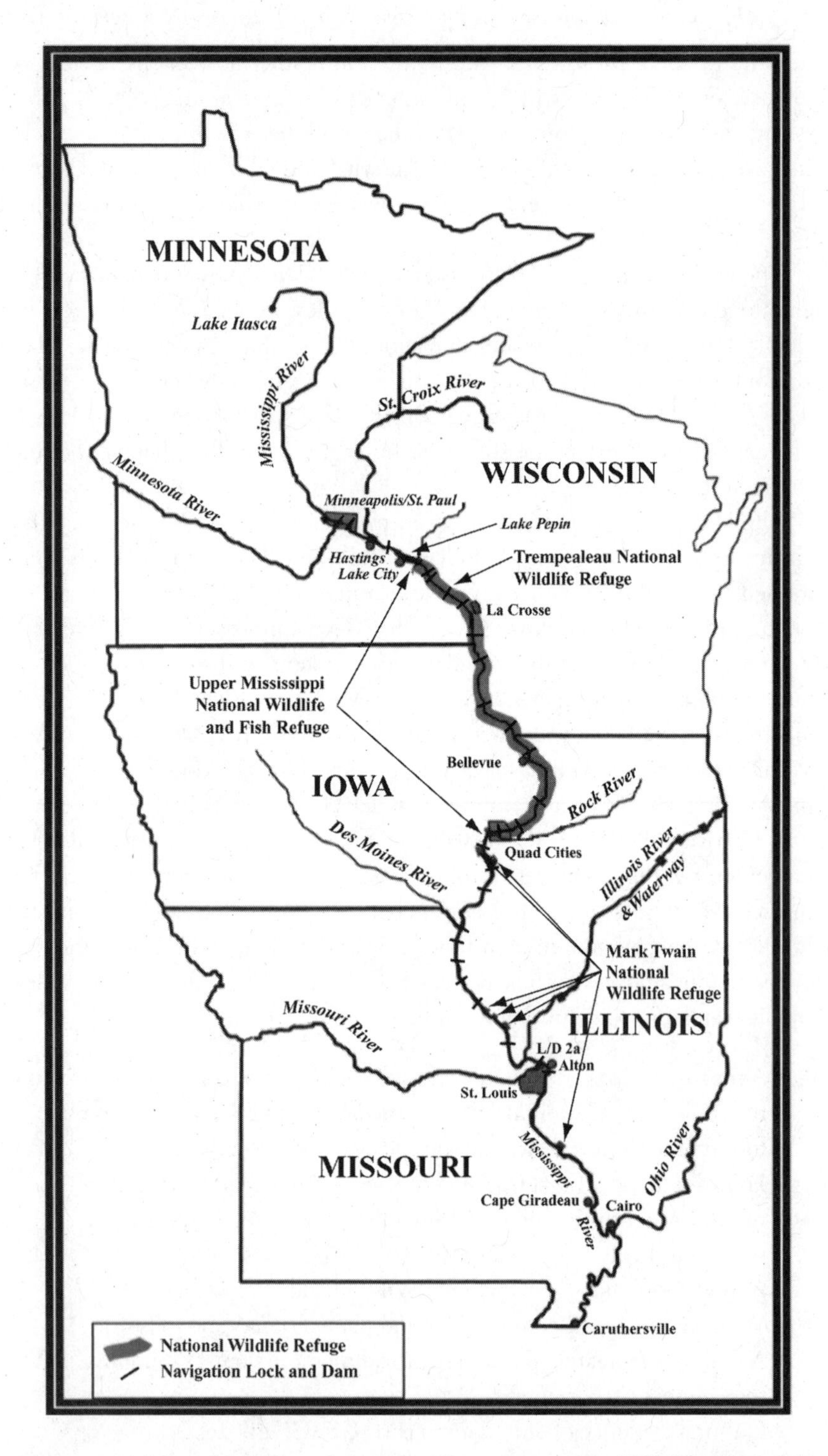

FIGURE 13.2. Map of the Upper Mississippi River Basin Area (USFWS 2002, 1-1).

Transportation Development

The UMR ecosystem's restoration challenges are the direct results of human activities, primarily from consequences of developing the river as a "nationally significant" commercial navigation system, an important role served by the Mighty Mississippi since early in America's history.[15] By 1830, the federal government had removed obstacles to navigation and confined flow to the UMR's main channel (USACE 2004a, 6). By 1880, Congress authorized the first of three comprehensive UMR channel projects to benefit navigation: a 4 1/2-foot channel (Act 1878); a 6-foot channel (Act 1907); and the present 9-foot channel, largely finished in 1940 (Act 1927). The Melvin Price Lock and Dam near St. Louis was completed in 1990 to replace Lock and Dam 26, which had 600-foot locks, and to provide a 1,200-foot lock to reduce commercial traffic congestion below the confluence of the Illinois Waterway (IWW) and the UMR.

In 1848, an important navigation improvement for the Illinois River, a series of rivers and dug canals, joined that river to Lake Michigan via the Illinois and Michigan Canal. Although in 1900 the diversion of water through the ensuing Chicago Sanitary and Ship Canal increased Illinois River levels by 3 feet, low summer flows continued to make year-round navigation unpredictable. Nor did the first locks and dams on the river (built by Illinois 1871–1899) guarantee a 9-foot channel. Only USACE's 1927 channel project did that. In 1960, the Thomas J. O'Brien Lock and Controlling Works was completed on the Calumet River near Chicago (USACE 2004a, 7). The current UMR system of thirty-five locks at twenty-nine locations on the mainstem river and eight locks at eight IWW locations is depicted in figure 13.3.

As increases in waterborne commerce and human activities, such as floodplain development and flood control efforts, caused complex effects within the UMRB ecosystem, the United States also strove to protect UMR natural resources. See figure 13.4, a UMRS time line of navigation and environmental initiatives, and figure 13.5, a summary of national wildlife refuge lands.[16]

Nonetheless, fish and wildlife habitats throughout the UMRB "have been severely degraded by navigation activities, floodplain development, poor water quality, and exotic species" (USFWS 2002, xi).

Ecosystem-Wide Challenges

There are still "gaps in knowledge of the large, complex Upper Mississippi ecosystem" (NRC 2001, 80). Yet we know definitively that human development of the Mississippi and Illinois rivers caused "an altered ecosystem that, depending on river reach, is in various stages of evolution towards a new quasi-stable

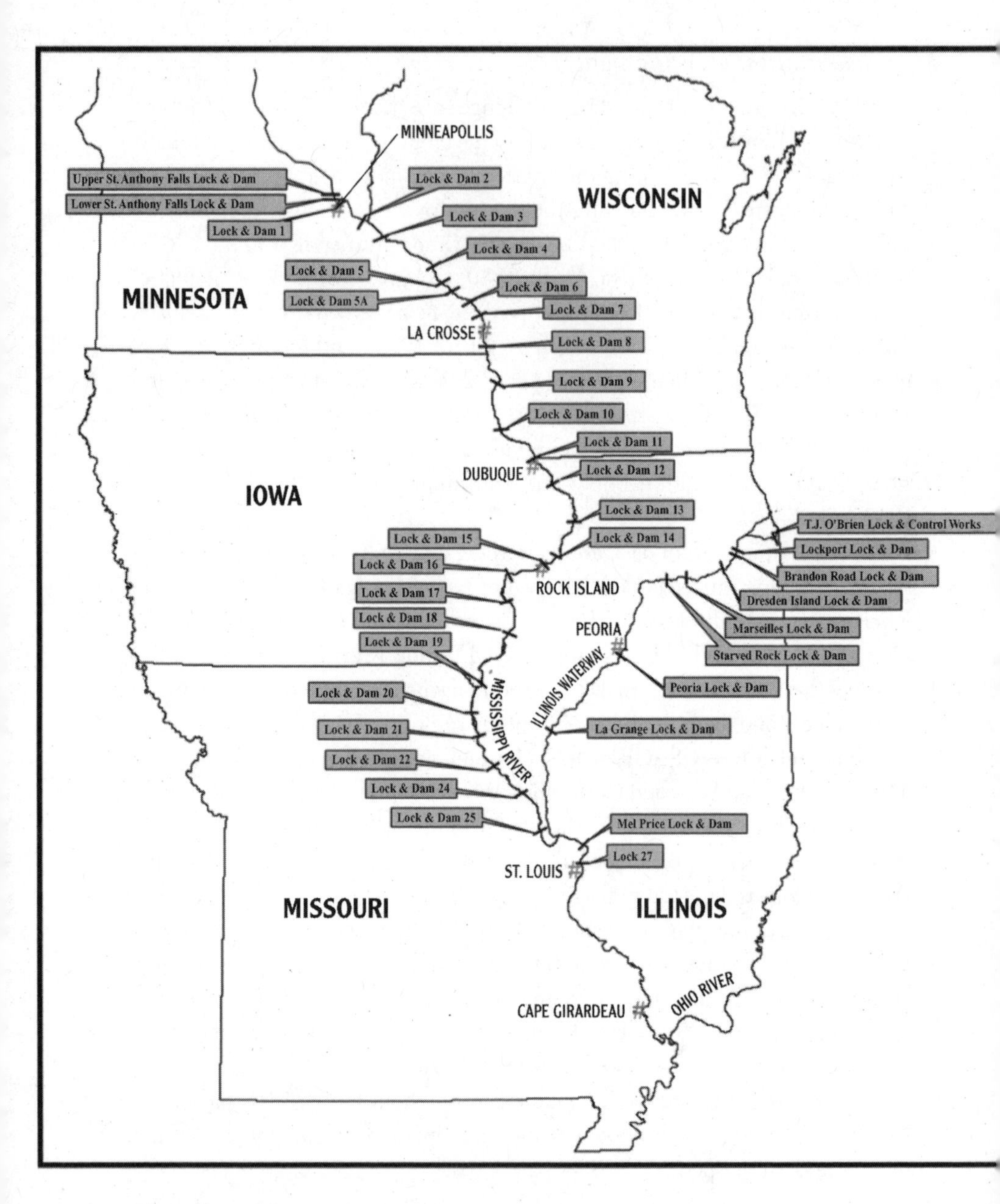

FIGURE 13.3. Present Navigation Features (Locks and Dams) of the Upper Mississippi River System (USACE 2004, 4).

UMR TIME LINE

1866	Snagging and clearing authorized by Congress
1872	US Fish Commission stocks American shad
1878	4-1/2 ft navigation channel authorized
1889	US Fish Commission begins fish rescue
1907	6 ft navigation channel authorized
1908	US Bureau of Fisheries establishes Fairport
1924	Upper Mississippi River Refuge established
1930	9 ft navigation channel using dams authorized
1934	Trempealeau National Wildlife Refuge established
1938	Guttenburg Fish Hatchery authorized for navigation compensation
1958	Mark Twain National Wildlife Refuge authorized
1965	12 ft channel proposed
1974	Wisconsin sues Corps over dredging
	GREAT programs funded Research Station, one of the first in the U.S.
1975	Lock and Dam 26 lawsuit
1978	New Lock and Dam 26 authorized
	UMRS Master Plan authorized
1982	Master Plan completed
1986	Lock and Dam 26 Second Lock authorized
	UMRS EMP authorized
1992	6-yr navigation expansion study authorized

FIGURE 13.4. UMRS Time Line Identifying Key Federal Navigation and Environmental Initiatives on the Upper Mississippi River System from 1866 to 1992 (USFWS 2002, 1-3).

Management Unit	Acres	Location
Upper Mississippi River National Wildlife and Fish Refuge		
Winona District	43,389	Pools 4-6
La Crosse District	46,469	Pools 7-8
McGregor District	90,678	Pools 9-11
Savanna District	52,973	Pools 12-14
Trempealeau NWR	5,733	Pool 6
Mark Twain National Wildlife Refuge Complex		
Port Louisa NWR	8,375	Pools 17-18
Great River NWR	10,037	Pools 20-24
Clarence Cannon NWR	3,751	Pool 25
Two Rivers NWR	2,660	Pools 25-26
Middle Mississippi NWR	4,400	Open River
Total Mississippi Acres	**268,465**	
Illinois River National Wildlife and Fish Refuges		
Cameron-Billsbach Unit	1,709	Peoria Pool
Chautauqua NWR	4,488	La Grange Pool
Emiquon NWR	1,303	La Grange Pool
Meredosia NWR	2,883	Alton Pool
Mark Twain National Wilfdlife Refuge Complex		
Two Rivers NWR	5,840	Alton Pool
Total Illinois Acres	**16,223**	

FIGURE 13.5. Summary of National Wildlife Refuge Lands in the Upper Mississippi River Basin (USACE 2004, 67).

condition." Some river reaches "may be close to this environmental stability, while it may take centuries for other reaches to achieve stability" (USACE 2002c, 46).

The "infrastructure supporting commercial navigation and the ecological response to it" differ along the length of the UMR (USACE 2002c, 42). After impoundments created reservoirs behind dams in the northernmost reach, aquatic habitat increased significantly. From the Quad Cities (Davenport and Bettendorf, Iowa; Moline and Rock Island, Illinois) south to Alton, Illinois, such impoundment effects were not so marked, but river levels were stabilized within channels. South of the Missouri River, the channel has been restored from "previously degraded" conditions, but off channel habitats away from the main river "are degraded and the floodplain is isolated from the mainstem in all but the worst floods" (USACE 2002c, 42). Illinois River aquatic habitats still fluctuate with low and high periods.

Channel training structures (wing dams) are limestone rock structures built perpendicularly from the shore to regulate water flow, reduce dredging, and fix the navigation channel in place. As to negative impacts, channel training structures, which act as ecological stressors, have caused "flow concentration, increased current velocity, increased structure and flow diversity" (USACE 2002c, 47). The ecological stressor "altered hydrology" additionally has caused "loss of low river stage, [and] altered water table" (USACE 2002c, 47). As to positive effects, wing dikes and other submerged structures in the navigation channel nonetheless have furnished substrate, aquatic habitat, and refuge important to many fish and aquatic invertebrates (USACE 2002c, 41).

The Corps' operation and maintenance of the 9-foot navigation channel significantly disrupted the UMRS's ecological integrity:

> The dams which were constructed to hold minimum water levels at low flows prohibit the natural river processes. This results in sediment accumulation above the dams and reduces transport of bedload to the open river. The high water levels maintained by the dams degrade bottomland hardwood forests. Channel training structures isolate backwaters and side channels from the mainstem river, alter natural sediment transport processes, and can change aquatic habitat to terrestrial habitat in accreted dike fields. Dredging for channel maintenance removes point bars, which provide important habitat for fish and wildlife. Additional impacts include facilitating the spread of exotic species and hindering fish passage at dams (USFWS 2002, xii).

Although these long-term effects of impoundment continue today, immediately following lock and dam construction, an initial biotic productivity boom occurred (USFWS 2002, 4–9). The extensive, shallow water expanses then created ideal conditions for some fish species and water birds. However, sedimentation in these habitats has decreased this benefit and remains the primary cause of habitat degradation (USFWS 2002, 5–25).

Impacts from Flood Protection

Levees pose a major conflict between land use in the built environment and the need to preserve the natural system. Determining how to mitigate the negative effects of decades of having built levees to provide significant levels of flood protection for numerous UMR communities (an important driver of UMRB-related land use) remains an essential task. The predominance of this land use is starkly exemplified in the 203-mile reach between Alton and Cairo, Illinois, where levees isolate 82 percent of the floodplain from the river (McGuiness 2000, 10).[17]

Levees are not evenly distributed in the UMRB. The proportion of leveed to total floodplain area "increases as the floodplain widens in a southerly direction" (USACE 2002c, 42). Levees today protect about 3 percent of the Mississippi River's floodplain north of Clinton, Iowa; 50 percent between Clinton and Alton, Illinois; 80 percent south of St. Louis to the Ohio River; and 60 percent of the Illinois River's floodplain south of Peoria (USACE 2002c, 42). Ironically, after the disastrous 1993 flood, a Galloway review committee (named for its chair, retired Brigadier General (BG) Gerald Galloway, Jr.) determined that the "elimination of natural floodplains by levee construction had in some instances actually served to increase the threat of flood damage" (Babbitt 2005, 47).

Where levees are present, lateral connectivity "allowing movements into inundated floodplains is reduced" (USACE 2002c, 46). Although land that could be reflooded is occasionally acquired after disasters, resource managers typically have been "very constrained" from doing so by lack of funding (USACE 2002c, 48). Increasingly, the long-term consequences of such historic activities for ecosystem restoration efforts are also being analyzed throughout the Lower Mississippi River Basin—after decades of losses of bottomland delta to open water in coastal Louisiana south of New Orleans were exacerbated by Hurricane Katrina.[18] Insofar as it still remains possible in the permanently altered, natural and built UMRS environment, projects for a sustainable UMRB ecosystem must reconcile future water management decisions with ongoing negative effects of past local and regional land use practices, both for flood protection and navigation.

Conflicting Views of Stakeholders

Use of the UMR-IWW as a national shipping artery has increased greatly over the past fifty years. Between 1960 and 2000, commerce tripled on the UMR, from 27 to 84 million tons; quadrupled on the Middle Mississippi River, from 30 to 122 million tons; and doubled on the IWW, from 23 to 47 million tons (USACE 2002c, 6; USACE 2004c, 13). Because most locks on the system are 600 feet long and barges today travel in 1,200-foot-long units,[19] barges must undergo unwieldy uncoupling procedures to pass through UMR locks, a process inland waterway operators argue poses untenable problems of delay.[20]

Congestion is the impetus behind years of congressional authorizations for USACE to study the feasibility of making UMRS navigation improvements. As some South American countries have been significantly improving their inland waterway systems, U.S. industry users desire longer locks to remain competitive in international markets:

> [O]ur South American competitors are investing millions to improve their transportation infrastructure to make their grain more competitive in the global market. Argentina, for example, has invested over $650 million to improve their transportation system. Likewise, Brazil is reviving its water transport network to reduce shipping costs for soybeans by at least 75 percent. As a result, Brazil and Argentina have captured 50 percent of the total growth in the world soybean market during the past three years. Further inaction with respect to the locks and dams of the Mississippi and Illinois Rivers will only allow this figure to grow. (USACE 2002c, 178)[21]

Because of the long planning and construction period required for lock projects, navigation interests have urged Congress to authorize the Corps to begin work for increased UMR lock capacity now.[22] However, many environmental interests insist that the need for new locks is not established. At the Study's outset, they deplored USACE's considering expanding lock capacity, urging the Corps to mitigate past impacts from operating the present navigation system comprehensively.[23] Weighing in on the side of environmental interests, the NRC recommended that USACE make "a careful assessment of the prospects of nonstructural alternatives for decreasing waterway congestion" and "an improved assessment study of the system-wide and cumulative effects of the existing navigation system on river ecology" before recommending new lock and dam construction (NRC 2001, 86–87).

Within this debate, the Upper Mississippi River Conservation Committee (UMRCC), a sixty-year-old voluntary organization of UMRB federal and state natural resources managers, plays a key technical role. Although UMRCC's professional members accept the premise of a "working river," they have identified five human modifications causing the "most detrimental" UMRB changes:

> 1. Levee construction, resulting in a 50% reduction in floodplain area
> 2. Construction of 36 locks and dams, converting most of the free-flowing river into a series of slackwater "pools"
> 3. Channelization of the formerly meandering river, . . . maintain[ing] the 9-foot navigation channel
> 4. Human settlement and use of the river's watershed, . . . degrad[ing] water quality and increas[ing] the amount and alter[ing] the rate of sediment and nutrient flows in the system
> 5. Connecting Lake Michigan to the Illinois River, . . . creat[ing] a pathway for invasion of nonnative species (McGuiness 2000, 5)

Three of these modifications causing negative impacts to the UMRB ecosystem, lock and dam construction, channelization, and creation of a pathway for invasion of some destructive nonnative species, resulted from human use of the UMR for commercial navigation. As many habitats are "expected to degrade further in coming decades," the U.S. Fish and Wildlife Service (USFWS) found that management actions are needed now "to reverse this trend" (2002, 4-1).[24]

After False Starts, A Better Planning Approach

In 1991, after conducting appraisal and separate reconnaissance studies for the UMR and IWW, the Corps began work on a joint feasibility study focused around a systems approach to solve navigation problems on both water bodies. In the new scope of work, USACE planned to include environmental studies required to address navigation traffic impacts, proposed by an earlier uncompleted study, entitled *Plan of Study* (USACE 1991; USACE 2004a, 17).[25]

At a Chicago review conference the Corps convened in 1992, environmental stakeholders and representatives of the U.S. Environmental Protection Agency (USEPA), USFWS, and the five UMRB states (Minnesota, Wisconsin, Iowa, Illinois, Missouri) discussed their views on the continuing need to complete the studies and on the planned scope, cost, and schedule for the new Study (USACE 2004a, 17–18). In 1993, the Corps began work on its Navigation Study.

The Corps apparently sowed the seeds for ongoing disputes with stakeholders and agency partners when it did not modify its Study project management plan "in response to various concerns raised" at the review conference. Instead, the Corps accepted consensus where it existed and moved on, recognizing but deferring attempts to resolve remaining areas of disagreement (USACE 2004a, 18). In 1993, the Corps adopted Study recommendations, upon which agency representatives and stakeholders agreed,[26] but did not undertake the "broader multi-purpose environmental study proposed by a number of agencies and organizations" (USACE 2004a, 18). Not until 2004 did USACE produce a broader *Final Integrated Feasibility Report*.

Why did USACE initially resist what it finally undertook? In a word, funding—that is, the Corps did not deem it had the authority in 1993 to undertake a multipurpose study as a 100-percent federal effort. Because such a study would have "address[ed] issues beyond the scope" of federal navigation project improvements, USACE leadership believed that conducting the Study as proposed would have required cost-sharing by states or other sponsors (USACE 2004a, 18). At the Study's outset, the Corps took an approach Congress had indisputably authorized: that the focus of federally funded environmental studies must "assess the effects of incremental increases in traffic associated with any navigation improvements" (USACE 2004a, 18).

Throughout the 1990s, federal and state agencies and environmental NGOs continued to criticize the Study's narrow scope and the process by which USACE had decided not to use their feedback in formulating the environmental plan.[27] In 2001, motivated by NRC's strong recommendations for a different course of action,[28] the Corps evidently reevaluated its legal authorities, determining that they did allow USACE to conduct a multipurpose Study.[29] The Corps then restructured the Study's scope to be more consistent with NRC's recommendations, an approach similar to the one that the agency representatives and stakeholders had been advocating since 1992. In its 2002 *Interim Report*, the Corps finally revised project goals and objectives to undertake an expanded Study addressing the "cumulative environmental effects of navigation and the needs for ecosystem restoration as an integral part of the restructured navigation study with a goal of an environmentally sustainable navigation system" (USACE 2002c, iii). The UMRB states, USACE's sister agencies, and environmental interests supported this approach,[30] as did navigation interests.[31]

Thus, although the Study's initial attempt to balance environmental values with economic goals failed dramatically, in its Restructured Study and NESP, the Corps found a better path, recommending "an integrated dual-purpose plan" focusing on navigation and ecosystem sustainability (USACE 2004b, 2). UMR citizens have consistently said they value integrating all ecosystem functions—economic, environmental, recreational, and aesthetic (USACE 2002c, 78). Even so, holistic restoration will not take place unless Congress grants yearly appropriations to achieve such balanced, multiple use of all UMR resources.

Federal Principals Group Leadership

After NRC's Study review, USACE "solicited help" in "evaluat[ing] the [NRC] comments and determin[ing] a new course of action," forming a UMR task force, the Federal Principals Group, on which senior officials of the following agencies worked with USACE to reach consensus on the best way forward: U.S. Department of Agriculture (USDA); USEPA; U.S. Department of the Interior (USDOI), U.S. Fish and Wildlife Service (USFWS); and U.S. Department of Transportation, Maritime Administration (MARAD) (USACE 2004a, 20). Per the Corps, the Group, a "collaborative and collegial forum," met several times in mid-2002 "in order to develop a plan of action on how to address" NRC recommendations (USACE 2004a, 20). Each agency in the Group drafted an issue paper offering suggestions on how to solve the environmental and economic problems identified in figure 13.6. The Corps then stated how it would deal with each one.

ENVIRONMENTAL THEMES & ISSUES:

Theme 1a: Equal consideration for fish and wildlife resources.

Theme 1b: Environmental effects of the existing Nine-Foot Channel Project.

Issue 2: Incorporate a cause and effects cumulative effects analysis in the System Study.

Issue 3: Should the scope of the tow traffic effects analysis be expanded to include quantification of the impacts of existing traffic (including Second Lock traffic) and traffic increases expected to occur without navigation expansion, or should existing traffic impacts remain identified as the baseline condition?

Issue 4: Include an assessment of ongoing project operation and maintenance (O&M) impacts as an element of the System Navigation Study.

Issue 5: Include a comprehensive mitigation plan that addresses the total array of navigation effects (O&M impacts, baseline traffic, Second Lock traffic, avoid and minimize, and incremental traffic) as part of the Navigation Study.

Issue 6: Assessment of traffic effects due to the Second Lock, Melvin Price Lock and Dam.

Issue 7: Upper Mississippi River cooperating Federal and state agencies should develop and implement a comprehensive ecosystem management plan for the Upper Mississippi River System.

Issue 8: How will site-specific impacts be addressed and incorporated into the overall environmental impact assessment?

Issue 9: Inadequacy of incremental effects studies due to insufficient data.

ECONOMIC ISSUES:

Issue 1a: Calculation of Traffic Forecast: Relates to Issue 1, "Spatial Equilibrium Model and Data of the National Research Council (NCR) review report.

Issue 1b: Demand Elasticities. Relates to Issue 1, "Spatial Equilibrium Model and Data" of the National Research Council (NCR) review report.

Issue 1c: Use of ESSENCE Model (Benefit Model). Relates to Issue 1, "Spatial Equilibrium Model and Data" of the National Research Council (NCR) review report.

Issue 2: Consider nonstructural options for improving traffic management as a baseline condition for the study. This relates to issue 2 of the National Academy of Sciences Review Report.

FIGURE 13.6. Environmental and Economic Themes and Issues Resolved by UMR Federal Principals Group (USACE 2004, 20-21).

In the Group's process of getting out on the table past disagreements with USACE's prior conduct of the Study, old disputes were resolved sufficiently to reach an initial consensus. Members of the Group were ultimately able to speak with one voice as to the best path forward, giving "equal consideration of fish and wildlife resources along with navigation improvement planning"; being "comprehensive and holistic" in considering the UMR system's multiple purposes; and seeking a "robust strategy that will work well under a variety of scenarios" (USACE 2002c, 1).

Considering how far apart Group members were when they began meeting, their eventual consensus was a remarkable achievement on two levels. First, the Group's success meant that USACE had brought its sister agencies into partnership roles rather than leaving them at loggerheads. Second, other agencies were able to use the Group as a forum for airing and resolving disputes early on, thus working effectively with USACE to help shape the ultimate plan, rather than finding themselves left with no opportunity to offer objections until formal consultation would ordinarily occur—on a completed plan.[32]

Federal Principals Group Chair James Johnson, USACE, created conditions needed for members to reach initial consensus and to deal equably with a wide range of stakeholders (pers. comm., Group members 2002; see note 32). Benjamin Tuggle, USFWS, attributed much of the Group's success to Johnson's ability to "keep everyone at the table, while not alienating anyone." Johnson was the key person who could have "done business as usual," but instead "made the effort" to get a new process "off the ground" and "sold" the benefits of this approach to fellow Corps colleagues and stakeholder groups, "trying to help them understand this new way of doing business" (pers. comm., Tuggle 2002). At a 2002 Group meeting with regional stakeholder representatives in St. Louis, for example, Tuggle observed navigation and environmental interests beginning to talk together about "how to make things happen," or, he noted, at least "screaming across the barricades" about it (Tuggle 2002).

This federal Group could not have achieved the success it did without the concomitant work of its counterpart regional group. The members of the regional group conveyed scientific and policy information back and forth among regional UMR stakeholders and agency officials and task force members in Washington, D.C. (pers. comm., Federal Principals Group members 2002; USACE 2002c, 16).

Reaching Consensus

In its Final Report, USACE identified many key participating UMR stakeholders, governmental and nongovernmental (USACE 2004a, xvi). The federal agencies included USDA, U.S. Geological Survey (USGS), USEPA, USFWS, and MARAD (all members of Federal Principals Group but USGS). All but MARAD also participated in the regional group. Other agencies and organizations that participated in Study efforts through the 2004 Final Report—and now continue to engage with the Corps' NESP as USACE plans to implement the Study's recommendations—are listed in figure 13.7. Those

UMR-IWW System Navigation Feasibility Study

Participating Agencies and Organizations

American Rivers
American Waterway Operators
Audubon Society
Illinois Department of Natural Resources
Illinois Department of Transportation
Illinois State Water Survey
Illinois Stewardship Alliance
Iowa Department of Agriculture
Iowa Department of Natural Resources
Iowa Department of Transportation
Iowa Institute of Hydraulic Research
Midwest Area River Coalition 2000
Minnesota Department of Agriculture
Minnesota Department of Natural Resources
Minnesota Department of Transportation
Mississippi River Basin Alliance
Missouri Department of Conservation
Missouri Department of Natural Resources
Missouri Department of Transportation
National Corn Growers Association
The Izaak Walton League of America
The Nature Conservancy
U.S. Army Corps of Engineers
U.S. Department of Agriculture
U.S. Department of Transportation, Maritime Administration
U.S. Fish and Wildlife Service
U.S. Geological Survey
U.S. Environmental Protection Agency
Upper Mississippi, Illinois and Missouri River Association
Upper Mississippi River Conservation Committee
Upper Mississippi River Basin Association
Wisconsin Department of Natural Resources
Wisconsin Department of Transportation

FIGURE 13.7. UMRB Agencies and Organizations Participating in USACE'S Restructured Study (USACE 2004, xvi).

from UMRB states generally include departments of natural resources, transportation, and agriculture.

The UMRB Association (UMRBA) is a significant participating entity that continues to represent state views. Formed to fill the void left when the UMRB Commission (UMRBC) was terminated by presidential executive order in

1981,[33] UMRBA includes representatives appointed by UMRB state governors. Given its origins, UMRBA was keenly interested in the evolving nature of USACE's Study, as it continues to be in the initial NESP activities (USACE 2004a, 2).

Navigation interests participating from the Restructured Study efforts through its heir, NESP, include American Waterways Operators; MARC 2000; National Corn Growers Association; and the Upper Mississippi, Illinois, and Missouri River Association (UMIMRA)[34] (USACE 2004a, xvi). Participating environmental interests include American Rivers, Audubon, Mississippi River Basin Alliance, Izaak Walton League of America, Nature Conservancy, and UMRCC (USACE 2004a, xvi). Other entities joined in or offered similar comments as these NGOs on the Study's 2002 draft Interim Report (see USACE 2004a, 476-487; USACE 2002c, 159-186).[35]

On the local level, the Quad City Riverfront Council noted that USACE's efforts to maintain a balance between the "commercial and economic value of the river" and the "need to sustain a vibrant and lasting ecological and recreational system" were "paramount" to the Study's success (USACE 2002c, 179). UMRB states agreed.[36]

USACE's Continuing Work with Stakeholders

Boundaries of the UMRB study area fall across three Corps Districts, north to south, St. Paul, Rock Island, and St. Louis. All report to the Mississippi River Division (MRD) Commander in Vicksburg, Mississippi. Rock Island had the "lead" on the Study; it now directs planning activities for NESP. The study area contains thirty-eight major river communities and seventy-seven counties (USACE 2004a, 24).

Within these bounds, the Corps organized five main stakeholder groups to consult with and provide feedback on UMR work. First, USACE worked through a Governors' Liaison Committee (GLC) to "build consensus" among UMRB states and to provide USACE each governor's position on Study issues (USACE 2004a, 28). The five governors each designated a representative to GLC, to "assure that study recommendations would merit the support of the people of each state" (USACE 2004a, 28). GLC often met with UMRBA, successor to the defunct UMRB Commission. When GLC was disbanded after the Study's 2004 completion, UMRBA assumed the role as collective voice of the UMRB states.[37]

Second, through a Navigation Environmental Coordination Committee (NECC), USACE works to facilitate interagency coordination for compliance with the National Environmental Policy Act, Endangered Species Act, Fish and Wildlife Coordination Act, and other statutes (USACE 2004a, 28). Representatives from state natural resource agencies, USFWS, and USEPA sit on the NECC, which met forty-five times between 1992 and 2005, and, since 2005, eight times concurrently with the Economic Coordinating Committee (pers. comm., S. Whitney, USACE 2007).

The goals of NECC are to help "refine environmental modeling procedures and to provide comments on environmental studies conducted as part of the overall study" (USACE 2004a, 28). The Corps is fortunate that so many NECC members are the same state and federal agency liaisons who have worked with USACE on UMRB issues for years. Many such long-time "river rats" had a history of coordinating and cooperating with each other through the UMRCC, which has held annual meetings on river issues for decades (see UMRCC 1993)[38] and with the Corps since the Plan of Study days.

Third, through an Economics Coordinating Committee (ECC), USACE works via NESP to provide partners and stakeholders with the opportunity "to share their views on economic matters pertaining to the study, to facilitate efforts to arrive at a consensus on those matters among the members, and to engender a shared set of goals and expectations for the economic portion of the study among all committee members and the public" (USACE 2004a, 28). Representatives of UMRB states and MARAD, USDA, and MARC 2000 sit on ECC, which met twenty-nine times on its own before beginning to meet concurrently with NECC (pers. comm., S. Whitney, USACE 2007). In 2007, for example, ECC met in St. Louis to discuss issues continuing to be assessed in the Corps' Economic Reevaluation Study, including the impact on the economics of proposed lock expansions if greater amounts of corn now shipped down the Mississippi for export are used instead for future ethanol production in the Midwest (NESP ECC 2007) (see chapter 15, this volume).

Fourth, through an Engineering Coordinating Committee (EnCC), USACE works on "key engineering assumptions and findings" (USACE 2004a, 28). Representatives of UMRB states sit on EnCC, which also meets with navigation industry technical experts and representatives to "review the practical and logistical application of both small-scale and large-scale engineering alternatives" (USACE 2004a, 28).[39]

Fifth, through a Public Involvement Coordinating Committee (PICC), USACE works "to create a shared set of goals and expectations regarding public involvement . . . among all committee participants, the navigation industry, and the public" (USACE 2004a, 28). Representatives from UMRB states sit on PICC.

Since USACE has for years conducted meetings of these groups in an open manner, stakeholder representatives are always invited participants. Moreover, in its Restructured Study (2001 through its 2004 Final Report) and now in its beginning NESP processes, the Corps has striven to conduct all UMR research and decision making not only transparently but also collaboratively. Thus an important achievement of the Restructured Study has been to change the Corps' traditional way of doing UMR business. In other words, the Corps' expanded outreach to stakeholders before its 2004 Final Report did not stop when the Final Report was completed but is being continued as the Corps moves forward with its new sustainability program. Nevertheless, it remains to be seen whether this ease of access to USACE's ongoing UMR planning will lead to greater continuing congressional appropriations for the envisioned multipurpose project.

Public Support for Balanced Multiple Use of UMR Resources

The NRC commended USACE's active public involvement efforts (NRC 2004, 19).[40] The public's long-standing appreciation for UMRB resources remains steady and manifests broad support for their balanced, multiple use. A 1996 Corps public expectations survey lists the following priorities:

- 99% valued the rivers for future generations
- 70% wanted to control industrial pollution
- 55% wanted improved water quality
- 45% wanted improved fish and wildlife habitat
- 25% wanted improved sport fishing
- 15% wanted less barge traffic (USACE 2002c, 78)

In the same survey, the public highly valued five future UMRS interest areas:

- More fish and wildlife in general (habitat diversity, species diversity, and abundance)
- Clean and abundant water
- Reduction of sediment and siltation
- Balance between the competing uses and users of the river
- Restoration of backwaters, side channels, and associated wetlands (USACE 2002c, 78)

In public meetings to solicit stakeholder input during work on the Restructured Study, survey results reflected the following:

- 79% agreed with taking a balanced, sustainable approach to navigation and the environment; 4% disagreed
- 77% agreed with improving the efficiency of the navigation system; 11% disagreed
- 75% agreed with sustaining a healthier ecosystem; 11% disagreed
- 66% agreed with restoring River habitat; 5% disagreed (USACE 2002c, 78–79)

Sustainability Planning Progress

The consensus achieved thus far regarding restoration goals is reflected in the collaboratively developed vision statement and definition of UMRS sustainability, resulting from USACE's enhanced coordination begun during the Restructured Study and ongoing throughout its planning for NESP:

> *Vision statement*: To seek long-term sustainability of the economic uses and ecological integrity of the [UMRS].
>
> *Definition of Sustainability*: The balance of economic, ecological, and social conditions so as to meet the current, projected, and future needs of the [UMRS] *without compromising the ability of future generations to meet their needs* (emphasis added). (USACE 2004a, 3)

The goal for NESP continues to be to "modify the way the Corps operates and maintains the [UMR] system to strive for economic, environmental, and social sustainability" (USACE 2004a, 2). The post-Study NESP has thus become "the mechanism to define the baseline ecosystem sustainability goals and objectives to be used across [existing] Federal management activities within the spatial limits" depicted in figure 13.8 (USACE 2002c, 20).

The following congressionally authorized management activities (jurisdictional areas shown in fig. 13.8) will also help determine NESP implementation requirements:

> The Illinois River Basin Restoration initiatives will define management for sustainability outside the navigation project limits on the Illinois Waterway and throughout the Illinois River Basin.[41] The Environmental Management Program[42] and Environmental CAP [Continuing Authorities Program] (Sections 204, 206, and 1135)[43] will integrate the baseline sustainability goals and continue to operate throughout the river floodplain system. The U.S. Fish and Wildlife Service Refuge Comprehensive Conservation Plans[44] will incorporate the baseline sustainability goals and objectives (USACE 2002c, 20).[45]

The Corps' planning process for "establishing UMRS ecological sustainability" focused on developing five areas: (1) objectives, (2) management actions, (3) anticipated costs and expected outcomes, (4) environmental alternatives analysis, and (5) integrated alternatives and tradeoffs analysis (USACE 2002a). By late 2002, the Corps had already presented and invited stakeholder feedback on a preliminary ecosystem conceptual model and set UMR goals and objectives

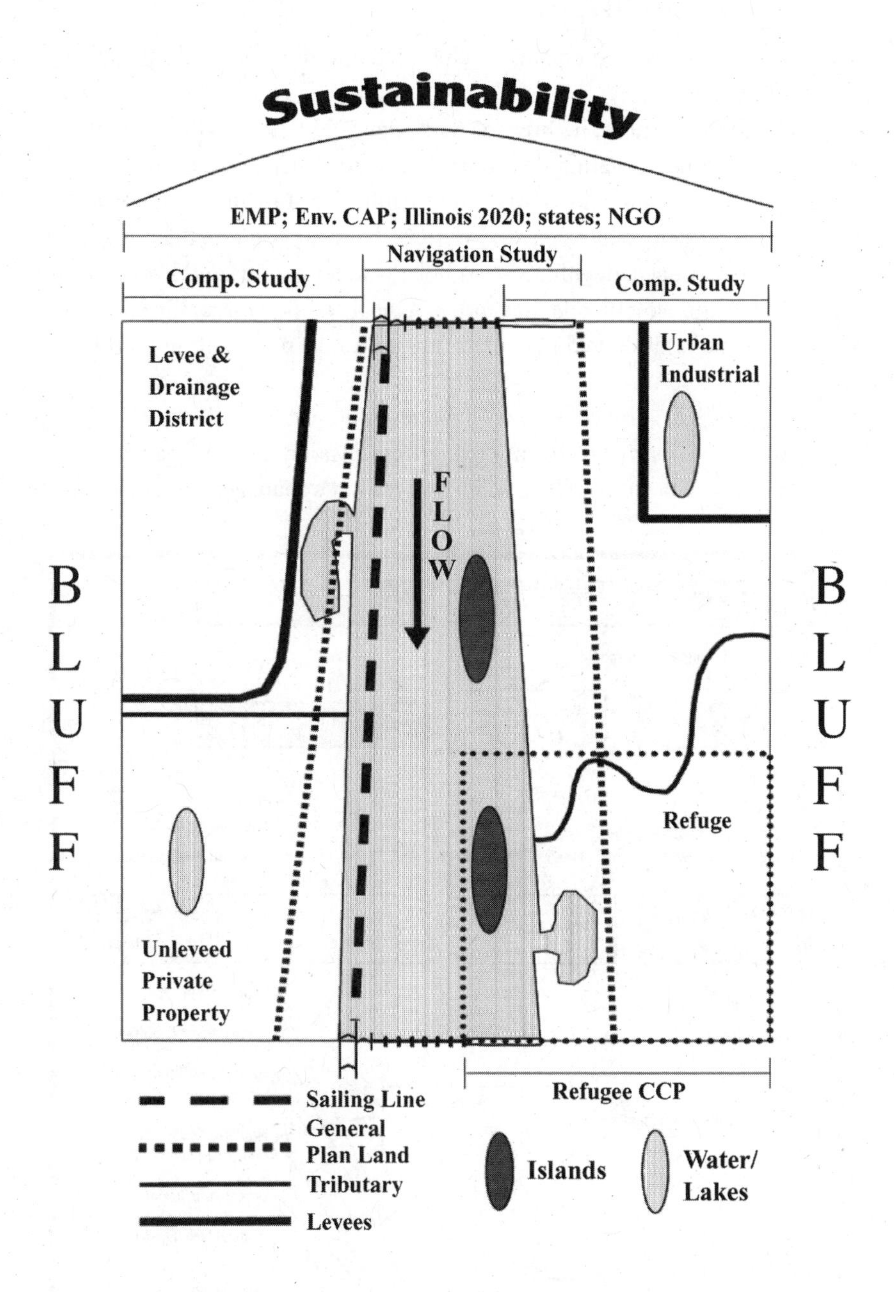

FIGURE 13.8. Schematic Representation of a River Reach Illustrating General Types of Land Uses and Ownership and the Approximate Extent of River Management Authorities (USACE 2004, 517).

"collaboratively, with participation of the full community of river stakeholders" (USACE 2002c, 74).

According to Chuck Theiling, USACE, figure 13.9 depicts the current conceptual model incorporating these elements: "Ecosystem drivers and stressors, like climate and extreme floods or droughts, influence functional and structural Essential Ecosystem Characteristics (EEC). Physical and chemical functions and processes associated with geomorphology, water quality, and hydrology are significant factors affecting habitat and biological components of the ecosystem. Outcomes can be evaluated by monitoring indicators representing each EEC" (pers. comm., 2007).

The Corps had achieved early collaboration on sustainability issues by organizing workshops with invited participants from a variety of area organizations to (1) solicit feedback from UMRS scientists, resource managers, engineers, and

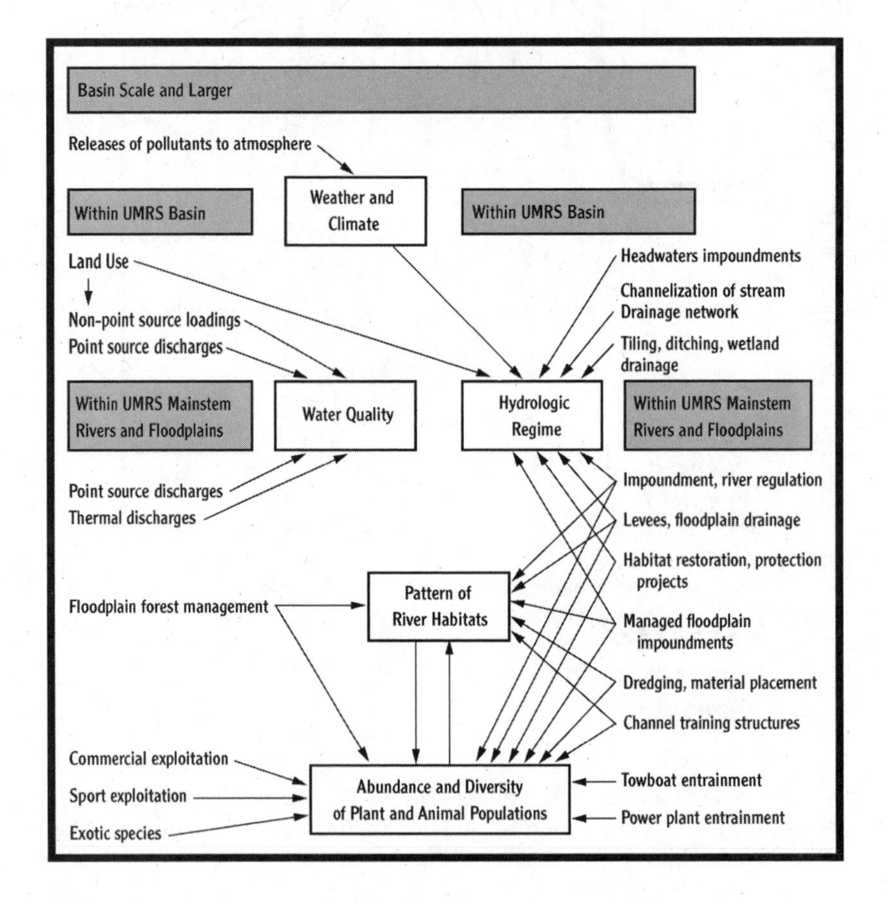

FIGURE 13.9. UMR Ecosystem Conceptual Model (WEST Consultants, Inc. 2000, 3, for USACE).

NGOs regarding desirable management actions in particular river reaches; and (2) ask the resulting stakeholder groups to formulate alternative environmental plans, linking the recommended actions with UMR objectives. The workshops also provided an initial forum for USACE to evolve its concepts of adaptive management needed to guide UMR restoration (pers. comm., A. Fenedick, USEPA 2002).

Initial Response to New Plan

After the Restructured Study, the UMRB states, UMRBA, UMRCC, and navigation and environmental interests supported the concept of "A River That Works and a Working River" (McGuiness 2000). Yet some environmental interests withheld support for NESP because they did not believe USACE had justified lock expansion proposals. According to Mark Beorkrem, of both the National Corps Reform Network Steering Committee and the Sierra Club, a separate environmental movement to reform Corps planning processes was driven by the abuses of USACE's late 1990s UMR efforts—which exemplified how not to plan a major Corps project (pers. comm. 2007).[46]

In July 2004, governors of the five UMRB states endorsed the Corps' plan for navigation efficiency and ecosystem restoration (USACE 2004b, 6). In a second review of the Study planning process, NRC supported the Corps' recommendation to adopt an adaptive management approach to implementing the ecosystem and navigation portions of a multipurpose project (NRC 2004, 28-29). Acknowledging that NRC still analyzed "limitations" in USACE's existing economic models (NRC 2004, 2, 24–25), Chief of Engineers Lieutenant General (LG) Carl Strock committed the Corps to continuing research "to develop a new suite of peer reviewed models and to refine future traffic projections" (USACE 2004b, 5).

Noting that the Mississippi River Commission agreed with the recommended plan "for an integrated 50-year framework" to achieve "navigation efficiency and environmental sustainability and to add ecosystem restoration as an authorized project purpose," LG Strock concurred with these proposals (USACE 2004b, 5–6). In transmitting the Final Report to the Secretary of the Army on December 15, 2004, Strock recommended that Congress "approve the integrated plan" (USACE 2004b, 6). As of fall 2007, the Secretary of the Army had not forwarded his recommendation on the plan to Congress.

Nonetheless, Congress began appropriating funds in fiscal year (FY) 2005 ($13.5 million) for USACE to conduct preconstruction engineering and design (PED) studies for navigation improvements and habitat restoration (Consolidated Act 2005). Authorized studies include field and planning investigations (pers. comm., D. L. Hays, USACE 2007). As depicted in figure 13.10, USACE

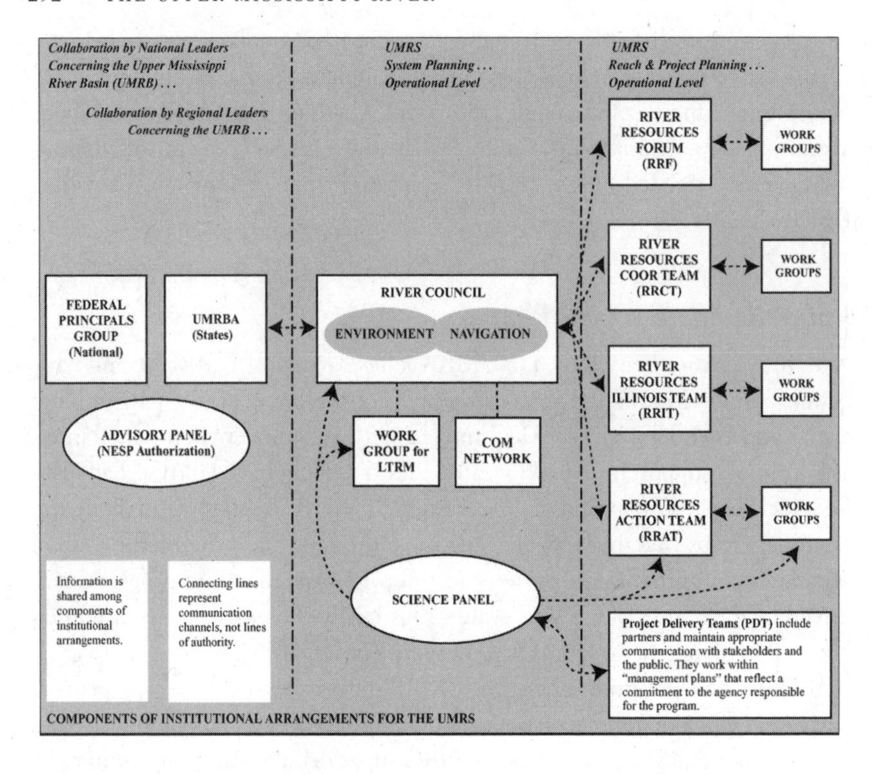

FIGURE 13.10. Components of Institutional Arrangements for UMRS (USACE 2005, 16).

has also proposed an institutional framework to oversee the recommended sustainability program's activities with the help of other river management agencies.

The Role of Science

By 2005, USACE had thus begun collaborating with partners and stakeholders during a PED phase to formulate an institutional framework for an integrated dual-purpose project using adaptive management of UMRS ecosystem resources. The intent is that new scientific information will be developed and communicated in a manner resulting in revised management decisions (USACE 2004a, 196–97). The primary way scientific and technical information has generally been communicated to project managers and policy makers is through USACE's and other agencies' normal chains-of-command and now NECC (USACE 2004a, 32–33).

After the Study's 2004 completion, the Corps organized an expert Science Panel to provide guidance for implementing an adaptive management approach to UMRS under NESP (Barko et al. 2006, ES-1). The panel's report highlights the need for further work to improve six functional areas of UMR ecosystem restoration:

1. Refining and clarifying ecosystem objectives
2. Developing evaluation criteria outcomes, including for ecosystem services
3. Evaluating and sequencing proposed ecosystem restoration projects
4. Monitoring, including selection of response variables appropriate to different scales
5. Evaluating relevant ecological indicators, metrics, and outcomes for a UMRS ecosystem report card
6. Integrating ecological models and using information technology to facilitate the adaptive management process (Barko et al. 2006, 1)

Cooperation between the scientists on the panel and Corps' team members on planning for NESP projects continues, albeit more slowly than some would prefer. According to Jon Duyvejonck, USFWS, the Science Panel's ability to address these objectives has been limited by lack of funding and time; most panel members have other full-time positions. The Science Panel is also responsible to assist NESP planning design teams (PDTs) as they develop high-priority projects, such as fish passage at Lock and Dam 22 and Melvin Price Locks and Dam (pers. comm., J. Duyvejonck 2007). The scientists' work on these six objectives also has been delayed because their task has proved more difficult than originally envisioned (pers. comm., J. Duyvejonck 2007).

Most ecosystem restoration programs have a goal to return the environment to "some semblance of a natural condition that existed before whatever perturbation caused its degradation" (pers. com., J. Duyvejonck 2007). However, it would be difficult to restore preimpoundment conditions on the UMRS because existing UMR locks and dams will not be removed. The political reality for the foreseeable planning horizon is that they will remain in place. The only real dispute is how many, if any, additional locks and dams may also be constructed in what locations. The Science Panel is still "wrestling with the implications of having to plan for such an atypical restoration goal" (pers. com., J. Duyvejonck 2007).

Initial Experiences with Adaptive Management

Some twenty-six different NESP activities were funded for FY 2007 under PED (pers. comm., S. Whitney 2007). While awaiting more specific input from the Science Panel, USACE and its partners working in PDTs on NESP projects have begun to implement basic adaptive management principles. Thus far, success has been "mixed," largely depending upon study team members' previous experience with NESP (pers. comm., J. Duyvejonck 2007). Some study team members who have yet to learn adaptive management principles are "still following traditional Corps planning guidance and policies and are reluctant to stray very far." However, PDT members who have had more experience in NESP planning "show an enthusiasm for adaptive management" (J. Duyvejonck 2007). The Corps and its state and federal partners addressed this issue by conducting a workshop presenting adaptive management principles to all NESP study team members. According to Rick Nelson, field supervisor, Rock Island field office, USFWS, "Certainly, for NESP to be successful, Corps managers and partners need to be on the same page about the role of adaptive management in NESP" (pers. comm. 2007).[47]

Although the Corps' ecosystem sustainability program is envisioned ultimately to encompass 1,009 individual projects and 43 overall objectives for a huge river system (USACE 2004a, xii–xiii, 208), each individual project needs to address only a few of the objectives to build a systems approach over time. An example of how well individual projects can work can be seen by results from the EMP and other partnership activities, under which phased island construction, seed island construction, dredging, and drawdowns are all coordinated to achieve desired physical conditions and biological outcomes in Pool 8 near La Crosse, Wisconsin (pers. comm., G. L. Benjamin 2007).

COL Robert Sinkler, USACE, recounts another excellent example of ecosystem improvement on the rivers. He grew up near the Illinois River and remembers seeing "huge chunks of brown suds, over ten feet in diameter and at least six feet high floating down the river, and decaying fish that had entered the river from a tributary at their peril. I vividly remember stories about sections of the river and waterway actually burning from petroleum products that had collected on the surface of the water." Now that he is back in the area, Rock Island District Commander Sinkler reported, "I am very pleased to see fishing tournaments in portions of the river basin, bald eagles nesting by the river in the winter and feeding on fish, and pelicans on the river, which our biologists tell me can consume up to six pounds of fish a day. Waterfowl numbers also have significantly increased" (pers. comm. 2007).

The Integrated Dual-Purpose Plan

The Study's Final Report included these recommendations: (1) "a fifty-year framework plan for small-scale structural and nonstructural navigation measures at a first cost of $2.4 billion; and (2) a fifty-year ecosystem restoration plan comprised of 1,009 individual projects with a first cost of $5.3 billion" (USACE 2004a, xii–xiii, 208). Navigation improvements will be funded from the Inland

Waterways Trust Fund per the fifty-fifty cost-sharing requirements of WRDA 1986, section 102. Ecosystem improvements will be funded through a combination of 100-percent federal funding and 65-percent federal/35-percent non-federal funding (USACE 2004a, 521–523).

The report also recommended constructing new 1,200-foot locks at Lock and Dams (L/D) 20–25, LaGrange and Peoria ($1.46 billion). During the first fifteen years of the plan, the Study's Final Report recommended implementing small-scale structural and nonstructural measures (for example, respectively, mooring cells and switchboats) at a cost of $207 million. Mitigation for impacts associated with increased tow traffic would cost $200 million. The total cost of this first fifteen-year navigation improvement plan would be $1.66 billion. The need for additional 1,200-foot lock extensions at L/D 14–18 would also be investigated but would require additional reporting and congressional authorization (USACE 2004a, xiii–xiv).

The Study's Final Report recommended an initial fifteen-year plan for habitat restoration, to be followed by a supplemental report considering additional ecosystem needs for the next thirty-five years. Components of the initial $1.463 billion restoration plan include the following:

- Fish passage at dams 4, 8, 22, and 26 and initial Engineering & Design at L/D 19 ($209 million total)
- Dam point control at L/D 25 and 16 ($41 million total)
- Programmatic authority to implement the specific measures listed below (#1–#9) under an adaptive management program ($935 million)
- Authorization to acquire 35,000 acres for increasing floodplain connectivity, and protecting and restoring wetland and riparian ecosystems ($277 million) (subject to finding willing sellers and 65/35 (federal/nonfederal) cost-sharing) (USACE 2004a, xiv–xv)

As of fall 2007, the Corps had sought, but not yet received from Congress, authority to implement these ecological restoration projects:

- Water level management in 12 pools
- 23 island building projects
- 33 backwater restoration projects
- 29 side channel restoration projects
- Wing dam/dike alteration at 19 locations
- Island/shoreline protection at 73 locations
- Topographic diversity improvement at 9 locations
- 13 dam embankment lowering projects
- Reduction of Illinois River's water level fluctuation (USACE 2004a, xiv–xv)

Will Congress Fund the Plan?

Acknowledging both types of primary stakeholders' interests as equally important, in the mid-1980s Congress proclaimed that the UMR is both a "nationally significant ecosystem" and "a nationally significant commercial navigation system" (WRDA 1986). Appearing almost an attempt to broker a "truce" between the two most prominent UMRB stakeholder groups' main interests in the UMR, this congressional declaration indisputably established a conceptual framework for the Corps' proposed integrated dual-purpose project.[48]

Nevertheless, navigation interests still generally want increased lock capacity now and believe it is already economically justified. Environmental interests still generally want no increased lock capacity ever, an opinion they hold more fervently now that midwestern ethanol production has increased and may continue to increase to meet U.S. alternative energy needs.

The five UMRB states, however, are "pleased" the Restructured Study incorporated a "balanced consideration of both navigation and ecosystem needs" (USACE 2002c, 144).[49] Like the federal government, the states wish to "thread the needle" to meet all UMRB stakeholders' needs.[50] Yet, stating the obvious, that the ultimate UMR solution at federal and state levels will be a political one, does not make the obvious any easier to achieve.

Despite environmental NGOs' recurring criticism of USACE's UMR activities, Congress, which retains ultimate power to say yea or nay to any proposed plan, has for years consistently funded USACE projects in general and the Study in particular. Even reporter Grunwald (critical of both USACE and its Study)[51] remained aware of realpolitik considerations, noting that Senator Christopher Bond (R-MO) "vowed to make sure the lock expansions get funded *no matter what the Corps study ultimately concludes*, saying the projects would make it easier and cheaper to move grain, fertilizer and other materials up and down the Mississippi" (emphasis added) (Grunwald 2001).

Although the views of one senator would not necessarily carry the day,[52] the question for environmental stakeholders remains: why should we put all our ecosystem restoration "eggs" in the basket of opposing *any* increased lock capacity? Moreover, considering the enhanced congressional gridlock on domestic matters that has reigned for years inside the post-9/11 Beltway, the comparable question for navigation stakeholders is: why should we assume that equal consideration of UMR ecosystem needs can be achieved without "revisioning" what possibly lesser transportation improvements may now be "needed"? And the ultimate question may yet be: if future studies should demonstrate little "need" for new locks (chapter 15, this volume), what priority, in a new fiscal environment of looming deficits, will citizens insist that Congress place on UMRB ecosystem restoration?

By June 2002, when USACE circulated its Study's draft Interim Report for stakeholder comment, the era a scant three years before—when URMB environmental interests had still conceived of justifying new ecosystem restoration "investments" because the United States had budget surpluses—already seemed but a quaint and remote anachronism.[53] Already then, even before the Iraq war, as President George W. Bush made homeland security and the military his top budget priorities, special interest groups and lawmakers were "fiercely competing for the limited funds up for grabs" in Congress's 2002 appropriations season: "But with so much of current budget dedicated to the war on terrorism and homeland security, advocates for national parks, education, water projects, local law enforcement and scores of other domestic programs are faced with shrinking domestic funds" (Pianin and Eilperin 2002). Three years after the 2004 Study's completion, America's burgeoning fiscal shortfall had soared to staggering deficits of hundreds of billions of dollars. In straitened economic circumstances, how will Congress generate the political will to fund UMRB ecosystem restoration?[54]

Only a groundswell of continuing support from committed UMRB constituents—such as the 99 percent of those surveyed who affirmed their appreciation of the UMR's value for future generations—would likely be required to make it happen (see USACE 2002c, 78). Citizens who care passionately about the UMRB's resources have worked for years with myriad groups and agencies that claim jurisdiction or authority to manage these resources.

To the extent any significant UMRB ecosystem restoration or rehabilitation is ever achieved, it will be due to the synergistic efforts of all these individuals, too numerous to name. The words of Dan McGuiness of Audubon perhaps best illustrate the common UMRB vision shared by many: "We need to move forward and get beyond old ways of doing business. We need to look for common ground, be honest and, as we develop a final plan for the Upper Mississippi River, seek opportunities to build trust and support for one another by working together side by side on projects and programs that are mutually beneficial."[55]

Acknowledgments

The author gratefully acknowledges the work of reference librarian Mark Plotkin, research assistants Heather Gogola, Sarah Schooley-Davidoff, Sarah Wood Borak, Trey Jordan, and Lauren Petrash; thanks USACE's and USFWS's Rock Island offices for permission to reproduce figures 13.1–13.10; and appreciates the diligence of all agency and stakeholder representatives whose commitment to UMRB ecosystem restoration included helping clarify various points in this chapter.

For their time, the author warmly thanks Dr. James Johnson, Dr. Benjamin Tuggle, Margaret Blum, Barbara Robinson, Robert Wayland, Federal Principals Group; Major General Don Riley, COL William Bayles, COL Robert Sinkler, Larry Barnett, David Hays, Rian Hancks, Denny Lundberg, Ken Barr, Scott Whitney, Chuck Theiling, Nicole McVay, Hank DeHaan, Richard Worthington, USACE; Al Fenedick, USEPA; Gretchen Benjamin, WDNR; Chris Brescia, MARC 2000; Mark Beorkrem, Sierra Club; Holly Stoerker, UMRBA; Rick Nelson, Bob Clevenstine, and Jon Duyvejonck, USFWS. On behalf of all chapter authors, the author acknowledges an immense debt of gratitude to copy editor Kathleen Hamman for her professionalism and cheerfulness.

Notes

1. The lock and dam expansions then expected to be recommended by the *Upper Mississippi River–Illinois Waterway System Navigation Feasibility Study* (ongoing since 1993).

2. Michael Grunwald, 2000, "Pentagon Probes Lock Proposal; Corps Officials Accused of Distorting Economist's Study," *Washington Post*, Feb. 15. The inspector general's investigation substantiated failures by Corps officials to meet established standards and one instance of showing preferential treatment to the barge industry. See U.S. Army Inspector General Agency Report of Investigation (Case 00-019) (n.d.), at www.mvr.usace.army.mil.

3. See NRC 2001, 87, recommending that USACE's draft Navigation Study "better integrate environmental and social considerations" into its ultimate decision as to whether UMR locks should be extended.

4. Group members included senior officials of USACE; U.S. Department of Agriculture; U.S. Department of Interior, U.S. Fish and Wildlife Service (USFWS); U.S. Department of Transportation, Maritime Administration (MARAD); and U.S. Environmental Protection Agency. See note 32.

5. On March 26, 2002, Chief of Engineers Lieutenant General (LG) Robert Flowers, USACE, announced these new principles, at www.hq.usace.army.mil/cepa/envprinciples.htm, charging Corps employees to do the following:

- Strive to achieve environmental sustainability
- Seek balance and synergy among human development activities and natural systems by designing economic and environmental solutions that support and reinforce one another
- Seek ways and means to assess and mitigate cumulative impacts to the environment; bring systems approaches to the full life cycle of [Corps] processes and work
- Respect the views of individuals and groups interested in Corps activities, listen to them actively, and learn from their perspective in the search to find innovative win-win solutions to the nation's problems that also protect and enhance the environment (USACE 2002a)

6. See also USACE, 2000, Planning Guidance ER 1105-2-100, C-12: "It is national policy that ecosystem restoration, particularly that which results in the conservation of fish and wildlife resources, be given equal consideration with other study purposes in the formulation of alternative plans."

7. The term "navigation interests" includes stakeholders interested in expanding the UMR system's navigation capacity: barge operators and agricultural NGOs.

8. This Final Report is the end product of one and the same Study with two relevant time frames: (1) 1993–2001, before the Study was restructured to seek authority for a new mission of ecosystem sustainability, that is, when it was a traditional Corps of Engineers' navigation study of the feasibility of adding locks and dams (considering mitigation mainly of environmental impacts relating to proposed construction), and (2) 2001–2004, after this same Study was restructured also to pursue a wholly different focus from prior USACE navigation feasibility studies—ecosystem sustainability. During this latter period the Corps produced both the Study's 2002 Interim Report and 2004 Final Report.

9. Letter, July 11, 2002, to Corps Study Manager Denny Lundberg from Scott Faber, Environmental Defense; Tim Sullivan, Mississippi River Basin Alliance; Melissa Samet, American Rivers; David Conrad, National Wildlife Federation; Mark Muller, Institute for Agriculture and Trade Policy; Bob Perciasepe, Audubon; Richard X. Moore, Izaak Walton League of America; Debbie Sease, Sierra Club; and Sol Simon, Mississippi River Revival ("July 11, 2002 Environmental NGO's letter") (commenting on Study's draft 2002 Interim Report circulated by Corps for stakeholder review) (USACE 2002c, 162).

10. The Corps uses the term "Illinois Waterway" (not Illinois River) because the Illinois Waterway (IWW) includes all or some of the Illinois, Des Plaines, Little Calumet, and Calumet rivers, the Chicago Sanitary and Ship Canal, and the Cal-Sag Channel (USACE 2002c, 5).

11. The Corps considers the Upper Mississippi River reach to include the 663 river miles from Upper St. Anthony Falls Lock in Minneapolis–St. Paul to its confluence with the Missouri River north of St. Louis, and the Middle Mississippi River reach to include the 195 river miles between its confluences with the Missouri River and with the Ohio River near Cairo, Illinois (USACE 2002c, 2). USACE's "study area" for the Restructured Study included lands drained by both the Upper and the Middle Mississippi River reaches. References throughout to "UMR" thus also include the Middle Mississippi River reach. The term is more broadly used to distinguish the Upper Mississippi River north of St. Louis from the Lower Mississippi River south of St. Louis down to New Orleans.

12. In addition to federal refuge lands, Minnesota, Wisconsin, Illinois, Iowa, and Missouri manage another 140,000 UMRB acres for conservation purposes. These lands are owned in fee by those states or the Corps, which leases states the lands it was required to buy for the 9-foot navigation channel. The states manage these "general plan lands" under cooperative agreements with USFWS and USACE (USFWS 2002, 3–13).

13. Half of UMRB's 30 million residents rely on water from the river and its tributaries for municipal and industrial water supplies (McGuiness 2000, 2). Over 650 manufacturing facilities, terminals, and docks are located within the system (USACE 2002c, 2).

14. See Black et al. 1999, ii. This exceeds annual visitation at most national parks, including Yellowstone (USACE 2002c, 7). The UMRB region received over $6.6 billion annually from over 12 million visitor-days by hunters, fishers, boaters, sightseers, and other visitors to the river and its scenic bluffs (USACE 2004c, 15).

15. For example, in 1673, French explorers Marquette and Joliet wrote about their experiences as they traveled along the UMR:

> The Mississippi takes its rise in several lakes in the North. Its channel is very narrow at the mouth of the Mesconsin [Wisconsin], and runs south until it is affected by very high hills. Its current is slow, because of its depth. In sounding we found nineteen fathoms of water. A little further on it widens nearly three-quarters of a

> league and the width continues to be more equal. Here we perceived the country change its appearance. There were scarcely any more woods or mountains. The islands are covered with fine trees, but we could not see any more roebucks, buffaloes, bustards, and swans. We met from time to time, monstrous fish, which struck so violently against our canoes, that at first we took them to be large trees, which threatened to upset us. We also saw a hideous monster; his head was like that of a tiger, his nose was sharp, and somewhat resembled a wildcat; his beard was long; his ears stood upright; the color of his head was gray; and his neck black (Marquette and Joliet [1673] 1850, 284).

16. Although UMRB also includes several national wildlife refuges (see fig. 13.5), when the first UMR National Wildlife and Fish Refuge (NWFR) was created in 1924, Congress specified that its management was not to interfere with using the river for commercial navigation: "Nothing in this Act shall be construed as exempting any portion of the Mississippi River from the provisions of Federal laws for the improvement, preservation, and protection of navigable waters, nor as authorizing any interference with the operations of the War Department in carrying out any project now or hereafter adopted for the improvement of [the] river" (UMR NFWR Act 1924).

17. Concurrently with the Study (in response to a different congressional authorization, WRDA 1999), USACE completed a UMR Comprehensive Plan that analyzed floodplain issues, particularly problems occurring after the devastating flood of 1993. See Knapp: "The 1993 flood on the Mississippi River has been recognized as one of the greatest natural disasters experienced in the Midwest, and has generated intense debate on the role of human impacts in the flood and future directions in floodplain management" (1994, 199). See also Babbitt: "It was the worst flood in the Mississippi River Basin since the great disaster of 1927, when the lower Mississippi broke through the levee, inundating nearly twenty million acres in the Delta region of Arkansas, Mississippi, and Louisiana" (2005, 46–47).

18. Compare, per Senator David Vitter (R-LA), that southern Louisiana is "losing a football field every 38 minutes and lost 217 square miles of coastal wetlands and land in merely two days, the days after Hurricane Katrina and Rita hit our coast" (Alpert 2007). See note 32 (below).

19. Of the 43 locks on the UMR-IWW, only three are over 600 feet long. Lock 19 has a 1,200-foot lock; Melvin Price Locks and Dam and Lock 27 each have one 1,200-foot lock and one 600-foot lock.

20. Depending on "many variables," the entire lockage process may take over one and one-half hours (USACE 2002c, 31).

21. Letter, June 7, 2002, to BG Edwin J. Arnold, Jr., Mississippi River Division (MRD), USACE, from Bart Ruth, President, American Soybean Association (commenting on [see note 22] Study's 2002 draft Interim Report circulated by USACE for stakeholder review).

22. See letter, June 14, 2002, to BG Edwin J. Arnold, Jr., MRD, USACE, from Gary D. Marshall, Missouri Corn Growers Association (commenting on Study's 2002 draft Interim Report circulated by Corps for stakeholder review): "If lock expansions are not made by 2020, the average cost of transporting corn and soybeans to export could increase by 17 cents per bushel. This seemingly modest cost increase will have a devastating impact on the agricultural economy in Missouri. These impacts include reduced production, lost export sales, and a drop in farm income. This precipitous drop in farm income would reduce agricultural employment[,] resulting [in] a decline in state and local tax receipts" (USACE 2002c, 175).

23. See July 11, 2002, Environmental NGO's letter (commenting upon Study's 2002 draft Interim Report, circulated by Corps for stakeholder review):

> Our organizations are deeply concerned that the Draft Interim Report for the [UMR]-[IWW] Navigation Study exaggerates expected barge traffic growth. . . . [W]e strongly oppose the . . . efforts to tie such a reevaluation to any potential lock and dam expansion. The Corps is already required by law to prepare a supplemental environmental impact statement on its operations and maintenance of the 9-foot channel. . . . [T]he Corps does not need additional legal authority or new internal policies to examine mitigation or to take immediate steps to improve the health of the Upper Mississippi and Illinois Rivers. . . . Under the most optimistic scenarios, locks on the Upper Mississippi and Illinois rivers are not likely to reach capacity . . . until at least 2015 (USACE 2002c, 162–65).

24. Compare undated letter from Gretchen L. Benjamin, Wisconsin Department of Natural Resources (WDNR), to Jon Duyvejonck, Rock Island Field Office, USFWS (commenting on *Preliminary Draft Fish and Wildlife Coordination Report for the UMR-IWW System Navigation Study through August 1, 2001* (circulated by USFWS for state review): "We still have a long way to go to achieve a river system we would be happy simply to sustain. . . . Integrated management of the UMR system to balance project purposes for both natural resources and commercial navigation is necessary to simply restore, rehabilitate and maintain the ecosystem over the next 50 years" (USFWS 2002, 11-6).

25. In 1978, Congress authorized replacing the existing 600-foot locks and dam at Alton, Illinois, with a single 1,200-foot lock and new dam (Inland Waterways Act [1978]). This act directed the UMRB Commission (UMRBC) to prepare a master plan to consider future navigation needs, specifically the size of a proposed second lock at Alton, in light of "broad mandates" including "impacts of navigation, and operation and maintenance on fish and wildlife, water quality, recreation, potential wilderness areas, and the carrying capacity of the UMR system" (USFWS 2002, 1–3). The last act of the UMRBC, terminated by presidential executive order on December 31, 1981, was to file with Congress the *Comprehensive Master Plan for the Management of the Upper Mississippi River System* (Master Plan)(UMRBC 1982).

The 1982 Master Plan recommended a program of habitat enhancement and long-term monitoring, which Congress authorized in 1986 as the EMP (see note 42). As stated by the interagency team that developed it, the *Plan of Study's* objective was to help fill data gaps: "[to] develop studies which identify and quantify navigation traffic impacts to significant [UMRS] resources where such impacts are currently poorly defined due to a lack of scientific data. Where possible, studies will quantify the impacts associated with that increment of traffic caused by the second lock" (USFWS 2002, 1–4).

26. All agreed to expand the Study's boundary to the Ohio River's mouth and to include additional studies on impacts: to mussels and fish spawning habitat; of recreational craft; and of sediment resuspension on plants.

27. The USFWS noted several undesirable results of USACE's early Study approach:

> As planning for implementation of the Plan of Study ensued, planning schedules for the proposed project dictated that the three-year navigation studies be reduced to one year; numbers of samples taken be reduced, areas to be sampled be reduced, and the number of studies be scaled back. In addition, the overall objective of examining traffic impacts to natural resources of the UMR system was eliminated. A new objective was set. That objective was to determine impacts only from the increased traffic as a result of the proposed navigation improvements.

> Data relevant to the total traffic picture would not be sought, or if available, would not be provided by the Corps of Engineers to the resource agencies or the public (USFWS 2002, 1–4).

28. The Corps "did not conduct the environmental analyses necessary" in considering only "incremental traffic effects" without "a more comprehensive analysis of system-level consequences," and should conduct its Study "under an adaptive management paradigm" (NRC 2001, 54, 87).

29. The 1970 Flood Control Act's plain meaning supports NRC's position: "The Secretary of the Army . . . is authorized to review the operation of projects the construction of which has been completed and which were constructed by the Corps of Engineers in the interest of navigation, . . . and to report thereon to Congress with recommendations on the advisability of modifying the structures or their operation, *and for improving the quality of the environment in the overall public interest*" (emphasis added) (Sec. 216).

30. See June 20, 2002, letter from Governor of Missouri Bob Holden to Colonel (COL) William J. Bayles, Rock Island District (RID), USACE (commenting on Study's 2002 draft Interim Report circulated by Corps for stakeholder review):

> The State of Missouri is pleased that the restructured study incorporates a balanced consideration of both navigation and ecosystem needs. The Missouri Departments of Agriculture, Economic Development, Conservation, Transportation, and Natural Resources have reviewed the Draft Interim Report and support the direction the Corps is going in addressing the many needs and uses of this great river system. We were especially pleased to see the inclusion of the Middle Mississippi River Reach (St. Louis to Cairo) in the study area. While this reach of the river has suffered some of the greatest ecosystem losses, it has only received limited support for restoration from the [EMP] (USACE 2002c, 144–45).

See also July 11, 2002, Environmental NGO's letter: "[W]e are encouraged that the Corps has recognized the need to reexamine and modify its operations and maintenance activities and to implement much needed restoration and mitigation" (USACE 2002c, 163).

31. See July 11, 2002, Navigation Industry Perspectives on Refocused UMR–IWW Study (commenting on Study's 2002 draft Interim Report circulated by Corps for stakeholder review): "The navigation community embraces a balanced approach to meeting the economic and ecological needs of the river basin" (USACE 2002c, 186).

32. Members of the UMR–IWW Federal Principals Group were James F. Johnson, Chief, Planning and Policy Division, USACE; Margaret D. Blum, Associate Administrator for Port, Intermodal and Environmental Activities, MARAD; Barbara C. Robinson, Deputy Administrator, Transportation and Marketing Programs, Agriculture Marketing Service, USDA; Benjamin N. Tuggle, Chief, Division of Federal Program Activities, USFWS; and Robert H. Wayland, III, Director, Office of Wetlands, Oceans, and Watersheds, USEPA. These members later formed a second group to deal with pressing problems posed by (pre–Hurricane Katrina) coastal Louisiana ecosystem restoration efforts (pers. comm., J. Johnson 2003).

33. See note 25.

34. The UMIMRA represents levee districts. MARC 2000 is "committed to modernizing" the UMR–IWW "while preserving their natural resources." See www.marc2000.org.

35. See also note 9.

36. See, for example, note 30.

37. See note 25.

38. Founded in 1943, UMRCC published a history of its UMR activities for its fiftieth anniversary.

39. NRC recommended that USACE use a kind of "adaptive cost-estimating" approach: that is, continually revisit estimated costs considering subsequent construction experience and increase the percentage amount estimated for contingency costs (NRC 2001, 59).

40. As Chief of Engineers LG Carl Strock informed the Secretary of the Army, USACE held "seven series of public meetings and open houses involving 54 meeting locations and over 5,000 attendees. The latest round of public hearings on the draft feasibility report included over 1,200 participants and produced 367 oral statements and over 4,000 written communications" (USACE 2004b, 6).

41. Post-Study, USACE completed a "2007 Final Draft of the Illinois River Basin Restoration Comprehensive Plan," establishing "the vision, goals, objectives, and recommended plan to restore the ecological integrity of the Illinois River Basin System," including "a restoration program; a long-term resource monitoring program; a computerized inventory and analysis system; and a program to encourage innovative dredging technology and beneficial use of sediments." NRC has identified the Illinois River (per early explorers, a "boundless marsh") as one of three "large-floodplain river systems in the lower 48 states with the potential to be restored to an approximation of their outstanding biological past" (USACE 2007a, ES-1).

According to COL Robert Sinkler, USACE, Rock Island District Commander, "The person in charge of the Illinois River [restoration program] is not an engineer; he's a biologist. . . . That tells you something about the direction we're going" (Bibo 2007).

42. Authorized by WRDA 1986 (see note 25) and reauthorized by WRDA 1999 under two continuing authorities, EMP, a "premier" ecosystem restoration authority, includes the following:

- A long-term resource monitoring, computerized data inventory and analysis and applied research (LTRMP)
- Planning, construction, and evaluation of fish and wildlife habitat rehabilitation and enhancement projects (USACE 2004a, 62–63)

By 2006, EMP habitat improvement projects covering 80,000 acres (3 percent of UMR floodplain acreage) had been completed, with another 54,000 acres slated for improvement. LTRMP is "the first comprehensive large river natural resources monitoring network in the world with six scientific stations in five states" (USACE 2006, 6).

43. The Environmental CAP are nationwide environmental restoration authorities limited to "smaller individual projects":

> Section 204, authorized in WRDA 1992 (Public Law 102-580), provides authority for projects for the protection, restoration, and creation of aquatic and ecologically related habitats, including wetlands, in connection with dredging for construction, operation, or maintenance of an authorized navigation project.
>
> Section 1135, authorized in WRDA 1986 (Public Law 99-662), provides authority to review and modify structures and operation of water resource projects completed by the Corps prior to 1986 for the purpose of improving the quality of the environment when it is determined that such modifications are feasible, consistent with the authorized project purposes, and will improve the quality of the environment in the public interest.

Section 206, authorized in WRDA 1996 (Public Law 104-303), provides authority for the development of aquatic ecosystem restoration and protection projects that improve the quality of the environment, are in the public interest, and are cost effective (USACE 2004a, 199).

44. These plans will "guide management decisions on the refuges for 15 years" (USACE 2002c, 24).

45. The Corps' "programmatic" Institutional Arrangements Project seeks to integrate these and all relevant federal management programs with operating and maintaining the 9-foot navigation channel (USACE 2007b, 1). See figure 13.10, depicting a proposed broad "umbrella" for such collaborative management. In this regard, Holly Stoerker, UMRBA, had noted that, while the Corps' 2004 Final Report "called for creation of a new River Management Council to address a broad array of river management issues," UMRBA and "the 5 state Governors' comments on [this] Report urged that existing institutions be used" (Stoerker 2005).

46. The particulars of arguments regarding "Corps reform" are beyond the scope of this paper. But compare McGuinness's speech: "The Corps of Engineers, the navigation/agricultural industries and the environmental organizations have all scored what we have each typically defined as 'victories' of one kind [or] another over the years. But in a 'We versus Them' paradigm, one group's victory is another group's defeat. When was the last time you saw the outcome of a war called 'win–win'?" (2002).

47. The USDOI (of which USFWS is part) also produced guidelines for implementing adaptive management (Williams 2006).

48. It also set the stage for inevitable conflicts among those agencies to which Congress had already delegated primary responsibility for one or the other, but not both, such functions: USACE for navigation; USEPA for environmental protection; USFWS for fish and wildlife; MARAD for maritime transportation; and USDA for agriculture.

49. See also USACE 2004a, 478. The governors of the five states "that share stewardship of the project area (Illinois, Minnesota, Iowa, Wisconsin, and Missouri) . . . jointly endorse the Corps' proposed plan . . . [and] support an integrated, balanced, adaptive, collaborative, and fairly funded plan."

50. Including those stakeholders who may not "frame" the ultimate UMR issue as choosing between the navigation and environmental interests' positions, but may, like local levee districts, wish to maintain existing levels of flood protection, or may, like recreational users, wish to ensure easy access with fewer fees for "using" UMR natural resources.

51. See the first page and note 2 of chapter 13 (this volume).

52. Nonetheless, according to Chris Brescia of MARC 2000, by 2002 some twenty-five congressional representatives had signified commitment to increasing lock capacity (pers. comm. 2002).

53. See Moore 1999: "In an era of budget surpluses, an investment in a national treasure [UMR NFWR] that serves the needs of 3.5 million Americans and contributes to a recreational economy valued at $1 billion is more than warranted" (12).

54. This question is especially apt after a 2007 Government Accountability Office (GAO) review of Everglades work found costs skyrocketing, increasing by 28 percent in six years and noting that the "true cost 'could be significantly higher'" (Clark 2007; see chapter 1, this volume). Florida officials "underscored the 'pressing need' for federal money [which since 2000 Congress has not appropriated as anticipated] in their response to GAO. . . . 'Until projects are authorized and appropriated through the Congress, the

full potential of the restoration program and the environmental benefits derived will remain limited,' their letter said" (Clark 2007).

By December 2000, according to USACE COL Sinkler, "Under Title VIII–Upper Mississippi River and Illinois Waterway System," the Water Resources Development Act of 2007 has now "authorized construction of navigation and ecosystem restoration projects. The $1.9 billion navigation component includes moorings, switchboats, nonstructural traffic management, environmental mitigation, and new locks. The $1.7 billion ecosystem restoration component includes systemic cultural stewardship, forestry management, fish passage, pool water level management, island shoreline protection, side channel restoration, and wing dam/dike alteration" (pers. comm. 2008).

But the question remains: Given the poor track record of congressional funding thus far for proposed Everglades restoration work, how can UMR stakeholders count on the UMR's receiving the years of consistent congressional funding needed to achieve ecosystem sustainability?

The remarks of a Florida federal district court judge who ruled in December 2006 on an issue regarding backpumping of water containing pollutants into Lake Okeechobee would not appear to give UMR stakeholders much cause for optimism that the answer to this question will be an unqualified "yes." As this Florida judge emphasized, (1) after six years, "very little" progress had been made "in constructing any of the proposed CERP projects," and (2) "although Florida has recently authorized SFWMD [South Florida Water Management District] to issue $1.8 million in bonds to 'accelerate' eight of these projects, the court emphasized that there is 'no statute' that 'mandates completion of the Accerler8 projects'"—and thus "'no guarantee that the federal-state-cost-shared CERP projects will ever be implemented'" (Drew 2007, 35).

55. See McGuinness speech 2002. Like many environmental NGOs, Audubon features UMR news on its Web site: www.audubon.org.

References

Act of June 18, 1878 ch. 264, 20 Stat. 152 (1878). (Acts of 1878, 1907, and 1927 authorized the construction, repair, and preservation of certain public works on rivers and harbors, and for other purposes.)

Act of March 2, 1907, ch. 2509, 34 Stat. 1073 (1907) (same).

Act of Jan. 21, 1927, ch. 47, 44 Stat. 1010 (1927) (same).

Alpert, B. 2007. "Bill Could Aid La. Flood Protection." *Times-Picayune*, July 28.

Audubon. At www.audubon.org.

Babbitt, B. 2005. *Cities in the Wilderness: A New VISION of Land Use in America*. Washington, DC: Island Press.

Barko, J. W., B. L. Johnson, and C. H. Theiling. 2006. *Environmental Science Panel Report: Implementing Adaptive Management*. Navigation and Ecosystem Sustainability Program (NESP) ENV Report 2. Districts in Rock Island (RI), IL; St. Louis (SL), MO; and St. Paul (SP), MN: USACE.

Benjamin, G. L. 2007. Personal communication.

Beorkrem, M. 2007. Personal communication.

Bibo, T. 2007. "River Has 'Come a Long Way.'" *Peoria Journal-Star*, October 4.

Black, R., B. McKenney, A. O'Connor, E. Gray, and R. Unsworth. 1999. "Economic Profile of the Upper Mississippi River Region." *Industrial Economics*. (March).

Brescia, C. 2002. Personal communication.

Clark, L. 2007. "Costs Rise as Glades Plan Lags." *Miami Herald*, July 3.

Clark, L., and C. Morgan. 2007. "State Lawmakers Fighting for Glades Bill." *Miami Herald*, October 3.

Consolidated Appropriations Act, 2005, Pub. Law 108-447, sec. 4, 118 Stat. 2827.

Drew, C. A. 2007. "Storm Water and the Consent Decree: The Life or Death of the Everglades." *Natural Resources and Environment* 21 (4): 30–35.

Duyvejonck, J. 2007. Personal communications.

Federal Principals Group (Group) members. 2002. Personal communications.

Fenedick, A. 2007. Personal communication.

Flood Control Act of 1970, Pub. L. No. 91-611, Title II, sec. 216, 84 Stat. 1830 (1970).

Grunwald, M. 2000. "How the Corps Turned Doubt into a Lock; In Agency Where the Answer Is 'Grow,' A Questionable Project Finds Support." *Washington Post*, February 13.

Grunwald, M. 2000. "Pentagon Probes Lock Proposal; Corps Officials Accused of Distorting Economist's Study." *Washington Post*, February 15.

Grunwald, M. 2001. "Public Works Study Halted; Army Corps' Analyses of Mississippi River Projects Faulted." *Washington Post*, March 1.

Hays, D. L. 2007. Personal communication.

Inland Waterways Revenue Act of 1978, Pub. L. No. 95-502, 92 Stat. 1693 (1978).

Johnson, J. 2002. Personal communication.

Knapp, H. 1994. "Hydrologic Trends in the UMR Basin." *Water International* 19:199–206.

MARC 2000. At www.marc2000.org.

Marquette, J., and L. Joliet. [1673] 1850. *An Account of the Discovery of Some New Countries and Nations in North America, in 1673*. In *Historical Collections of Louisiana, Embracing Translations of Many Rare and Valuable Documents Relating to Natural, Civil, and Political History of That State*. Vol. 2. Ed. B. F. French. Translated from French. New York: J. Sabin & Sons.

McGuiness, D. 2000. "A River That Works and a Working River." January. Rock Island, IL: UMRCC.

McGuiness, D. 2002. Speech presented at a U.S. Corps of Engineers Conference, "Balancing Economy and Environment: The Upper Mississippi River and Its Watershed: Seeking a Sound Economy and a Healthy Environment." New Orleans. July 18.

Michigan v. U.S. Envtl. Prot. Agency, 268 F.3d 1075 (D.C. Cir. 2000).

Moore, R. 1999. *Refuge at the Crossroads: The State of the Upper Mississippi River National Wildlife and Fish Refuge*. St. Paul, MN: Izaak Walton League of America.

National Research Council (NRC). 2001. *Inland Navigation System Planning: The Upper Mississippi River Illinois Waterway*. Washington, DC: National Academies Press.

NRC. 2004. *Review of the USACE Restructured UMR-IWW Feasibility Study*. Washington, DC: National Academies Press.

Navigation and Ecosystem Sustainability Program (NESP) Economics Coordinating Committee (ECC). 2007. Minutes from conference call, July 10.

Nelson, R. 2007. Personal communication.

Pianin, E., and J. Eilperin. 2002. "In a War Budget, Congress Is Dueling for Dollars: Spending Constraints Intensify the Scramble for Funds." *Washington Post*, June 4.

Sinkler, R. 2007, 2008. Personal communications.

Stoerker, H. 2006. Letter to Chuck Sptizack, U.S. Army Corps of Engineers. June 15.

Supplemental Appropriations Act of 1985, Pub. L. No. 99-88, ch. 4, 99 Stat. 293 (1985).

Theiling, C. 2007. Personal communication.

Tuggle, B. 2002. Personal communication.

Twain, M. 1863. *Life on the Mississippi*. Boston: James R. Osgood and Co.

Upper Mississippi River (UMR) Basin Commission (UMRBC). 1982. *Comprehensive Master Plan for the Management of the Mississippi River System*. January 1. Minneapolis: UMRBC.

UMR Conservation Committee (UMRCC). 1993. *50 Years of Conservation through Cooperation*. Rock Island, IL: UMRCC.

UMR Management Act of 1986, Pub. L. No. 99-662, Title XI, 100 Stat. 4082 (1986).

UMR NFWR Act of 1924, ch. 346, sec. 13, 43 Stat. 650, 652 (1924).

U.S. Army Corps of Engineers (USACE or Corps). 1991. *The Plan of Study: Navigation Effects of the Second Lock, Melvin Price Locks and Dam*. February. SL District. St. Louis, MO: USACE.

USACE. 2000. *Planning Guidance ER 1105-2-100*. April. Washington, DC: USACE.

USACE. 2002a. "Environmental Objectives Planning Workshop." Slideshow presented at 37th Meeting of NECC. December 10. Davenport, IA: USACE.

USACE. 2002b. *Environmental Operating Principles*. March. At www.hq.usace.army.mil/cepa/envprinciples.htm.

USACE. 2002c. *Interim Report for the Restructured UMR–IWW System Navigation Feasibility Study*. July. RI District. Rock Island, IL: USACE.

USACE. 2004a. *Final Integrated Feasibility Report and Programmatic Environmental Impact Statement for the UMR-IWW SNFS*. September. Rock Island, IL: USACE.

USACE. 2004b. Report of Lieutenant General Carl A. Strock, USACE. December.

USACE. 2005. "Components of Institutional Arrangements for UMRS." In *October 20–21, 2005 Workshop Handbook*. St. Paul District. St. Paul, MN: USACE.

USACE. 2006. "Environmental Successes Celebrated on Program's 20th Anniversary." *UMR-IWW System Navigation Study Newsletter*, Sept. 1. RI District. Rock Island, IL: USACE.

USACE. 2007a. *Illinois River Basin Restoration Comprehensive Plan with Integrated Environmental Assessment*. March. Rock Island, IL: USACE.

USACE. 2007b. *Institutional Arrangements Information Paper*. March. Rock Island, IL: USACE.

U.S. Army Inspector General Agency Report of Investigation (Case 00-019). n.d. At www.mvr.usace.army.mil.

U.S. Fish and Wildlife Service (USFWS). 2002. "Draft USFWS Coordination Act Report for the UMR-IWW System Navigation Study through August 1, 2001." April. RI Field Office. Rock Island, IL: USFWS.

Water Resources Development Act (WRDA), Pub. L. No. 99-662, sec. 1103, 100 Stat. 4082 (1986).

WEST Consultants, Inc. 2000. *Upper Mississippi River and Illinois Waterway Navigation Feasibility Study–Cumulative Effects Study*, Vol. 2. Prepared by WEST Consultants, Inc., for the U.S. Army Corps of Engineers, RI District. Rock Island, IL: USACE.

Whitney, S. 2007. Personal communication.

Wiener, J. F., C. R. Fremling, D. E. Korschgen, K. P. Kenow, E. M. Kirsch, S. J. Rogers, Y. Yin, and J. S. Sauer. 1998. "Mississippi River." In *Status and Trends of the Nation's Biological Resources, 1* ed. M. J. Mac, P. A. Opler, C. E. Puckeet Haeker, and P. D. Doran. Reston, VA: U.S. Geological Survey, 351–84.

Williams, B. K., R. S. Szaro, and C. D. Shapiro. 2007. *Adaptive Management: The U.S. Department of Interior Technical Guide*. Washington, DC: USDOI, Adaptive Management Working Group.
WRDA, Pub. L. No. 106-53, sec. 509, 113 Stat. 269 (1999).

Chapter 14

Upper Mississippi Restoration Ecology

Putting Theory into Practice

THOMAS L. CRISMAN

The United States government, in the Water Resources Development Act of 1986, defined the Upper Mississippi River System (UMRS) as extending from Minneapolis, Minnesota, to Cairo, Illinois, the confluence with the Ohio River. Also included in the UMRS are the Illinois Waterway from Chicago to Grafton, Illinois, and the navigable portions of the Minnesota, St. Croix, Black, and Kaskaskia rivers. Notably absent in this definition is the segment of the Mississippi River from its origin at Lake Itasca in northern Minnesota south to Minneapolis, the true Mississippi River headwaters.

The overall management approach for this segment of the Mississippi River historically has taken two parallel tracks: (1) move water efficiently and quickly downstream to avoid possible flooding of farms and cities and (2) facilitate navigation on the river by building and maintaining locks and dams. Sometimes the two management tracks appear to be in conflict: one discharges water to prevent flooding, whereas the other retains water for navigation. Currently, there are twenty-nine lock and dam facilities, with thirty-five locks on the UMRS and an additional eight locks on the Illinois Waterway (USACE 2005).

Given the multiple purposes governing river use and the myriad stakeholders, it is critical that a firm theoretical foundation be developed against which goals of sustainability and restoration and actions to achieve both can be evaluated. Any management plan must also consider multiple temporal and spatial scales to be successful. Temporally, biotic and abiotic factors display pronounced daily, seasonal, and interannual variations that must be addressed. Management scenarios must incorporate data on projected, long-term climate change and the significant variances of structure and function along the length of the river. Ecologists, water managers, and environmentalists would agree on one point: it would be extremely difficult to conceive of a single management model that could be applied uniformly along the length of the UMRS and its tributaries.

Integrating Riparian Areas into Management

The riparian zone is loosely defined as that portion of the watershed immediately adjacent to the river that interacts with it on a regular basis. Precise delineation of the landward extent of the zone is problematic, given that direct linkages with the channel reflect the extent and duration of flooding and change with the seasons and from year to year. The riparian zone functions in the transformation and storage of chemicals and sediments from the upper watershed and the stream channel; both types of material can be transferred into the channel. Biologically, the riparian zone can support high levels of primary production that can be the main energy source for stream food webs, especially in upper reaches and headwaters. Primary production is the biomass of energy created by autotrophic photosynthesis.

Vegetated riparian buffer zones can provide effective treatment options for nonpoint chemical and sediment export from agricultural areas in the Midwest and an overall economic gain for the region (Qiu and Prato 2001). The benefit comes from these buffer zones' retention of nutrients and sediments within the local area rather than exporting them downstream, then having to replace their importance via fertilizers. While maximum water delivery efficiency is seen where land slope from field to stream is minimized, ditching and stream dredging in headwaters throughout the Midwest have resulted in steep-sided lotic (moving water) systems, promoting minimum uptake by the riparian zone, regardless of width. Given that approximately 66 percent of the Upper Mississippi River Basin (UMRB) is used for agricultural purposes, with values ranging from 25 percent in the St. Croix and Wisconsin river basins to 95 percent in the Minnesota River Basin (Allan 2004), the potential of riparian buffer zones for treating chemical and sediment export from agricultural lands should be a key component of any UMRS management plan.

The riparian zone also transforms and stores sediments and nutrients originating from the stream proper, ultimately exporting a proportion downstream for biotic utilization (Webster and Patten 1979). The potential importance of the riparian zone as a pollution control management tool for the Mississippi River Basin was highlighted by Mitsch et al. (2001), who calculated that restoring from 78,000 to 200,000 square kilometers (km^2) (from 19 to 48 million acres) of riparian bottomland forest (from 2.7 to 6.6 percent of total basin area) would be sufficient to reduce nitrogen loading to the Gulf of Mexico significantly, by 300,000 to 800,000 metric tons per year.

Forested riparian zones export large quantities of particulate and dissolved organic carbon, derived from the growth and decomposition of riparian vegetation (Sweeney 1993a). Such allochthonous (derived from terrestrial sources) organic matter enters the stream via direct vertical or lateral input of leaves and wood or during flood events. Lower order streams act as heterotrophic ecosystems, with metabolisms driven primarily by such detrital material, while higher order streams are increasingly autotrophic (Vannote et al. 1980). Heterotrophic organisms are those that must obtain organic matter and compounds to consume for energy and cannot produce their own food. Autotrophic organisms, algae and macrophytes (visible aquatic plants), are able to make their own food via photosynthesis when increased levels of nutrients are in the stream.

Macroinvertebrates, aquatic creatures lacking a backbone, have developed various feeding strategies, such as filter feeder, collector–gatherer, predator, scraper, and others (feeding guilds), to process the various forms and particle sizes of organic matter entering a stream (Allan 1995). Many invertebrates have evolved life cycles in parallel with predictable seasonal patterns of organic matter input from individual plant communities, especially the timing of leaves falling. Pronounced differences exist in both the quantity and quality of detritus produced by individual species; thus the successional status (series of changes taking place in an ecosystem after cessation of a disturbance) of the riparian forest can have a direct impact on the structure and function of stream food webs (Roberts 2001). Replacement of mature riparian forests with high biotic diversity by early successional herbaceous species or monocultures of planted or invasive species, such as the exotic purple loosestrife (*Lythrum salicaria*), can lead to alteration in the timing, quantity, and quality of detrital input to streams to disruption of stream food webs from invertebrates to fish (Sweeney 1993b).

Increasingly, the importance of the riparian zone for biological conservation is recognized. Sabo et al. (2005) found that riparian zones support distinct species pools from adjacent uplands and, in all cases, increase species richness regionally by more than 50 percent. In addition, the riparian zone serves as a conservation corridor for species to disperse along the length of the river to reach patches of suitable habitat in fragmented watersheds, is critical for the breeding and foraging of amphibians, and provides refugia (places where organisms can exist or be protected) in isolated water bodies for species lost from the river channel due to contamination (Boon et al. 1992).

While physically and biologically diverse riparian zones are critical for the functioning of stream ecosystems, two key questions remain. The first regards the importance of the riparian zone as a controlling factor for the quality and quantity of water exchange along the hyporheic corridor (area beneath the stream channel that interacts vertically and laterally with groundwater) (Stanford and Ward 1993). The second relates to development of methodology to determine the aerial (overall) extent of the riparian zone needed to maintain stream structure and function.

Headwater Streams

The importance of headwater streams as a control for river structure and function is usually overlooked. This is unfortunate, given that first- and second-order streams comprise approximately 95 percent of all streams and 73 percent of total channel length in North America (Leopold et al. 1992). None of the theories developed for streams consider headwater portions of lotic ecosystems immediately below their point of origin. Even whole watershed studies have focused on higher streams exiting the system.

Throughout the upper Midwest, most of the original headwater stream reaches were a finely dissected network of nondistinct channels or swales, but these have been replaced with highly efficient drainage systems, composed of slotted pipes designed to remove even base flow from the soil. Thus the upper watersheds have lost any capacity to retain water, resulting in a spiked hydrograph in receiving streams after each rain event (single rain episode). Dredging of receiving streams to facilitate increased flows usually accompanies such drainage "improvements," resulting in box cut systems; that is, channel walls are essentially vertical, showing cross-sectional views, with no riparian zone. Mitsch et al. (2001) suggested that much of the natural function of headwater areas for water retention could be replaced by constructing wetlands at the margins of agricultural fields, to collect water drained by the slotted pipe systems and to allow both percolation into the ground and biological transformation of nutrients prior to their discharge into receiving streams. Mitsch et al. (2001) proposed that creating or restoring from 21,000 to 53,000 km^2 (from 5 million to 13 million acres) of wetlands throughout the Mississippi River Basin (from 0.7 to 1.8 percent of total basin area) could reduce nitrogen loading by 300,000 to 800,000 metric tons per year.

The structure and function of headwater streams of the upper Midwest have been seriously impaired by a combination of channelization and increased intensity of spates (flood events). Headwater streams usually have a great degree of habitat heterogeneity. Riffles (light rapids where water flows across a shallow section of river) and pools promote vertical heterogeneity of the channel, and the sediment matrix appears in a broad array of sizes. Allochthonous organic matter from the riparian zone collects in headwater stream channels as wood, mixed debris dams, and leaf packs and serves as a habitat and a food source (leaves) for invertebrates. The residence time of such organic matter at a channel location and the configuration and size of individual detrital patches are reflections of the frequency and intensity of spates. Enhanced drainage networks within the upper watershed and stream channelization have resulted in simplification of stream channels through the scouring action of spates, leading to removal of all organic debris structure and a smoothing of the stream bottom.

Benstead and Pringle (2004) found that agricultural streams had only 13 percent of the leaf litter storage exhibited by comparable forested streams. The patch dynamics concept (Townsend 1989) stresses that habitat patchiness (especially detrital components) are likely to display a predictable seasonality for a given stream reach, related to the frequency of spates and organic matter input, and to vary significantly along the length of the stream. Patches are considered essential to invertebrate colonization and provide critical refugia between successive spates. Elimination of such habitat patchiness will likely alter the food webs of headwater streams profoundly.

Destruction of the riparian zone along headwater streams via maximization of tillable area and/or stream channelization increases sunlight availability in the stream channel and promotes a shift in ecosystem metabolism from being predominantly heterotrophic, driven by allochthonous organic matter from the riparian zone, to dominance by autotrophic production, contributed by macrophytes and attached algae (Hartman and Scrivener 1990). Such a shift in system metabolism typically results in increased density, biomass, and diversity of macroinvertebrates, as well as a shift in dominance from shredders to filter feeders and grazers (Jackson et al. 2001). Dominance by autotrophs also leads to hypoxia (concentrations of oxygen stressful to organisms) and anoxia (absence of oxygen) of streams as a result of algal metabolic processes.

Upper Mississippi and Major Tributary Rivers

The major tributary systems included in the UMRS are the Illinois Waterway from Chicago to Grafton, Illinois, and the navigable portions of the Minnesota, St. Croix, Black, and Kaskaskia rivers. As mentioned, there are twenty-nine lock and dam facilities and thirty-five locks on the UMRS and eight locks on the Illinois Waterway (USACE 2005). Management of the major tributaries and the main channel of the Mississippi River has focused on efficient movement of water downstream while maintaining prescribed water levels in the channel to facilitate navigation. As a result, channels were dredged and control structures (dams and locks) built for navigation, and levees were constructed to isolate the channel from the adjacent riparian/floodplain zone in the name of flood protection. Elimination of lateral connectivity in combination with efficient drainage of headwater farmland and water level stabilization has resulted in increased sedimentation within the main channel and loss of habitat diversity.

Ward and Stanford (1983) developed the serial discontinuity concept to explain the role of constructed control structures, especially dams, on longitudinal structure and function of streams, and later (1995) they extended it to floodplain rivers. Dams disrupt longitudinal patterns in the evolution of stream structure and function, but expected patterns can often return below the channel segment affected by human-altered structure (the discontinuity distance). Poole (2002) noted, however, that each stream confluence within the overall network also acts to promote discontinuities in stream organization. The spacing of natural and human-induced discontinuities downstream determines both the discontinuity distance and whether a stream or river as a whole can recover from those influences.

Connell (1978) developed the intermediate disturbance hypothesis to explain the positive influence of a moderate level of disturbance on ecosystem biotic diversity, and Townsend et al. (1997) quantified its importance for the biotic structure of lotic systems. Of all the potential disturbance factors, bed disturbance accounted for the most variance in biotic diversity in the fifty-four streams studied. While intermediate levels of substrate disturbance can stimulate biotic diversity, hydrological alterations to increase the periodicity and intensity of spates and floods, plus active channel dredging, are likely posing severe threats to biotic diversity of the Upper Mississippi and its tributaries. The UMRB is home to 25 percent of fish species in the United States and 20 percent of mussel species (Weitzell et al. 2003). Currently, the basin has 286 state-listed or candidate species and 36 federally listed or candidate species of threatened or endangered biota (those species in danger of extinction as a result of human alteration of the environment) considered endemic. Many of the taxa (species) have intricate interactions with the substrate for both breeding and feeding, but few data exist relating levels of bed disturbance to species' responses. The ability of many native fishes to reproduce is threatened by the loss of periodic lateral connectivity of the river with large segments of its floodplain (Gutreuter et al. 1999) and by channel control structures hindering upstream movements (Weitzell et al. 2003).

Additionally, aquatic and riparian food webs have been altered throughout the basin. Growth of invasive plants, such as purple loosestrife and autumn olive (*Elaeagnus umbellata*), not only changes the quality and quantity of allochthonous organic matter entering streams, but plants that are nitrogen fixers, like the autumn olive, can actually enhance nitrate loading to streams (Church et al. 2004). Increased nutrient loading has promoted productivity of benthic (on river bottom or in bottom sediments) and epiphytic (nonparasitic growth on other plants) algae in small streams and phytoplankton (suspended microscopic algae) in larger rivers. The importance of these algal communities is that production has shifted to substrate attached forms; therefore, the structure of the animal community will also shift to those taxa capable of feeding on attached rather than suspended organic matter. When phytoplankton are dominant, the structure of the animal community shifts to those species that feed within the water column via filter feeding mechanisms.

Although macrophytes are threatened by phytoplankton production and suspended sediment in the main channel and major tributaries, the practice of altering flooding regimes to promote expansion of macrophytes in the Illinois River may be challenged by expanding populations of grass carp (*Ctenopharyngodon idella*), an exotic herbivorous fish native to Asia (Raibley et al. 1995). However, excessive macrophytes in rivers reduce the flow of water by clogging the channel and can stunt fish populations by altering the ability of predator fish to see their prey, resulting in many small and fewer large fish. Superfluous macrophytes also cause shifts in animal communities to favor forms that feed on substrates.

Putting Management into Proper Perspective

Two important questions must be addressed by any large-scale restoration effort of watersheds: (1) What is the reference condition against which to evaluate the success of individual and overall management and restoration efforts? (2) Where in the watershed should such efforts be conducted to maximize the cost to environment benefit ratio?

Identification of a reference or control against which to evaluate management and restoration actions is a vexing problem in all stream and river investigations. At the large scale, no two river basins are similar enough for direct comparison, while intrabasin comparisons are plagued by the facts that streams and rivers evolve significantly along their stream length and that sampling, even within short reaches of streams, can lead to statistical problems associated with pseudoreplication (Karr and Chu 1999). The classical approach of comparing river reaches above (for control) versus below (for treatment) pollution inputs is rarely valid at either small or large scales. Sadly, one assessment model does not fit all streams, and monitoring schemes are moving targets in time and space. Increasingly, scientists are questioning whether the concept of a reference or control condition is valid for either streams or rivers.

The second question regarding where in the watershed to focus management and restoration efforts for maximum cost benefit and environmental effectiveness is somewhat less complex. Without implementation of best management practices in and around headwater streams, there is little likelihood for sustainability of management and restoration schemes downstream. Unfortunately, water managers often overlook headwaters in their equations, not recognizing their additive effects on observed conditions lower in the watershed, especially sediments, nutrients, and hypoxia/anoxia. The old adage that the solution to pollution is dilution, sending pollutants downstream, has been debunked in favor of containment, keeping potential contaminants as near to their sources as possible.

Those who oversee the ecology of streams and large rivers, such as managers of wetlands and lakes, must consider vertical (channel to groundwater) and horizontal (channel to riparian zone) interactions in their management equations and recognize that these relationships, especially the importance of surface versus groundwater sources, evolve along the course of the system. We know that dams and other channel alterations can disrupt broad-scale evolving patterns in downstream structure and function, as outlined in the serial discontinuity concept (Ward and Stanford 1995), and require costly management. Water managers must address environmental contaminants at their sources throughout the UMRB, which can often be accomplished at a fraction of the cost of in-channel management. Responsible, informed decision making is essential for sustainable management of this critically important river system.

References

Allan, J. D. 1995. *Stream Ecology: Structure and Function of Running Waters*. London: Chapman and Hall.

Allan, J. D. 2004. "Landscapes and Riverscapes: The Influence of Land Use on Stream Ecosystems."*Annual Review of Ecology, Evolution and Systematics* 35:257–84.

Benstead, J. P., and C. M. Pringle. 2004. "Deforestation Alters the Resource Base and Biomass of Endemic Stream Insects in Eastern Madagascar."*Freshwater Biology* 49:490–501.

Boon, P. J., B. R. Davies, and G. E. Petts. 1992. *Global Perspectives on River Conservation*. New York: Wiley and Sons.

Church, J. M., K. W. J. Williard, S. G. Baer, J. W. Groninger, and J. J. Zaczek. 2004. "Nitrogen Leaching below Riparian Autumn Olive Stands in the Dormant Season." In *Proceedings, 14th Central Hardwood Forest Conference*, eds. D. A. Yaussy, D. M. Hix, R .P. Long, and P. C. Goebel. General Technical Report NE-316. Newtown Square, PA: USDA Forest Service, Northeastern Research Station, 211–16.

Connell, J. H. 1978. "Diversity in Tropical Rain Forests and Coral Reefs." *Science* 199:1302–10.

Crisman, T. L., C. Mitraki, and G. Zalidis. 2005. "Integrating Vertical and Horizontal Approaches for Management of Shallow Lakes and Wetlands." *Ecological Engineering* 24:379–89.

Gutreuter, S., A. D. Bartels, K. Irons, and M. B. Sandheinrich. 1999. "Evaluation of the Flood-pulse Concept Based on Statistical Models of Growth of Selected Fishes of the Upper Mississippi River System." *Canadian Journal of Fisheries and Aquatic Sciences* 56:2282–91.

Hartman, G. F., and J. C. Scrivener. 1990. "Impacts of Forestry Practices on a Coastal Stream Ecosystem, Carnation Creek, British Columbia." *Canadian Bulletin of Fisheries and Aquatic Sciences* 223:1–148.

Jackson, C. R., C. A. Sturm, and J. M. Ward. 2001. "Timber Harvest Impacts on Small Headwater Stream Channels in the Coast Ranges of Washington." *Journal of the American Water Resources Association* 37:1533–49.

Karr, J. R., and E. W. Chu. 1999. *Restoring Life in Running Waters*. Washington, DC: Island Press.

Leopold, L. B., M. G. Wolman, and J. P. Miller. 1992. *Fluvial Processes in Geomorphology*. Mineola, NY: Dover Publications.

Mitsch, W. J., J. W. Day, Jr., J. W. Gilliam, P. M. Groffman, D. L. Hey, G. W. Randall, and N. Wang. 2001. "Reducing Nitrogen Loading to the Gulf of Mexico from the Mississippi River Basin: Strategies to Counter a Persistent Ecological Problem." *BioScience* 51:373–88.

Poole, G. C. 2002. "Fluvial Landscape Ecology: Addressing Uniqueness within the River Discontinuum." *Freshwater Biology* 47:641–60.

Qiu, Z., and T. Prato. 2001. "Physical Determinants of Economic Value of Riparian Buffers in an Agricultural Watershed." *Journal of the American Water Resources Association* 37:295–303.

Raibley, P. T., D. Blodgett, and R. E. Sparks. 1995. "Evidence of Grass Carp (*Ctenopharyngodon idella*) Reproduction in the Illinois and Upper Mississippi Rivers." *Journal of Freshwater Ecology* 10:65–74.

Roberts, C. R. 2001. "Riparian Tree Associations and Storage, Transport, and Processing of Particulate Organic Matter in a Subtropical Stream." Ph.D. Dissertation. Gainesville, FL: University of Florida.

Sabo, J. L., R. Sponseller, M. Dixon, K. Gade, T. Harms, J. Heffernan, A. Jani, G. Katz, C. Soykan, J. Watts, and A. Welter. 2005. "Riparian Zones Increase Regional Species Richness by Harboring Different, Not More, Species." *Ecology* 86:56–62.

Stanford, J. A., and J. V. Ward. 1993. "An Ecosystem Perspective of Alluvial Rivers: Connectivity and the Hyporheic Corridor." *Journal of the North American Benthological Society* 12:48–60.

Sweeney, B. W. 1993a. "Streamside Forests and the Physical, Chemical and Trophic Characteristics of Piedmont Streams in Eastern North America." *Water Science and Technology* 26:2653–71.

Sweeney, B. W. 1993b. "Effects of Streamside Vegetation on Macroinvertebrate Communities of White Clay Creek in Eastern North America." *Proceedings of the Academy of Natural Sciences of Philadelphia* 144:291–340.

Townsend, C. R. 1989. "The Patch Dynamics Concept of Stream Community Ecology." *Journal of the North American Benthological Society* 8:36–50.

Townsend, C. R., M. R. Scarsbrook, and S. Doledec. 1997. "The Intermediate Disturbance Hypothesis, Refugia and Biodiversity in Streams." *Limnology and Oceanography* 42:938–49.

U.S. Army Corps of Engineers (USACE). 2005. "Upper Mississippi River System Navigation and Environmental Sustainability Program Implementation Plan for First Increment (15-year)." Fiscal year 2005. Working Draft.

Vannote, R. L, G. W. Minshall, K. W. Cummins, J. R. Sedell, and C. E. Cushing. 1980. "The River Continuum Concept." *Canadian Journal of Fisheries and Aquatic Sciences* 37:130–37.

Ward, J. V., and J. A. Stanford. 1983. "Serial Discontinuity Concept of Lotic Ecosystems." In *Dynamics of Lotic Systems*, eds. T. D. Fontaine and S. M. Bartell. Ann Arbor, MI: Ann Arbor Science, 29–42.

Ward, J. V., and J. A. Stanford. 1995. "The Serial Discontinuity Concept: Extending the Model to Floodplain Rivers." *Regulated Rivers: Research and Management* 10:159–68.

Webster, J. R., and B. C. Patten. 1979. "Effects of Watershed Perturbations on Stream Potassium and Calcium Dynamics." *Ecological Monographs* 49:51–72.

Weitzell, R. E., J. L. Khoury, P. Gagnon, B. Schreurs, D. Grossman, and J. Higgins. 2003. *Conservation Priorities for Freshwater Biodiversity in the Upper Mississippi River Basin*. Arlington, VA: NatureServe and the Nature Conservancy.

Chapter 15

Comparing Apples and Oranges?

Costs and Benefits of Upper Mississippi River System Restoration

STEPHEN POLASKY

The Upper Mississippi River flows through the heart of the upper Midwest, draining the majority of the surface area of Illinois, Iowa, Minnesota, Missouri, and Wisconsin. A means for transporting bulk items, including grain, coal, and other materials, the Mississippi River links Minneapolis and other cities along its path to the Atlantic Ocean through the port of New Orleans on the Gulf of Mexico. The river provides recreation benefits, such as boating, fishing, and wildlife viewing. The Upper Mississippi is the heart of a vibrant and dynamic ecological system that provides habitat for numerous terrestrial and aquatic species. Naturally, many segments of the public have a great interest in all proposed plans that affect the river in any significant way.

While the Upper Mississippi River System (UMRS) provides many things to many people in the upper Midwest, it cannot provide all things to all people simultaneously. Many of the demands placed on the river system conflict at least to some degree. For example, one cannot simultaneously have a natural river restored to conditions pre-1900 and accommodate barge transportation on the river at current levels of demand. Myriad trade-offs require countless difficult decisions.

The U.S Army Corps of Engineers (USACE or the Corps) has the unenviable task of trying to reconcile the often conflicting environmental and economic objectives for the UMRS, which includes the Mississippi River from the mouth of the Ohio River to Minneapolis and the Illinois Waterway from the Mississippi River to Chicago. In 1988, the Corps began to investigate the need to expand locks on the Upper Mississippi and in 1993 initiated a study of the benefits and costs of expanding locks and dams and making other improvements to transportation to meet transportation needs from 2000 to 2050. After more than a decade of controversy and outside review, including two reviews by the National Research Council, in 2004 the Corps released the *Upper Mississippi River–Illinois Waterway System Navigation Feasibility Study: Final Integrated Feasibility Report and Programmatic Environmental Impact Statement* (USACE 2004a, hereafter UMR–IWW Feasibility Study). With the release of the report, Lt. Gen. Carl A. Strock of the Corps said, "I am confident that our plan balances the need for economic growth and environmental sustainability" (USACE 2004b). However, striking such a balance and getting interested parties to agree that the plan is properly balanced can be more easily said than done.

The challenge facing the Corps as it performed the analysis for the *UMR–IWW Feasibility Study* and recommended a preferred alternative has several dimensions. How to balance the value of improved transportation benefits, typically measured in dollar terms, versus the value of ecosystem restoration, typically not measured in dollar terms, is far from clear. This involves a comparison of apples and oranges, and different observers may draw quite different conclusions. Even with a clear way to balance economic and environmental benefits, decision makers still had the difficult tasks of estimating economic benefits on the one hand and environmental benefits on the other. Each of these tasks raises complex questions. Corps members' and observers' understandings of ecological systems or economic markets could be inadequate, or data on which to base empirical estimates could be insufficient.

Estimating Transportation Benefits of Expanding Locks

The Corps, responsible for transportation infrastructure on the UMRS, built and operates twenty-nine locks and dams and maintains a 9-foot channel from Minneapolis to the mouth of the Ohio River. The Corps also maintains the Illinois Waterway from the Mississippi River to Chicago. In 2002, the UMRS carried 50 percent of U.S. corn exports and 40 percent of U.S. soybean exports, as well as coal, other energy products, and other goods (USACE 2004a). The Corps and others whose livelihoods depend upon the river are concerned that the UMRS is inadequate to meet current and future transportation needs because the locks and dams are too old and the system's capacity is insufficient.

Many of the locks and dams, built in the 1930s, have outlasted original estimates of their useful life spans. Locks were built to accommodate 600-foot barges. Many barges today are 1,200 feet long. Barge operators have to uncouple longer barges to go through the locks one-half at a time and recouple them after both halves have gone through; this process adds time and cost to barge transport. Delays due to traffic congestion at the locks add to trip times and costs. The Corps estimates that during the period from 1990 to 1999, delays averaged between three and four hours at Locks 20 through 25, with lesser delays farther upriver (USACE 2004a). Expanding the locks and modernizing the system, however, costs money. From an economic point of view, one would want to know whether the benefits of expanding the locks exceed the costs of doing so. A central focus of the *UMR–IWW Feasibility Study* was to answer this question.

The Corps has established procedures for conducting analyses of its projects set forth in the *Economic and Environmental Principles and Guidelines for Water and Related Land Resource Implementation Studies* (USWRC 1983). As set out by the Principles and Guidelines, the Corps is to estimate a project's contribution to national economic development (NED), which measures the monetary impact of the project in terms of benefits and costs. In the case of expansion of the locks, the benefits would include reduction in transportation costs, while the costs would include construction costs and increased operations and maintenance expenses. Among alternatives considered by the Corps, the alternative that maximizes NED, subject to environmental constraints, must be included. Other alternatives that may score lower on NED but do better on any of three other accounts, environmental quality (EQ), regional economic development (RED), or other social effects (OSE), can also be considered. Under the Principles and Guidelines, the benefits of environmental improvement are considered in the environmental quality account, but no attempt is made to estimate monetary values of these benefits and include them in NED.

The *UMR–IWW Feasibility Study* considered six alternative plans: alternative 1, no action; alternative 2, which considered nonstructural changes (such as congestion fees); and four alternatives that considered various levels of structural changes involving moorings and lock expansion. The Corps estimated NED under several different assumptions for each alternative, using five different scenarios for future transportation demands and three different methods for figuring demand elasticity, resulting in fifteen estimates of NED for each alternative. The results of the analysis are summarized in figure 15.1.

Compared with the no action alternative (alternative 1), four alternatives generated positive NED at least under some assumptions. Of these, the only alternative that generated positive NED under all assumptions was alternative 2, which made no structural changes to the system but changed river traffic management through congestion fees. Alternatives 4, 5, and 6 all make structural changes to the system, with alternative 6 being the most aggressive. For each of the structural change alternatives, a negative NED results when transportation demand does not grow or a high elasticity of demand is present. When transportation demand growth is high and a low elasticity of demand is assumed, positive NED results.

These results illustrate that the answer to the question of what is the most preferred alternative, even when considered strictly on national economic

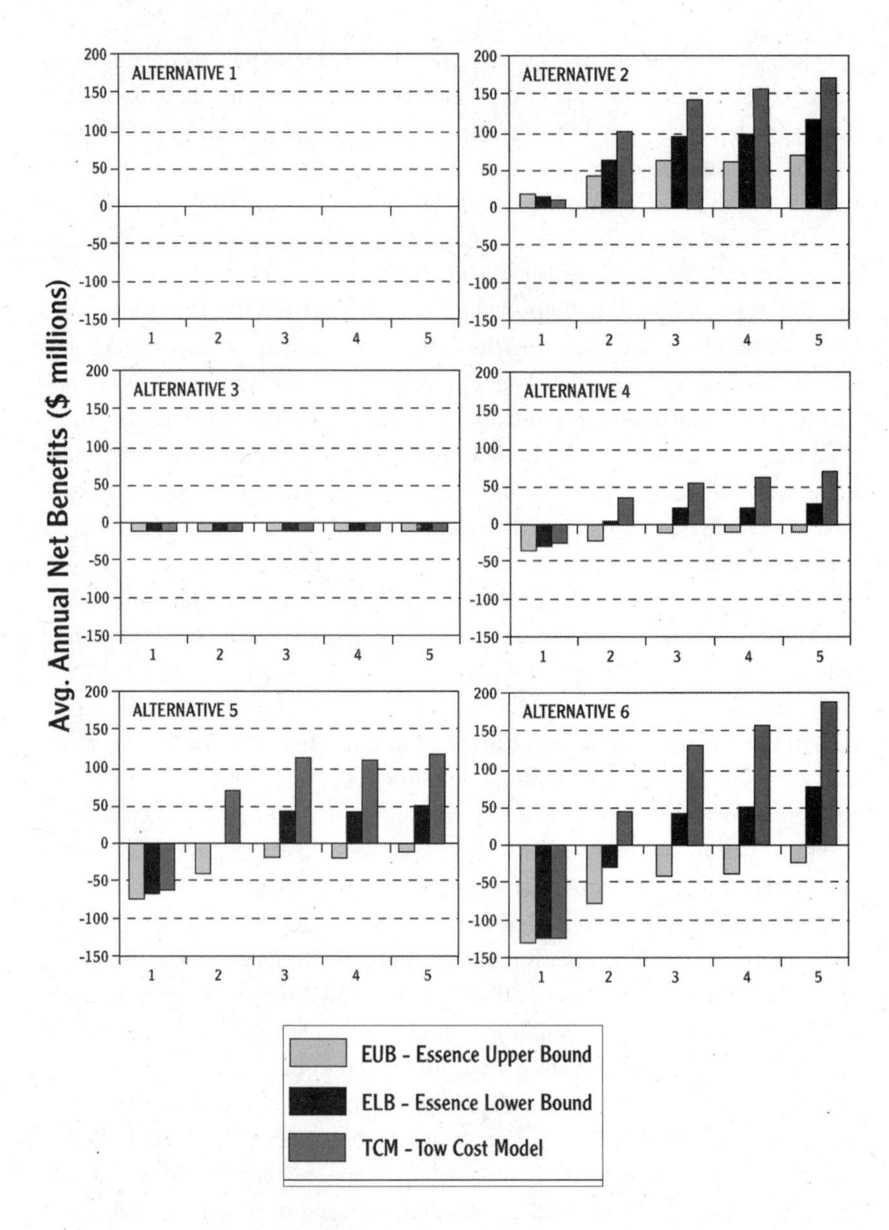

FIGURE 15.1. Average Annual Net Benefits ($ millions) for Navigation Efficiency Alternatives across the Range of 15 Possible Economic Conditions Created by the Use of Five Scenarios and Three Economic Models (USACE 2004c, vi).

development terms—ignoring environmental, distributional, and other social concerns considered in other accounts—is, It depends. The preferred alternative depends upon what the likely growth in traffic on the river will be; how easy it will be to switch to alternative modes of transportation, such as rail or truck; and whether one assumes that nonstructural policy changes can be made to manage congestion on the river.

To go further than this somewhat unhelpful answer of "it depends" requires a more detailed look at each of three issues: (1) projected growth in demand for river transportation, (2) degree of substitution among alternative means of transportation, and (3) availability of nonstructural congestion management options such as congestion fees. Each of these three issues was highlighted in the National Research Council (NRC) studies that reviewed the initial and final drafts of the Corps' feasibility study (NRC 2001; NRC 2004a). The NRC was highly critical of the Corps' approach on each issue in the initial draft of the feasibility study (NRC 2001). The Corps responded to these criticisms by expanding its analysis to include alternative 2, which focused on nonstructural management changes, and by including a greater range of uncertainty about transportation demand growth and demand elasticity.

One difficulty in estimating benefits of infrastructure improvement projects is that such projects have long lives. The benefits of a project continue to accrue over the life of the project, necessitating estimates of conditions well into the future. Estimating the benefits of expanding the locks on the UMRS requires estimates of demand for river transportation decades into the future. Estimates of future demand can be derived in several ways, including statistical time series analysis of past trends or specifying a structural model that predicts demand, based on changes in fundamental conditions. Estimating the future based on past trends is subject to the criticism that the future may not be like the past. Technological changes or shifts in production or consumption patterns can cause large shifts in transportation patterns. A cautionary tale along these lines is illustrated by the decision to expand the airport at Gander, Newfoundland, in the early 1950s (NRC 2001). At that time, Gander served as an important refueling station for airplanes flying between North America and Europe. Traffic was growing rapidly, and the decision was made to expand capacity. However, the expansion occurred at about the same time that Boeing introduced jets capable of flying between cities in the United States and Europe without refueling. Gander was left with a large facility and virtually no demand.

Building a structural model and using it to forecast future conditions can potentially avoid such pitfalls but involves more work and more data. A structural model of grain shipments on the UMR would, at minimum, need to include components capable of predicting grain production in the upper Midwest, local grain demand (including demand for ethanol), foreign grain demand, and alternative means of transporting grain from the upper Midwest. If data or understanding about any of these components is lacking, the resulting structural model may not predict well.

A third alternative is for engineers simply to use a range of scenarios that they hope span the spectrum of possible future conditions. In their analysis, members of the Corps chose this approach. They analyzed five potential demand scenarios for grain shipments, ranging from a slight downturn in demand to several cases with moderate to large increases. The inclusion of a range of scenarios was an improvement over the initial draft of the feasibility study (USACE 2000), in which only a single prediction was used. Even with this improvement, what is still lacking is confidence that these scenarios really span the range of potential conditions and some sense about which scenarios are more likely. The NRC report commented that important changes in world grain markets would be likely to affect grain shipments on the UMRS and should be reflected in the analysis of demand (NRC 2004). Income growth in China could lead to rapid growth in demand for grain exports, or alternatively, it could lead to growth of demand for beef exports and a decline in grain exports. Competition from South American producers might reduce grain exports. Without further data on these issues, the NRC found it difficult to have much confidence in the Corps' opinion regarding the trend in grain shipments on the UMRS. The NRC was also critical of the assumption that nongrain transport would grow by 20 percent over a fifty-year period, without consideration of the underlying uncertainty in this estimate.

Demand for ethanol has led to a major shift in midwestern agricultural markets. The rapid expansion of ethanol production has increased demand for corn within the upper Midwest and decreased demand to ship grain on the UMRS. Annual ethanol production increased from 1.9 million barrels in 2000 to 4.9 million barrels in 2006 (RFA 2007). Production capacity on an annual basis was 5.6 billion gallons in early 2007, but an additional 6.2 billion gallons of capacity were planned or under construction (Baker and Zahniser 2007). In July 2006, the Food and Agricultural Policy Research Institute projected that the amount of corn for ethanol would surpass the amount of corn exported in 2006–07 and that corn exports would decline in coming years (IATP 2006, fig. 6,7).

The degree to which benefits accrue to society when an expansion in capacity for river transportation occurs also depends upon the availability of substitute modes of transportation. If the costs of transporting grain or other commodities on the Mississippi River and by other alternatives, such as by rail transport, are virtually identical, then almost no benefit derives from expanding river transport capacity. In its modeling efforts to date, the Corps has not attempted to include rail or other substitute transportation methods. The Corps has collected little evidence on the elasticity of substitution among alternative transportation methods. Therefore, it is not clear what degree of substitution exists between river transport and alternative means of transport.

The lack of information about substitutes combined with a lack of information about demand growth means that a wide range of estimates regarding transportation benefits from UMRS lock expansion is possible. This range includes benefit values that are less than the costs of lock expansion. In other words, considering NED by itself, without trying to factor in environmental or other social values, lock expansion might generate negative net benefits and, hence, not be justified. To make a case for building locks, analysts need to have a sense that cases giving rise to negative net benefits are unlikely, so that net benefits on average are positive. Given the current state of knowledge and evidence, it is hard to know whether this is true.

The other major area of contention in the economic analysis of lock expansion is the treatment of nonstructural policy changes. Congestion fees or slot allocations could shift use to less congested time periods and reduce congestion with the existing infrastructure. In its second report, the NRC stated, "Implementing some nonstructural measure for managing waterway congestion could decrease congestion, reduce shipping costs, and use the existing waterways more efficiently. Because the costs of implementing nonstructural measures are low and because some have positive net benefits, implementation of these measures should be of the highest priority" (NRC 2004a, 6). In fact, alternative 2, which considered nonstructural changes such as congestion fees, was the *only* alternative that generated positive NED estimates under the complete range of demand and elasticity estimates used by the Corps (see fig. 15.1). Despite these results, the Corps did not consider the alternative of nonstructural changes because "it fails to fully meet the planning objectives of economic sustainability by limited growth on the system. In addition, current law prohibits congestion fees, and the current national policy makes institutional acceptability of this alternative doubtful" (USACE 2004a, x).

One area of agreement between the Corps and the NRC is the need to use adaptive management; using this system, the Corps would make decisions in stages, carefully monitor the consequences of its decisions, and modify decisions based upon new information learned from monitoring. Some form of adaptive management is important in cases where potentially costly mistakes would be difficult to reverse and could be avoided through gathering more information. Within economics and management sciences, the combination of avoiding irreversible decisions and preserving options in the face of uncertainty is known as "option value" (Arrow and Fisher 1974; Dixit and Pindyck 1994; Freeman 1993).

Given the current state of uncertainty over demand growth and elasticity, decision makers would be prudent to postpone making costly and largely irreversible investments in infrastructure. Nonstructural policy changes should be pursued immediately because they are not costly, appear to have positive net benefits, and can be changed relatively easily. But structural changes are costly, appear to have negative net benefits for a range of plausible conditions, and are difficult to change. If it turns out in the future that demand for river transport grows, the locks can be expanded, with the only loss being a temporary increase in congestion. Alternatively, if demand does not grow, decision makers can avoid the expense of expanding the locks and prevent an outcome similar to the airport at Gander. What is prudent and what comes from political processes, however, may not align. According to Senator Christopher (Kit) Bond (R-MO), "We cannot wait for all the economists to agree. We need to modernize our antiquated river transportation system before we get whipped by foreign competition" (Grunwald 2001).

Estimating Ecosystem Restoration Benefits

Over time, the Corps' responsibilities have expanded beyond maintaining the transportation infrastructure on the UMRS to include ecological restoration. Because changes in transportation infrastructure can have major impacts on ecosystems, it is appropriate that the Corps explicitly consider ecological restoration as one of its goals. In fact, the larger part of requested funding recommended in the *UMR–IWW Feasibility Study* is for ecosystem restoration, $5.3 billion initial expenses for ecosystem restoration versus $2.4 billion initial expenses for transportation infrastructure improvement (USACE 2004a).

The Corps faces two major challenges in considering the effects of its actions on the ecosystem. First, analysis of ecological restoration is hindered by a lack of scientific understanding and evidence about the dynamics of complex river ecosystems (see chapter 14, this volume). Given this lack of understanding, it is hard to predict how a restoration project or combinations of projects would contribute to attaining specified ecosystem restoration objectives. With large current uncertainties over the potential effects of projects, it makes sense for scientists to adopt adaptive management of ecosystem restoration efforts, which the Corps has endorsed.

Second, even if uncertainties about the dynamics of complex river ecosystems could be overcome and clear predictions of the effects of alternatives on ecosystems were in hand, how should we compare ecological end points with economic end points? Suppose, for example, that alternative 1 has greater ecosystem restoration benefits but costs more than alternative 2. Ranking alternative

1 versus alternative 2 requires a judgment of whether the ecosystem restoration benefits are worth the additional cost. Comparing ecological objectives, measured in biophysical terms, and economic objectives, measured in monetary terms, is something of an apples and oranges comparison. One could try to measure the value of ecosystem restoration benefits in monetary terms, thereby comparing apples and apples. Active, ongoing research programs exist to estimate the value of ecosystem services (Daily 1997; Millennium Ecosystem Assessment 2005; NRC 2004b). Such efforts are not without difficulties, and it is not clear that a credible comprehensive estimate of the value of ecological benefits for a system as large and complex as the UMR can be obtained (NRC 2004b).

The Corps analyzes ecosystem restoration projects, judged primarily on meeting ecosystems' objectives, separately from its analysis of transportation improvements projects, judged primarily on meeting national economic development goals. These differing objectives are not directly compared. Nevertheless, in reality, ecological and economic issues cannot be separated. Transportation improvement projects done for economic reasons have negative environmental consequences, which, in turn, necessitate ecological restoration projects. Ecological restoration projects cost a lot of money. Implicitly, the Corps makes trade-offs between environmental quality and economic returns in its decisions. These trade-offs are not made explicit; thus it is hard to judge whether trade-offs implied in transportation improvement plans and ecological restoration plans are consistent or are working at cross purposes.

Ideally, the Corps would develop an integrated assessment showing how all of its activities, including transportation improvement and ecological restoration projects, affect the suite of ecological and economic objectives. With an integrated assessment from the Corps, scientists and water managers could more clearly see whether the Corps' package were internally consistent, as well as more clearly determine the implicit or explicit weights placed on ecological versus economic objectives. As Lt. Gen. Carl A. Strock said after the *UMR–IWW Feasibility Study* was released: "We recognize the need to improve our evaluation of economic and ecosystem restoration matters" (USACE 2004b).

Conclusions

Given the large scale of the analysis involved in the *UMR–IWW Feasibility Study*, it is perhaps not surprising that some questions cannot be resolved fully with our current state of understanding and data. Given the high stakes involved on both ecological and economic issues, it is also perhaps not surprising that the Corps' plans for the UMRS have been enveloped in controversy. To the Corps' credit, its members have broadened the scope of their mission to include ecological restoration and have broadened the process of reaching decisions to include more input from interested stakeholders (see chapter 13, this volume). Whether these changes will prove to be sufficient to blunt criticism of the process or the substance of decisions made by the Corps on the Upper Mississippi remains to be seen.

Certainly, the potential for further controversy remains. The economic analysis contained in the *UMR–IWW Feasibility Study* gives ammunition to both critics and boosters of lock and dam expansion on the UMRS. Critics can point out that the Corps continues to downplay or ignore the potential for nonstructural changes in policy to address congestion on the river, as well as to reasonable projections of future demand for river transport that do not justify expansion. Boosters can counter by pointing to other reasonable projections of future demand that would justify expansion. Ecosystem restoration projects could also prove to be controversial because they potentially would cost much more than lock and dam expansion, and the potential benefits of such projects at this point are unproven and largely unknown.

Whatever initial decision is made, the Corps will need to reevaluate decisions and may need to shift course should new information show that its initial plans are ineffective or inefficient. Such changes are possible in ecosystem restoration projects. However, it is difficult to shift course without incurring great expense once infrastructure is in place. This fact places a greater burden of proof on the Corps for lock and dam expansion. Not only does the Corps need to present evidence that the expected net benefits of expansion, which should include ecological as well as economic effects, are positive, but also that waiting to gain more information is not justified. At present, evidence is lacking to justify the case for lock and dam expansion on the UMRS.

References

Arrow, K., and A. Fisher. 1974. "Environmental Preservation, Uncertainty, and Irreversibility." *Quarterly Journal of Economics* 88:312–19.

Baker, A., and S. Zahniser. 2007. *Ethanol Reshapes the Corn Market. Amber Waves: The Economics of Food, Farming, Natural Resources and Rural America*. U.S. Department of Agriculture, Economic Research Service. At www.ers.usda.gov/AmberWaves/May07SpecialIssue/Features/Ethanol.htm.

Daily, G. 1997. *Nature's Services: Societal Dependence on Natural Ecosystems*. Covelo, CA: Island Press.

Dixit, A., and R. Pindyck. 1994. *Investment under Uncertainty*. Princeton, NJ: Princeton University Press.

Freeman, A. M. III. 1993. *The Measurement of Environmental and Resource Values: Theory and Methods*. Washington, DC: Resources for the Future.

Grunwald, M. 2001. "Public Works Study Halted; Army Corps' Analyses of Miss. River Projects Faulted." *Washington Post*, March 1.

Institute for Agriculture and Trade Policy (IATP). 2006. *Staying Home: How Ethanol Will Change U.S. Corn Exports*. Minneapolis: IATP.

Millennium Ecosystem Assessment. 2005. *Living Beyond Our Means: Natural Assets and Human Well-Being*. At www.millenniumassessment.org.

National Research Council (NRC). 2001. *Inland Navigation System Planning: The Upper Mississippi River–Illinois Waterway*. Washington, DC: National Academies Press. At www.nap.edu/books/0309074053/html/R1.html.

NRC. 2004a. *Review of the U.S. Army Corps of Engineers Restructured Upper Mississippi River–Illinois Waterway Feasibility Study: Second Report*. Washington, DC: National Academies Press.

NRC. 2004b. *Valuing Ecosystem Systems: Toward Better Environmental Decision-Making*. Washington, DC: National Academies Press.

Renewable Fuels Association (RFA). 2007. *Historic U.S. Fuel Ethanol Production*. At www.ethanolrfa.org/industry/statistics/#A.

U.S. Army Corps of Engineers (USACE). 2000. "Upper Mississippi River–Illinois Waterway System Navigation Study Feasibility Report." Draft. Rock Island, IL: USACE.

USACE. 2004a. *Upper Mississippi River-Illinois Waterway System Navigation Feasibility Study: Final Integrated Feasibility Report and Programmatic Environmental Impact Statement* (*UMR-IWW Feasibility Study*). Vicksburg, MI: Mississippi Valley Division Corps of Engineers.

USACE. 2004b. "Chief of Engineers Recommends Ecosystem Restoration and Navigation Improvements for Upper Mississippi River and Illinois Waterway." News Release PA-04-27. At www.hq.usace.army.mil/cepa/releases/uppermissrpt.htm.

USACE. 2004c. "Average Annual Net Benefits ($millions) for Navigation Efficiency Alternatives across the Range of 15 Possible Economic Conditions Created by the Use of Five Scenarios and Three Economic Models Figure EX1." Rock Island, IL: USACE. In Final Integrated Feasibility Report (USACE 2004a).

U.S. Water Resources Council (USWRC). 1983. Economic and Environmental Principles and Guidelines for Water and Related Land Resource Implementation Studies. At www.iwr.usace.army.mil/iwr/pdf/p&g.pdf.

Conclusion

Assessing Ecosystem Restoration Projects

MARY DOYLE

The five fascinating ecosystem restoration projects that are the subjects of this book share many of the qualities of the natural systems to which they are dedicated. They sprawl across broad geographical and political landscapes, embracing many component governmental entities and interest groups, which interact in complex ways that are sometimes not well understood. They are designed to endure over long periods of time. And like the natural systems with which they are concerned, these projects are bound to experience almost continuous, often profound, change. Scientific uncertainties are resolved or gain in importance. As science casts more light on the intricacies of ecosystem function, restoration plans may be exposed as inadequate or ill-founded, and innovation will be required to discover new paths leading forward. The natural systems themselves alter with time, influenced by drought, flood and fire, global climate change, invasion of exotic species, and other forces.

Constant change also characterizes the political realm, as elected leaders come and go and public attention waxes and wanes. With the inevitable shifts in the political and economic climates, the priority accorded to environmental issues in general and to these large restoration projects in particular rises or falls, and the levels of public funding and support fluctuate accordingly. Thus, in the post–9/11 world, we have seen national priorities move overwhelmingly toward national security and away from other domestic concerns, including the environment.

The case studies presented in this book are like photographs, capturing views of the challenges facing those engaged in such enormous undertakings. But we know as the projects move forward, the pictures will be shifting continuously. The purpose of this conclusion is to offer a checklist for the interested observer—whether government executive, legislator, stakeholder, student, or concerned citizen—to employ in assessing the progress of a project beyond what is accounted for here. The checklist may also prove useful to those concerned with judging the viability and chances for success of early plans for restoration. The list focuses on six interrelated aspects of a restoration effort that, if fully explored, provide the basis for a reliable evaluation of how the project is structured and performing. It should also prove helpful in penetrating the welter of official and unofficial documents published by and about such projects.

Even with the checklist firmly in hand, however, a researcher may find these undertakings daunting. Simple questions, like how much public money was spent on the project in the last year, can be extremely difficult to answer accurately. This is because each of the responsible governmental entities follows different budgeting procedures, has a different fiscal year, and reports spending in terms of varying categories and accounts not necessarily identified with the project by name. The South Florida Ecosystem Restoration Task Force is one entity that has done exemplary work in developing techniques for researching and reporting cross-cut budgets, which account for public spending by all sources in the same time period, and for other efforts at comprehensive reporting on the progress of Everglades restoration. It would be a real step forward if responsible agencies were to agree on a common accounting and reporting system for a project, so that their activities and expenditures could be compiled and publicly reported on a regular basis in an understandable, comprehensive way.

Where an evaluation of the project reveals deficiencies or failures and the need arises for corrective action or new initiatives, the responsible decision makers must be called to account. This is easier to say than to do, however. Transparency and accountability are hard to achieve when responsibility is shared among many government entities with no hierarchy of authority and no single agency has the mandate and authority to carry out or oversee all aspects of the program. Recognizing this problem, the CALFED Bay-Delta Program and its successor, the California Bay-Delta Authority, for example, have struggled to create a hybrid federal–state entity with overall responsibility for the Delta effort. The CALFED goal remains elusive, and similar struggles for primacy are under way in the Everglades and Platte River projects.

Tracking Success Checklist

Judging whether a project is on track for success calls for inquiry into the following, interrelated aspects of the undertaking:

1. Timing and levels of funding
2. Setting and meeting interim and final goals
3. Nature and status of the federal–state partnership

4. Quality of the science and its integration in decision making
5. Conflict management and resolution
6. Building and maintaining public awareness and support

Timing and Levels of Funding

For a small minority of ecosystem restoration projects, like the Platte River effort, funding is not a central concern. The Platte program is directed to land and water management, not construction, so its cost is relatively small, and government entities and private water users share the price of land acquisition for habitat protection under the program. For most of these efforts, however, funding is the issue that receives the most attention from policy makers, stakeholders, the press, and the public, who find it easy to judge progress on the basis of funding levels alone. Depending on the circumstances, the issue may be whether the responsible governments are willing and able to commit the funds necessary to meet project goals on time. Where governments have made funding commitments, the question is whether they have fulfilled their promises on schedule.

As we have seen, the price tags on these large restoration projects can be enormous, running into the tens of billions of dollars. (To put these costs in context, 2006 spending on the Iraq war reached $8 billion per month, an amount that dwarfs the cost of ecosystem restoration.) In some areas, the will and capacity on the part of the state or federal government to appropriate the necessary funds have never been marshaled. Chesapeake Bay is an example. While in the 1990s Maryland's governor and legislature gave Bay cleanup goals high budget priority and posted impressive public expenditures on water quality and habitat protection, Virginia's efforts during those years lagged. Many people who work in government agencies and stakeholder groups on Chesapeake Bay restoration believe that the estimated total cost of $19 billion is beyond the current financial reach of the states involved, even with legislatures giving the matter the highest priority. A substantial funding commitment from the federal government is required to save the Bay. However, the federal government has never viewed the Chesapeake Bay as a matter of high national priority (which is ironic given that the bay lies in the watershed of the nation's capital), and a substantial federal funding commitment to save the bay seems unlikely at present.

In the California Bay-Delta, the political will to fund the CALFED project properly was evidenced at earlier stages of its development, only to recede after a change in political administrations with the result that commitments of federal funding have not been met. In the Everglades, promised federal funds have yet to be appropriated, due in part to Congress's refusal for the seven years since authorizing Everglades restoration to pass a Water Resources Development Act

(WRDA) bill. The federal government's inaction has caused a serious disruption in the federal–state partnership that was to be central to implementing CALFED and Everglades restoration. In frustration, Florida's legislators set aside a $1 billion fund to create a program, called Acceler8, under which the state will build eight component projects called for in the restoration plan, outside the federal partnership. Likewise, a California state bonding effort from 2000 to 2005 provided the preponderance of funding for CALFED in those years, and the governor has called for a long-term finance plan not reliant on future congressional funding.

In fiscal year 2005, Congress authorized $13.5 million in preconstruction engineering and design funds to allow the Corps of Engineers (Corps) to begin planning its first recommended fifteen-year increment of a proposed fifty-year UMR navigation and ecosystem sustainability plan. If constructed, the plan would redress many of the deleterious effects the UMR ecosystem has suffered since the Great Mississippi Flood of 1927, when present navigation and flood control projects were built along the river after congressional authorization of the 9-foot channel. However, Congress has not yet appropriated the $1.463 billion required to implement the first increment of the Corps' UMR ecosystem sustainability plan.

Under ideal circumstances, the state and federal government partners would each set aside a dedicated source of funds sufficient to meet their obligations to the project over time. So far the federal government has not done this, leaving its funding commitment to be met by individual appropriations bills or omnibus bills, such as WRDA or the Farm Bill, that come only once every few years. Few states have the budget resources of Florida and California to establish dedicated funding for ecosystem restoration.

Setting and Meeting Interim and Final Goals

Agreement among government representatives, scientists, and stakeholders on the overarching goal of an ecosystem restoration project is an essential condition of its creation, and adherence to this goal is required for its continuing viability. This ultimate objective is sometimes stated in hortatory language at a level of broad generality, so that all potentially disputing interest holders can feel comfortable signing on. The goal for Everglades restoration, for example, was stated by Congress in WRDA 2000 in a way intended to satisfy environmentalists and urban water users alike: "The overarching objective of the Plan is the restoration, preservation and protection of the south Florida ecosystem while providing for other water-related needs of the region, including water supply and flood protection." For other programs, the goal is couched in terms of continuing collaboration and deliberation among the interested parties and governments. An example is the Platte River project. When the parties could not come to agreement on specific target water flow numbers as the project's ultimate goal, they instead fixed on a series of incremental milestones to reduce water flow shortages and agreed to keep negotiating to reach a set of final objectives. In fact, the Platte River negotiations were able to move forward despite the state and federal parties' harboring different and competing visions of the final objective.

Because these projects are planned to carry on for long periods, sometimes many decades, interim goals and timetables (along with the baselines against which the goals are measured) need to be established with sufficient specificity to keep the work on track and moving in the desired direction. Interim goals may be substantive, as the *Chesapeake 2000* agreement's commitment to achieve a tenfold increase in native oysters in the bay (over 1994 levels) by 2010, or they may consist of process requirements such as monitoring and research, as in the Platte River agreement.

Although interim goals are necessary, once adopted, they can have the perverse effect of threatening the project's chances for success. Legislators and the public might see failure to meet certain intermediate goals in a timely way as a reason to withhold continued funding necessary for their achievement, a truly painful catch-22. Or government regulatory actions, such as Fish and Wildlife Service (USFWS) enforcement of water use reductions to meet Endangered Species Act (ESA) requirements, might be triggered by the program's failure to reach its interim goals. This, in turn, could impede the ongoing collaboration between the government and the regulated parties. Of course, looked at positively, regulatory certainty may be extended, along with funding and other types of government support, when the regulated parties satisfy agreed-upon interim measures.

In some projects, interim goals have been reformulated from time to time, as deadlines passed without their achievement. This has been true, for example, in the Chesapeake Bay program, when, in 1987, the partners committed to a 40-percent reduction in nutrients entering the main stem of the Bay by the year 2000. Soon after its adoption, this goal was deemed to be unachievable. Officials relaxed the goal but were still unable to meet the deadline; Chesapeake 2000 commits the signatories to continue to work toward the redefined 40-percent reduction target until 2010. This pattern deserves special scrutiny: Are the goals not being met because of a lack of funding or a faltering political will? Or were the goals unrealistic, perhaps based on faulty science, thus impossible to achieve? Or is litigation required before enforceable goals can be established? Which agencies bear responsibility for failures in setting and meeting goals? Answers to these questions are likely to be elusive or contested.

Nature and Status of the Federal–State Partnership

A robust federal–state partnership that is perceived by both levels of government to be mutually beneficial is one key to the success of large-scale, complex ecosystem restoration projects. Perhaps the most essential hallmark of a productive intergovernmental relationship is that both partners are actively engaged in carrying out the project as a matter of significant legislative and executive branch priority. This engagement requires visionary, capable leadership in both sectors. If leaders place the project's progress high on their public agendas, the impetus exists to find ways to bridge disagreements and to take the actions necessary to attain goals.

The change of administration in Washington, D.C., in 2001 and the events of September 11, 2001, weakened the federal–state partnerships in the California Bay-Delta and Everglades projects in similar ways: as discussed earlier, neither project continued to receive the high priority enjoyed before 2001 by the administration and Congress. The federal government largely withdrew its support of the projects, as evidenced by decreases in funding and in the attention of high-ranking officials. This history makes even more remarkable the accomplishment of the federal and state negotiators in the Platte River Basin, who persevered after 2001 and reached agreement, with stakeholder approval, in 2006.

Water quality lawsuits in the Chesapeake Bay and the Everglades resulted in consent decrees under which the states' conduct in enforcing water cleanup is mandated and monitored by federal judges. Litigation of this kind can engender resentment of the continuing federal court oversight and, thus, sour the relation between the federal and state governments, as it did in the Everglades in the final few years of Governor Jeb Bush's last term. Or, a federal court consent decree can be used in a positive way by a state to overcome internal political and legislative opposition to environmental regulation, as in Virginia's case.

Quality of the Science and Its Integration in Decision Making

Multiple teams of scientists often work on a restoration project simultaneously. Every significant federal agency involved in the project has its own cadre of scientists. This may also be true at the state and local levels, and academic scientists are likely to be at work in the ecosystem as well. One tall challenge is to coordinate the work of the governments' science teams, getting them to agree on what areas of scientific inquiry deserve priority. The watershed-wide nature of the project requires the specialized knowledge and application of numerous scientific disciplines—hydrology, biology, geology, botany, chemistry, and ecology—to plumb the intricacies of the interactions of natural systems. Finding ways to integrate the work of these disciplines is necessary and can be difficult.

Policy decisions in ecosystem restoration must be made on the basis of the best available science, but how can policy makers, stakeholders, and the public be assured as to the quality of the science being produced by the agencies? Each agency engaging in scientific research should have an established peer review process; similarly, science produced by interagency and intergovernmental teams must be peer reviewed by reputable independent scientists. Congress understood this when, in authorizing Everglades restoration in WRDA 2000, it mandated creation of an independent scientific panel to review "the Plan's progress toward achieving the natural system restoration goals" and required it to provide biennially to Congress "an assessment of ecological indicators and other measures of progress in restoring the ecology of the natural system" (WRDA 2000, Sec. 601(j)).

All five projects examined in this book embrace the concept of adaptive management, which means applying, in a systematic and orderly way, the results of scientific research to refine continuously or, if needed, redesign their plans and operations. This approach recognizes that current understandings and solutions may be supplanted by new scientific and technical knowledge and, as a consequence, the plans for restoration may have to be changed. Thus it is worth inquiring how each project's decision makers go about integrating science into their policy making. Among the five projects, only the Everglades project has put into place a set of procedures for the practice of adaptive management, though the procedures have yet to be tested by full-scale scientific reassessment.

While the governmental entities participating in the Bay-Delta project have committed to adaptive management, one critical report stated, "[A]daptive management has not become a way of doing business at CALFED." This section of the report concludes, "CALFED needs to embrace adaptive management if it wants to accelerate progress toward its goals" (Little Hoover 2005, 70). For the UMR, the Corps' ecosystem sustainability design teams have begun using adaptive management principles in their initial project planning.

Conflict Management and Resolution

Scientific and policy disputes and conflicts among water claimants are common in ecosystem restoration projects. A broad array of techniques for managing and resolving conflicts exists, ranging from simple jawboning, to elevating decisions to higher officials, to formal intervention using professional mediators or outside science panels, to litigation. Environmental disputes and conflicts come in many varieties: some are procedural, some substantive; some are scientific or technical, others political. Techniques for conflict resolution are more suitable than others for handling certain types of conflict appropriately; therefore, being able to analyze the type of dispute can be important in addressing it most effectively. Early in a project's development, those in charge need to consider conflict management and resolution issues and lay out procedures for classifying conflicts and directing them to the appropriate forum for resolution.

For example, the proposed fifty-year, $5.3 billion Upper Mississippi ecosystem sustainability program would never have seen the light of day had the Corps not taken seriously the National Research Council's peer-review criticisms of the Corps' prior environmental plan. In response, the Corps convened an innovative Federal Principals Group, committed to finding a new way of doing business that gave all involved agencies a forum to resolve differences—and, eventually, to come together to speak with one voice on a new path forward, ecosystem sustainability.

Building and Maintaining Public Awareness and Support

Most large ecosystem restoration projects have well-planned programs for public outreach. Everyone agrees that public education, understanding, and involvement are important to the success of these projects because education builds the public support necessary to secure the appropriation of taxpayer dollars they require. An example worth emulating is the Chesapeake Bay Foundation's program that takes 38,000 school children on boat trips on the bay every year. Public education and involvement, however, are often hard to accomplish. Understanding these projects' complexities and the technicalities of their issues and options is difficult for nonscientists, including most journalists and the general public.

Technology has advanced the project managers' ability to communicate with the public. Most restoration projects have well-designed Web sites that give access to copious information about their origins and purposes and allow public access to all official documents and deliberations. Decisions are not made behind closed doors but in venues that allow public access to the decision making. However, accessibility does not equal transparency in these efforts. With the multiplicity of government agency participants, the projects often lack common public accounting and reporting systems, which can hamper the success of public education efforts.

Seeing Success

This six-item checklist explained earlier can be put to use in several contexts. It may be used as a guide to researching and judging progress of the five large, complex projects covered in this book and others like them. The list can be employed to study and evaluate simpler, more regional or localized efforts as well, though modified to inquire into the productivity of the state–local, rather than the federal–state partnership. For projects in the inception stage, the checklist is designed as a guide for envisioning and putting in place concrete plans to achieve the elements of success in science-based, comprehensive, collaborative ecosystem restoration, regardless of scale.

The success of ecosystem restoration projects in setting sound watershed-wide, science-based goals; securing necessary funding; and meeting the goals is important not only to the future of their home regions. Their success is also important as a clear demonstration that this nation's citizens have the political will and capability to take on issues of environmental degradation in a scientific way. Well-managed projects account for all the interconnections and synergies in the natural system and are flexible enough to adapt as science and technology advance understanding. This work requires bridging many divides—between science and public policy, between federal government and the states, between regions and localities, among scientific disciplines, and between water users and habitat protectors. We need to inform ourselves and communicate with those charged with responsibility for ecosystem restoration to urge them to continue the active pursuit of success. Our future depends on it.

Reference

Little Hoover Commission. 2005. *Still Imperiled, Still Important: The Little Hoover Commission's Review of the CALFED Bay-Delta Program*. November. Report #183. Sacramento: Little Hoover Commission. At www.lhc.ca.gov/lhcdir/183/report183.pdf.

About the Authors

Alf W. Brandt, J.D., serves as the principal consultant for the Committee on Water, Parks & Wildlife for the California State Assembly. He served as counsel to the U.S. Department of the Interior, as federal agency coordinator for the CALFED Bay-Delta Program, and as a member of the Board of Directors for the Metropolitan Water District of Southern California.

Thomas L. Crisman, Ph.D., is Patel Professor of Environment at the University of South Florida and former professor of environmental engineering sciences and director of the Howard T. Odum Center for Wetlands at the University of Florida. He earned his Ph.D. in zoology from Indiana University.

Mary Doyle, LL.B., is professor of law, former dean of the University of Miami School of Law, and codirector of the Leonard and Jayne Abess Center for Ecosystem Science and Policy at the University of Miami. She has served as associate general counsel of the Environmental Protection Agency and acting assistant secretary for Water and Science, as well as counselor to the secretary of the U.S. Department of the Interior.

Cynthia A. Drew, Ph.D., J.D., is associate professor of law at the University of Miami School of Law and former program manager of the South Florida Water Management District's Surface Water Improvement and

Management Plan for the Indian River Lagoon, an Estuary of National Significance. She has published articles on the Everglades and Endangered Species Act regulations and practiced environmental law at Jenner & Block, Chicago, and the U.S. Department of Justice, Washington, D.C.

David M. Freeman, Ph.D., emeritus professor of sociology, Colorado State University, has researched, taught, and analyzed policy regarding the sociology of water organization in the United States and South Asia. He undertook systematic study of the Platte River Habitat Recovery negotiating process in 1997, continued as a participant-observer until the program was launched in January 2007, and is writing a book on mobilizing water users to implement the Endangered Species Act in the Platte River Basin.

Stuart Langton, Ph.D., has been a consultant to five hundred organizations, including the South Florida Ecosystem Restoration Task Force, which coordinates work on Everglades restoration. He is senior fellow at the Florida Conflict Resolution Consortium, Florida State University, and was the Lincoln Filene Professor of Citizenship and Public Affairs at Tufts University.

Fernando Miralles-Wilhelm, Ph.D., is associate professor of civil and environmental engineering at Florida International University and has served on the faculties of Northeastern University and the University of Miami. A graduate of the Universidad Simón Bolívar in Venezuela, he has a master's degree in engineering from the University of California, Irvine, and a Ph.D. in environmental engineering from the Massachusetts Institute of Technology.

David Nawi, LL.B., a mediator in the area of water and other environmental issues, served as regional solicitor for the U. S. Department of the Interior, Pacific Southwest Region, from 1993 to 2001. He has been involved extensively in California water issues, including the development of the CALFED Program and mediation of major litigation involving California's State Water Project.

Stephen Polasky, Ph.D., is Fesler-Lampert Professor of Ecological/Environmental Economics at the University of Minnesota. He has served as coeditor of the *Journal of Environmental Economics and Management*, senior staff economist for the President's Council of Economic Advisers, and a committee member on panels on ecosystem services and natural resource damage assessment for the Environmental Protection Agency, the U.S. Department of the Interior, and the National Research Council.

Colonel Terrence "Rock" Salt, U.S. Army, retired, is director of Everglades restoration initiatives for the U.S. Department of the Interior. He served as executive director of the South Florida Ecosystem Restoration Task Force and as Jacksonville district commander with the U.S. Army Corps of Engineers.

ABBREVIATIONS AND ACRONYMS

A

1983 Agreement	Chesapeake Bay Agreement of 1983
1987 Agreement	Chesapeake Bay Agreement of 1987
Accord, the (or Bay-Delta Accord)	Principles for Agreement on Delta Standards between the State of California and the Federal Government
ASA	assistant secretary of the army
ASR	aquifer storage and recovery
Authority, the	California Bay-Delta Authority

B

Bay-Delta Accord (or the Accord)	Principles for Agreement on Delta Standards between the State of California and the Federal Government
BBCAC	Bi-State Blue Crab Advisory Committee
BBTAC	Bi-State Blue Crab Technical Advisory Committee
BEA	Bureau of Economic Analysis
BG	brigadier general

C

C2K	Chesapeake 2000 (Chesapeake Bay agreement made in 2000)
C-38 Canal	a 90-kilometer-long, 9-meter-deep, 100-meter-wide canal, formerly the 166-kilometer-long winding path of the Kissimmee River
C&SF Project	Central and South Florida Project
CA	Cooperative Agreement (or Platte River Cooperative Agreement)
	Cooperative Agreement for Platte River Research and Other Efforts Relating to Endangered Species Habitats along the Central Platte River, Nebraska
CALFED (or the Program)	originally CALFED Bay-Delta Program; its successor is the California Bay-Delta Authority
CAP	Continuing Authorities Programs
CBC	Chesapeake Bay Commission
CBF	Chesapeake Bay Foundation
CBO	Congressional Budget Office
CBP	Chesapeake Bay Program
CBWBRFP	Chesapeake Bay Watershed Blue Ribbon Finance Panel
CDFG	California Department of Fish and Game
CDWR	California Department of Water Resources
CEQA	California Environmental Quality Act
CERP (or the Plan)	Comprehensive Everglades Restoration Plan
CESA	California Endangered Species Act
Club Fed	Federal Ecosystem Directorate

CNPPID	Central Nebraska Public Power and Irrigation District
COL	Colonel
Corps, the (or USACE)	U.S. Army Corps of Engineers
CRC	Chesapeake Research Consortium
CROGEE	Committee for the Restoration of the Greater Everglades Ecosystem
CSB	California State Budget
CSFP	Central and Southern Florida Project
CVP	Central Valley Project
CVPIA	Central Valley Project Improvement Act
CWA	Clean Water Act (Federal Water Pollution Control Act, 1948 and 1972, amended in 1977 as the Clean Water Act)

D

DEP	Department of Environmental Protection
DOI (or Interior or USDOI)	U.S. Department of the Interior
DWR	Department of Water Resources

E

EAA	Everglades Agricultural Area
ECC	Economic Coordinating Committee
EEC	essential ecosystem characteristics
EFA	Everglades Forever Act
8.5 SMA	8.5 Square Mile Area (in the Everglades)
EIR	environmental impact report
EIS	environmental impact statement
EMP	environmental management program

EnCC	Engineering Coordinating Committee
ENP (or the Park)	Everglades National Park
ENPPEA	Everglades National Park Protection and Expansion Act of 1989
EPA (or USEPA)	Environmental Protection Agency
EQ	environmental quality
ERP	Ecosystem Restoration Program
ERS	Economic Research Service
ESA	Endangered Species Act
EWA	Environmental Water Account

F

FACA	Federal Advisory Committee Act
FERC	Federal Energy Regulatory Commission
FWS (or USFWS)	U.S. Fish and Wildlife Service (an agency within the larger USDOI)
FY	fiscal year

G

GAO (or USGAO)	U.S. Government Accountability Office
GIS	geographic information systems
GLC	Governor's Liaison Committee
Governor's Commission	Governor's Commission for a Sustainable South Florida
Group, the	Federal Principals Group

H

ha	hectares
HAB	harmful algal blooms

I

IATP	Institute for Agriculture and Trade Policy
Interior (DOI or USDOI)	U.S. Department of the Interior
IPCC	Intergovernmental Panel on Climate Change
ISP	Integrated Science Plan
ITQs	individually transferable quotas
IWW	Illinois Waterway

K

km^2	square kilometers

L

LCA	Louisiana Coastal Area
L/D	locks and dams
LG	lieutenant general

M

MARAD	Maritime Administration
MARC	Midwest Area River Coalition
MCGA	Missouri Corn Growers Association
Mod Water	Modified Water Delivery Project
MRD	Mississippi River Division, USACE
MSCS	Multi-Species Conservation Strategy
MWD	Metropolitan Water District (of Southern California)

N

NAS	National Academy of Sciences
NASS	National Agricultural Statistical Service

NCCP	natural community conservation planning
NECC	Navigation Environmental Coordinating Committee
NED	national economic development
NEPA	National Environmental Policy Act (administered by EPA, requires federal agencies to prepare EISs)
NESP	Navigation and Ecosystem Sustainability Program
NGO	nongovernmental organization
NMFS	National Marine Fisheries Service
NOAA	National Oceanic and Atmospheric Administration
NPDES	National Pollutant Discharge Elimination System
NPPD	Nebraska Public Power District
NPS	National Park Service (an agency within the larger USDOI)
NPCA	National Parks Conservation Association
NRC	National Research Council (part of the National Academies)
NWFR	National Wildlife and Fish Refuge

O

OCE	Office of Chief of Engineers, USACE
OEEECT	Outreach, Environmental, and Economic Equity Coordination Team
O/M	operations and maintenance
OSE	other social effects

P

Panel, the	Chesapeake Bay Watershed Blue Ribbon Finance Panel

PCBs	polychlorinated biphenyls
PDT	Planning Design Teams
PED	preengineering design
PICC	Public Involvement Coordinating Committee
POS	plan of study
POSST	Public Outreach Steering and Support Team
ppb	parts per billion
PRESP	Platte River Endangered Species Partnership

Q

Quad Cities	Davenport and Bettendorf, Iowa; Moline and Rock Island, Illinois

R

RECOVER	Restoration, Coordination, and Verification team (interdisciplinary, interagency team of scientists established by USACE and SFWMD)
Reclamation (or USBR)	U.S. Bureau of Reclamation
RED	regional economic development
Restudy	Comprehensive Everglades Review Study
RFA	Renewable Fuels Association
RID	Rock Island (IL) District, USACE
ROD	record of decision

S

SAV	submerged aquatic vegetation
SCT	Science Coordination Team

SFERTF (or Task Force)	South Florida Ecosystem Restoration Task Force
SFWMD (or District)	South Florida Water Management District
SLD	St. Louis (MO) District, USACE
SMA	square mile area
SOE	Save Our Everglades program
SPCG	Science Panel Coordination Group
SPD	St. Paul (MN) District, USACE
STA	storm water treatment area
STAC	Scientific and Technical Advisory Committee
SWIM	Surface Water Improvement and Management Act
SWP	State Water Project
SWRCB (or State Board)	State Water Resources Control Board
SWSI	Statewide Water Supply Initiative

T

TMDL	total maximum daily load

U

UMIMRA	Upper Mississippi, Illinois, and Missouri Rivers Association
UMR	Upper Mississippi River
UMRB	Upper Mississippi River Basin
UMRBA	Upper Mississippi River Basin Association
UMRBC	Upper Mississippi River Basin Commission
UMRCC	Upper Mississippi River Conservation Committee

UMR-IWW	Upper Mississippi River Illinois Waterway
UMR NFWR	Upper Mississippi River National Fish and Wildlife Refuge
UMRS	Upper Mississippi River System
USACE (or Corps)	U.S. Army Corps of Engineers (an agency within the larger U.S. Army)
USBR (or Reclamation)	U.S. Bureau of Reclamation (an agency within the larger USDOI)
USBR/EIS	U.S. Bureau of Reclamation/ Environmental Impact Statement
USCB	U.S. Census Bureau
USDA	U.S. Department of Agriculture
USDOI (or DOI or Interior)	U.S. Department of the Interior
USEPA (or EPA)	U.S. Environmental Protection Agency
USFWS (or FWS)	U.S. Fish and Wildlife Service (an agency within the larger USDOI)
USGAO (or GAO)	U.S. Government Accountability Office
USGS	U.S. Geological Survey
USWRC	U.S. Water Resources Council

W

WCA	water conservation area
WDNR	Wisconsin Department of Natural Resources
WRAC	Water Resources Advisory Commission
WRDA	Water Resources Development Act

INDEX

Italicized page numbers refer to figures.

Island Press | Board of Directors